MACHINE INTELLIGENCE 13

MACHINE INTELLIGENCE

Machine Intelligence 1 (1967) (eds N. Collins and D. Michie) Oliver & Boyd, Edinburgh

Machine Intelligence 2 (1968) (eds E. Dale and D. Michie) Oliver & Boyd, Edinburgh

(1 and 2 published as one volume in 1971 by Edinburgh University Press) (eds N. Collins, E. Dale and D. Michie)

Machine Intelligence 3 (1968) (ed D. Michie) Edinburgh University Press, Edinburgh

Machine Intelligence 4 (1969) (eds B. Meltzer and D. Michie) Edinburgh University Press, Edinburgh

Machine Intelligence 5 (1970) (eds B. Meltzer and D. Michie) Edinburgh University Press, Edinburgh

Machine Intelligence 6 (1971) (eds B. Meltzer and D. Michie) Edinburgh University Press, Edinburgh

Machine Intelligence 7 (1972) (eds B. Meltzer and D. Michie) Edinburgh University Press, Edinburgh

Machine Intelligence 8 (1977) (eds E. W. Elcock and D. Michie) Ellis Horwood, Chichester/Halsted, New York

Machine Intelligence 9 (1979) (eds J. E. Hayes, D. Michie and L. Mikulich) Ellis Horwood, Chichester/Halsted, New York

Machine Intelligence 10 (1982) (eds J. E. Hayes, D. Michie and Y.-H. Pao) Ellis Horwood, Chichester/Halsted, New York

Machine Intelligence 11 (1988) (eds J. E. Hayes, D. Michie and J. Richards) Oxford University Press, Oxford

Machine Intelligence 12 (1991) (eds J. E. Hayes, D. Michie and E. Tyugu) Oxford University Press, Oxford

Machine Intelligence 13 (1994) (eds K. Furukawa, D. Michie and S. Muggleton) Oxford University Press, Oxford

MACHINE INTELLIGENCE 13

Machine Intelligence and Inductive Learning

edited by

K. FURUKAWA

Keio University, Tokyo

D. MICHIE

Turing Institute, Glasgow

and

S. MUGGLETON

Oxford University Computing Laboratory

CLARENDON PRESS · OXFORD

1994

Oxford University Press, Walton Street, Oxford OX2 6DP

Oxford New York
Athens Auckland Bangkok Bombay
Calcutta Cape Town Dar es Salaam Delhi
Florence Hong Kong Istanbul Karachi
Kuala Lumpur Madras Madrid Melbourne
Mexico City Nairobi Paris Singapore
Taipei Tokyo Toronto
and associated companies in
Berlin Ibadan

Published in the United States by
Oxford University Press Inc., New York

A catalogue record for this book is available from the British Library

Library of Congress Cataloging in Publiction Data available

ISBN 0 19 853850 2

Typeset by the authors using LaTeX
Printed in Great Britain on acid-free paper by
Bookcraft (Bath), Midsomer Norton

PREFACE

The founder of modern computational logic, J.A. Robinson, opens this volume with a chapter on the field's great forefathers John von Neumann and Alan Turing. Stephen Muggleton follows with an analysis of Turing's legacy in logic and machine learning, conceiving these not in generality, but as specific means of imparting knowledge to computers, a theme first articulated by Turing in the late 1940's.

The present volume records the Machine Intelligence Workshop of 1992, held at Strathclyde University's Ross Priory retreat on Loch Lomond, Scotland. Here the series entered not only its second quarter-century but a new phase. As can be seen in these pages, machine learning emerged to declare itself as a seed-bed of new theory, as a practical tool in engineering disciplines, and as material for new mental models in the human sciences.

Connections with behavioural and cognitive psychology are illuminated in Chapters 9 and 10. The pioneers always stressed these connections. In 1953 Claude Shannon had this to say:

> The problem of how the brain works and how machines may be designed to simulate its activity is surely one of the most important and difficult facing science ...Can we organise machines into a hierarchy of levels, as the brain appears to be organised, with the learning of the machine gradually progressing up the hierarchy? ...How can computer memory be organised to learn and remember by association, in a manner similar to the human brain?

Approaches to learning by association "in a manner similar to the human brain" have recently engendered unprecedented interest, one might almost say turbulence. Chapter 13 pre-views a joint European endeavour of six academic and six industrial laboratories to steer the topic towards clearer waters. The complete comparative study is now available as a book from Ellis Horwood (Simon and Schuster).

January 1994

Stephen Muggleton
Executive Editor
Donald Michie
Koichi Furukawa
Associate Editors

ACKNOWLEDGEMENTS

New beginnings *par excellence* also spring from an agreement concluded in 1991 between the Turing Institute, UK and the Japan Society for Artificial Intelligence, Tokyo, under the generous auspices of the Daiwa Anglo–Japanese Foundation. The Foundation provided funding, covering Workshops 13 and 14, to defray travel and attendance costs for six Japanese and six British scientists nominated by the respective parties. To this we owe the circumstance that this volume has been able properly to reflect something of the vigour with which the subject is being advanced in Japan.

We are also indebted to the Royal Society of London for facilitating Professor Enn Tyugu's participation from the Estonian Academy of Sciences. Strathclyde University, the Turing Institute, and Scottish Enterprise also contributed help and resource in the many small ways that go towards the making of a great occasion.

The Editors would also like to express their thanks to Ashwin Srinivasan for the many hours of effort involved in persuading LaTeX to produce the standard Machine Intelligence look-and-feel within this volume. Thanks are also due to the Oxford University Computing Laboratory for kindly allowing use of printing and document preparation facilities in the production of this volume.

CONTENTS

HISTORICAL PERSPECTIVES

1. Logic, Computers, Turing, and von Neumann 1
J. A. Robinson

2. Logic and learning: Turing's legacy 37
S. Muggleton

INDUCTIVE INFERENCE

3. A generalization of the least general generalization 59
H. Arimura, T. Shinohara, S. Otsuki and H. Ishizaka

4. The justification of logical theories based on data compression 87
A. Srinivasan, S. Muggleton, and M. Bain

5. Utilizing structure information in concept formation 123
K. Handa, M. Nishikimi and H. Matsubara

6. The discovery of propositions in noisy data 143
Hiroshi Tsukimoto and Chie Morita

7. Learning non-deterministic finite automata from queries and counterexamples 169
T. Yokomori

SCIENTIFIC DOMAINS

8. Machine Learning and biomolecular modelling 193
M. J. E. Sternberg, R. A. Lewis, R. D. King and S. Muggleton

9. More than meets the eye: animal learning and knowledge induction 213
E. J. Kehoe

10. Regulation of human cognition and its growth 247
C. Trevarthen

11. Large heterogeneous knowledge bases 269
E. Tyugu

EXPERIMENTAL MACHINE LEARNING

12. Learning optimal chess strategies 291
M. Bain and S. Muggleton

13. A comparative study of classification algorithms: Statistical, Machine Learning and Neural Network 311
R. D. King, R. Henery, C. Feng and A. Sutherland

LEARNING CONTROL

14. Recent progress with BOXES 363
C. Sammut

15. Building symbolic representations of intuitive real-time skills from performance data 385
D. Michie and R. Camacho

16. Learning perceptually chunked macro operators 419
M. Suwa and H. Motoda

17. Inductively speeding up logic programs 441
M. Numao, T. Maruoka, and M. Shimura

INDEX 459

HISTORICAL PERSPECTIVES

1

Logic, Computers, Turing, and von Neumann†

J. A. Robinson

Syracuse University
New York 13244-2010, U.S.A.

1 INTRODUCTION

The two outstanding figures in the history of computer science are Alan Turing and John von Neumann, and they shared the view that logic was the key to understanding and automating computation. In particular, it was Turing who gave us in the mid-1930s the fundamental analysis, and the logical definition, of the concept of 'computability by machine' and who discovered the surprising and beautiful basic fact that there exist universal machines which by suitable programming can be made to

†This essay is an expanded and revised version of one entitled *The Role of Logic in Computer Science and Artificial Intelligence*, which was completed in January 1992 (and was later published in the Proceedings of the Fifth Generation Computer Systems 1992 Conference). Since completing that essay I have had the benefit of extremely helpful discussions on many of the details with Professor Donald Michie and Professor I. J. Good, both of whom knew Turing well during the war years at Bletchley Park. Professor J. A. N. Lee, whose knowledge of the literature and archives of the history of computing is encyclopedic, also provided additional information, some of which is still unpublished. Further light has very recently been shed on the von Neumann side of the story by Norman Macrae's excellent biography *John von Neumann* (Macrae 1992). Accordingly, it seemed appropriate to undertake a more complete and thorough version of the FGCS'92 essay, focussing somewhat more on the interesting historical and biographical issues. I am grateful to Donald Michie and Stephen Muggleton for inviting me to contribute such a 'second edition' to the present volume, and I would also like to thank the Institute for New Computer Technology (ICOT) for kind permission to make use of the FGCS'92 essay in this way.

perform any computation whatsoever. This theoretical work of Turing's had direct practical influence on the designs of the first modern general-purpose electronic digital computers in the mid-1940s. Von Neumann also had a prominent role in the emergence of these first machines, whose logical design has changed remarkably little in its general features during the almost fifty years since it was first formulated. The conventional wisdom is that von Neumann is primarily responsible for this design, which is accordingly known as the 'von Neumann architecture'. Without detracting from von Neumann's enormous importance, we maintain that this attribution somewhat distorts reality. The birth of the general-purpose computer, and with it computer science, owes far more to Turing than is usually acknowledged. Turing's contribution was both direct (arising from his own activities in the 1940s) and indirect (because of his little-known but powerful intellectual influence on von Neumann, starting from the mid-1930s).

In his recent sympathetic biography of von Neumann, Norman Macrae characterizes his subject's propensity to develop new insights with incredible speed and brilliant virtuosity, even though they were often not his own:

> ...Johnny grabbed other people's ideas, then by his clarity leapt five blocks ahead of them, and helped put them into practical effect (Macrae 1992, p. ix).
>
> ...Johnny borrowed (we must not say plagiarized) anything from anybody, with great courtesy and aplomb. His mind was not as original as Leibniz's or Newton's or Einstein's, but he seized other people's original (though fluffy) ideas and quickly changed them in expanded detail into a form where they could be useful ...(Macrae 1992, p. 23).

Macrae might have added that neither was von Neumann's mind as original as Turing's. Not only was Turing highly creative and imaginative, but he was also capable of intellectual feats just as dazzling as those ascribed to von Neumann.

Nor was Turing's universal machine idea at all 'fluffy'. Yet it was one of the great ideas von Neumann 'grabbed' enthusiastically in this fruitful way. Von Neumann always freely acknowledged this intellectual debt to Turing, not merely 'with great courtesy and aplomb' but with quite open admiration and generous praise[2]. There was, however, no question of von Neumann's leaping 'five blocks ahead of' Turing in seeing that the general-purpose digital computer was implicit in Turing's idea. Turing had himself clearly seen virtually all the practical implications of his universal machine concept from the beginning[3]. Among them were the crucial role that logic would play in both the design and use of computers, and the alluring possibility of creating 'artificial intelligence' by programming a fast universal machine in suitable ways. In the course of exploiting Turing's idea, von Neumann also (when writing the EDVAC

[2]In a letter to Professor Brian Randell, Stanley Frankel (a Los Alamos physicist who, in collaboration with Nicholas Metropolis and at the instigation of von Neumann, had written the first major program to run on the ENIAC) writes: "I know that in 1943 or '44 von Neumann was well aware of the fundamental importance of Turing's paper of 1936 ... which describes in principle the 'Universal Computer' of which every modern computer (perhaps not ENIAC as first completed, but all later ones) is a realisation. Von Neumann introduced me to that paper and I studied it with care. Many people have acclaimed von Neumann as the 'father of the computer' ... but I am sure that he would never have made that mistake himself. He might well be called the midwife, perhaps, but he firmly emphasized to me, and to others I am sure, that the fundamental conception is owing to Turing. ... In my view von Neumann's essential role was in making the world aware of these fundamental concepts introduced by Turing and of the development work carried out in the Moore School and elsewhere. ... Both Turing and von Neumann, of course, also made substantial contributions to the 'reduction to practice' of these concepts but I would not regard these as comparable in importance with the introduction and explication of the concept of a computer able to store in its memory its program of activities and of modifying the program in the course of these activities." (Randell 1972, p.10)

[3]In (Carpenter and Doran 1986), p. 125, Mike Woodger writes: "During his stay at Princeton University shortly before the war Turing had discussed with John von Neumann the possibility of constructing a high-speed automatic computer with radio valves used essentially as switches. Both men were experts in the field of mathematical logic, and while the subject had very little bearing on the design or use of such machines it enabled them to see at once how the general problems of control and manipulation of sequences of binary digits could be affected."

Report in the spring of 1945) 'grabbed' the artifical neuron concept originated by McCullogh and Pitts in 1943, and the digital electronic techniques of Eckert and Mauchly, which he first heard about from Hermann Goldstine on the platform of the Aberdeen (Maryland) railway station in August 1944. Turing knew of the practical feasibility of large-scale digital electronic switching devices from his involvement in the early 1940s with the cryptanalytic 'Colossus' machines at Bletchley Park.

Mathematical logic was the primary professional field of both men. It gave them a view of computing which was more general than the view held by almost all of their colleagues, most of whom were either applied scientists interested in solving particular problems by numerical computation, or engineers interested in devising special tools to speed up traditional computing processes. The enormous practical significance of the idea of universality, or general-purposeness, seems at first to have been appreciated by hardly anyone other than Turing and von Neumann[4]. Indeed the conceptual power and significance even of the logical idea itself was at first understood only by the logicians and mathematicians who had followed the argument of (Turing 1936) and who were familiar with its background.

2 LOGIC AND THE COMPUTER

I expect that digital computing machines will eventually stimulate a considerable interest in symbolic logic ... [Turing 1947]. (Carpenter and Doran 1986, p. 122)

Logic and technology came together and gave us the computer. All modern general-purpose digital computers are phys-

[4]One of the few who noticed was, interestingly, J. R. Womersley, who as Superintendant of the Mathematics Division of the National Physical Laboratory was to recruit Turing there in 1945 in order to develop the ACE. He had read (Turing 1936) before the war, in 1937 or 1938, when he was employed at the Woolwich Arsenal on practical computation. He immediately saw its practical significance and was accordingly inspired to try to build a relay-based 'Turing Machine' (presumably a universal Turing Machine) with C. I. Norfolk, a telephone engineer. After some preliminary experiments the work was abandoned. See (Hodges 1983), p. 306 and note 5.30 on p. 556.

ical embodiments of the same logical abstraction Turing's universal machine. The practical exploitation of the universal machine concept, starting in the mid-1940s, owed much to Turing himself as well as to von Neumann. Turing and von Neumann not only played leading roles in the design and construction of the first working computers, but were also largely responsible for laying out the general logical foundations for understanding the computation process, for developing computing formalisms, and for initiating the methodology of programming: in short, for founding computer science as we now know it.

Today, logic continues to be a fertile source of abstract ideas for novel computer architectures for inference machines, dataflow machines, database machines, and rewriting machines. It provides a unified view of computer programming (which is essentially a logical task), and offers a systematic framework for reasoning about programs. Logic has been important in the theory and design of high-level programming languages. Indeed logical formalisms are the immediate models for two major logic programming language families: Church's lambda calculus (Church 1941) for functional programming languages such as LISP, SCHEME, ML, LUCID and MIRANDA, and the Horn-clause-resolution predicate calculus for relational programming languages such as PROLOG, PARLOG, and GHC. Peter Landin noted over twenty years ago (Landin 1966) that ALGOL-like languages, too, were merely 'syntactically sugared' only-slightly-augmented versions of Church's lambda calculus, and recently, another logical formalism, Martin-Löf's Intuitionistic Type Theory, has served (in, for example, Constable's NUPRL) as a very-high-level programming language, a notable feature of which is that a proof of a program's correctness is an automatic accompaniment of the program-writing process.

To design, understand and explain computers and programming languages; to compose and analyze programs and reason correctly and cogently about their properties; these are to practice an abstract logical art built on (in H. A. Simon's apt phrase) a 'science of the artificial' which studies rational artifacts in abstraction from the practical engineering details of their physical realization, yet maintains an analytical eye on their intrinsic efficiency. The formal logician has had to become also an abstract engineer. In earlier times, the logician could be, and often was, cavalier about matters of efficiency when offering an algorithm for carrying out some logical construction. The point was often merely to establish that an algorithm existed, and not necessarily to give one which might in fact be practically useful.

3 LOGIC AND ARTIFICIAL INTELLIGENCE

In artificial intelligence (AI) research, logic has been used (for example, by McCarthy and Nilsson) as a rational model for knowledge representation and (for example, by Plotkin and Muggleton) as a guide for the organization of machine inductive inference and learning. It has also been used (for example, by Wos, Bledsoe, and Stickel) as the theoretical basis for powerful automated deduction systems which have proved theorems of interest to professional mathematicians. Logic's roles in AI, however, have been more controversial than its roles in the theory and practice of computing. Until the difference (if any) between natural intelligence and artificial intelligence is better understood, and until more experiments have tested the claims both of logic's advocates and of logic's critics concerning its place in AI research, the controversies will continue.

4 LOGIC AND THE ORIGIN OF THE COMPUTER

Logic's dominant role in the invention of the modern computer is not widely appreciated. The computer as we know it today was invented by Turing in 1936, an event triggered by an important logical discovery announced by Kurt Gödel in 1930. Gödel's discovery (Gödel 1931) decisively affected the outcome of the so-called Hilbert Program. Hilbert's goal was to formalize all of mathematics and then give positive answers to three questions about the resulting formal system: is it consistent? is it complete? is it decidable? Gödel found that no sufficiently rich formal system of mathematics can be both consistent and complete. In proving this, Gödel invented, and used, a high-level symbolic programming language: the formalism of primitive recursive functions. As part of his proof, he composed an elegant modular functional program (a set of connected definitions of primitive recursive functions and predicates) which constituted a detailed computational presentation of the syntax of a formal system of number theory, with special emphasis on its inference rules and its notion of proof. This computational aspect of his work was auxiliary to his main result, but is enough to have established Gödel as the first serious programmer in the modern sense. Gödel's computational example inspired Turing a few years later, in 1936, to find an explicit but abstract logical model not only of the computing process, but also of the computer itself. Using these as auxiliary theoretical concepts, Turing disposed of the third of Hilbert's questions by showing (Turing 1936) that the formal system of mathematics is not decidable. Although his original computer was only an abstract logical concept, during the following decade (1937–1946) Turing became a leader in the design, construction, and operation of the first real computers.

The problem of answering Hilbert's third question was known as the Decision Problem. Turing interpreted it as the challenge either to give an algorithm which correctly decides, for all formal mathematical propositions A and B, whether B is formally provable from A, or to show that is there no such algorithm. Having first clearly characterized what an algorithm is, he found the answer: there is no such algorithm.

For our present purposes the vital part of Turing's result is his characterization of what counts as an algorithm. He based it on an analysis of what a 'computing agent' does when making a calculation according to a systematic procedure. He showed that, when boiled down to bare essentials, the activity of such an agent is nothing more than that of (as we would now say) a finite-state automaton which interacts, one at a time, with the finite-state cells comprising an infinite memory.

Turing's machines are plausible abstractions from real computers, which, for Turing as for everyone else in the mid-1930s, meant a person who computes. The abstract Turing machine is an idealized model of any possible computational scheme such a human worker could carry out. His great achievement was to show that some Turing machines are 'universal' in that they can exactly mimic the behavior of any Turing machine whatever. All that is needed is to place a coded description of the given machine in the universal machine's memory together with a coded description of the given machine's initial memory contents. How Turing made use of this universal machine in answering Hilbert's third question is not relevant to our purpose here. The point is that his universal machines are the abstract prototypes of today's stored-program, general-purpose computers. The coded description of each particular machine is the program which causes the universal machine to act like that particular machine.

Abstract and purely logical as it is, Turing's work had an obvious technological interpretation: there is no need to build a separate machine for each computing task. Only one machine is needed — a universal machine — because it can be made to perform any conceivable computing task simply by writing a suitable program for it. Turing himself set out to build a universal machine.

He began his detailed planning in 1944. After 1943 he had gradually transferred his activities from Bletchley Park to Hanslope Park, ten miles distant from Bletchley, in order to work on speech-encipherment devices. He had spent four months — from November 1942 to March 1943 — in the United States, during which he discussed his 1936 paper and its practical implications with Claude Shannon at the Bell Laboratories[5]. There has been much speculation concerning the plausible possibility that during his time in the United States Turing met with von Neumann for discussions about the possibility of practical general-purpose computers. The only evidence of any kind we have for this is the fact that Lord Halsbury in 1959 referred (in Ten Years of Computer Development, Computer Journal **1**, pp. 153–159) to:

> . . . a meeting of two minds which cross-fertilised one another at a critical epoch in the technological development which they exploited. I refer of course to the meeting of the late Doctors Turing and von Neumann during the war, and all that came thereof . . . .

[5] (Hodges 1983), 241 – 255.

But when Professor Brian Randell[6] later wrote to Lord Halsbury for more details he received the reply:

> I am afraid I cannot tell you more about the meeting between Turing and von Neumann except that they met and sparked one another off. Each had, as it were, half the picture in his head and the two halves came together in the course of their meeting. I believe both were working on the mathematics of the atomic bomb project.

I recently became aware of the existence of a photograph, inscribed *National Cash Register Co., Dayton, Ohio, July 1943*, showing Turing together with a group of naval officers, one of whom is identified[7] as William Friedman, the famous American code-breaker. This is the only evidence we have that Turing made at least one other visit to the United States after the November 1942 to March 1943 visit.

When the war ended in 1945 he moved to the National Physical Laboratory to pursue his goal full time. His real motive was already to investigate the possibility of artificial intelligence, a possibility he had frequently discussed with Donald Michie, I. J. Good, and other colleagues. He wanted, as he put it, to build a brain. By the end of 1945 Turing had completed his design for the ACE computer, based on his abstract universal machine. In designing the ACE, he was able to draw on his expert knowledge of the sophisticated new electronic digital technology which had been used at Bletchley Park to build special-purpose code-breaking machines (such as the Colossus). In the event, the ACE

[6]This account is taken from (Randell 1972).

[7]By Henry Tropp, a historian of computing, who has the photograph in his possession [personal communication, 25 June 1992]. It was Professor J. A. N. Lee who alerted me to the existence of this photograph. Tropp had earlier written (*The Effervescent Years*, IEEE Spectrum, February 1974, p. 76): "The National Cash Register Company offers a particularly intriguing 'might have been'. NCR actually had an electronic computing device constructed during the late 1930s. It was a high-speed arithmetic machine which could add, subtract and multiply electronically, and presumably this machine could have become the first commercial electronic computer had the company wished to pioneer in this field." Tropp was sent the photograph by the daughter of the late Joseph Desch, the NCR employee who designed this electronic device at Dayton in 1939.

would not be the first physical universal machine, for there were others who were after the same objective, and who beat NPL to it. Turing's 1936 idea had started others thinking. By 1945 there were several people planning to build a universal machine. One of these was John von Neumann[8].

Turing and von Neumann first met in 1935, when Turing was still an unknown 23-year-old Cambridge graduate student. Von Neumann was already world-famous for his work in many scientific fields, including theoretical physics, logic and set theory, and several other important branches of mathematics. Ten years earlier, he had been one of the leading logicians working on Hilbert's Program, but after Gödel's discovery he suspended his specifically logical researches and turned his attention to physics and to mathematics proper. In 1930 he emigrated to Princeton, where he remained for the rest of his life.

Turing spent two years (from mid-1936 to mid-1938) in Princeton, obtaining a doctorate under Alonzo Church, who in 1936 had independently solved the Decision Problem. Church's method (Church 1936) was quite different from Turing's and was not as intuitively convincing. During his stay in Princeton, Turing had many conversations with von Neumann, who was enthusiastic about Turing's work and offered him a job as his research assistant. Turing turned it down in order to resume his research career in Cambridge, but his universal machine had already become an important item in von Neumann's formidable intellectual armory. Then came the war. Both men were soon completely immersed in their absorbing and demanding wartime scientific work.

---

[8]Ironically, von Neumann's EDVAC Draft Report had already been completed in June 1945 and a copy of it soon afterwards reached J. R. Womersley at the National Physical Loboratory. Womersley immediately visited Turing and showed the EDVAC Report to him as part of a (successful) attempt to recruit him for his project to design and build a comparable machine. Turing joined the NPL on 1 October 1945, and within three months had completed the ACE Proposal. So after having freely broadcast the logical seeds almost a decade previously, Turing was at last preparing to harvest the results, but only in the knowledge that on the neighboring farm the ears had ripened even faster.

By 1943, von Neumann was deeply involved in many projects, a recurrent theme of which was his search for improved automatic aids to computation. In late 1944 he became a consultant to a University of Pennsylvania group, led by J. P. Eckert and J. W. Mauchly, which was then completing the construction of the ENIAC computer (which was programmable and electronic, but not universal, and its programs were not stored in the computer's memory). Although he was too late to influence the design of the ENIAC, von Neumann essentially supervised the design of the Eckert–Mauchly group's second computer, the EDVAC. Most of his attention in this period was, however, focussed on designing and constructing his own much more powerful machine in Princeton at the Institute for Advanced Study (IAS) computer (von Neumann 1946). The EDVAC and the IAS machine both exemplified the so-called von Neumann architecture, a key feature of which is the fact that instruction words are stored along with data in the memory of the computer, and are therefore modifiable just like data words, from which they are not intrinsically distinguished.

The IAS computer was a success (Aspray 1990). Many close copies were eventually built in the 1950s, both in US government laboratories (the AVIDAC at Argonne National Laboratory, the ILLIAC at the University of Illinois, the JOHNIAC at the Rand Corporation, the MANIAC at the Los Alamos National Laboratory , the ORACLE at the Oak Ridge National Laboratory, and the ORDVAC at the Aberdeen Proving Grounds), and in foreign laboratories (the BESK in Stockholm, the BESM in Moscow, the DASK in Denmark, the PERM in Munich, the SILLIAC in Sydney, the SMIL in Lund, and the WEIZAC in Israel); and there were at least two commercial versions of it (the IBM 701 and the International Telemeter Corporation's TC-1).

The EDSAC, a British version of the EDVAC, was running in Cambridge by June 1949, the result of brilliantly fast construction work by M.V. Wilkes following his attendance at a 1946 EDVAC course. Turing's ACE project was, however, greatly slowed down by a combination of British civil-service foot-dragging and his own lack of administrative deviousness, not to mention his growing preoccupation with AI. In May 1948 Turing resigned from NPL in frustration and joined the small computer group at the University of Manchester, whose small but universal machine started useful operation the very next month and thus became the world's first working universal computer. All of Turing's AI experiments, and all of his computational work in developmental biology, took place on this machine and its successors, built by others but according to his own fundamental idea.

Von Neumann's style in expounding the design and operation of EDVAC and the IAS machine was to suppress engineering details and to work in terms of an abstract logical description, based on the diagrammatic notation of McCullogh and Pitts' artificial neuron networks (McCullogh and Pitts 1943). He discussed both its system architecture and the principles of its programming entirely in such abstract terms. We can today see that von Neumann and Turing were right in following the logical principle that precise engineering details are relatively unimportant in the essential problems of computer design and programming methodology. The ascendancy of logical abstraction over concrete realization has ever since been a guiding principle in computer science, which has kept itself organizationally almost entirely separate from electrical engineering. The reason it has been able to do this is that computation is primarily a logical concept, and only secondarily an engineering one. To compute is to engage in formal reasoning, according to certain formal symbolic rules, and it makes no logical difference how the formulas are physically represented, or how the logical transformations of them are physically realized.

Of course no one should underestimate the enormous importance of the role of engineering in the history of the computer. Turing and von Neumann did not. They themselves had a deep and quite expert interest in the very engineering details from which they were abstracting, but they knew that the logical role of computer science is best played in a separate theater.

5 LOGIC AND PROGRAMMING

Since coding is not a static process of translation, but rather the technique of providing a dynamic background to control the automatic evolution of a meaning, it has to be viewed as a logical problem and one that represents a new branch of formal logics. [von Neumann 1947]

Much emphasis was placed by both Turing and von Neumann, in their discussions of programming, on the two-dimensional notation known as the *flow-diagram.* This quickly became a standard logical tool of early programming, and it can still be a useful device in formal reasoning about computations. The later ideas of Hoare, Dijkstra, Floyd, and others on the logical principles of reasoning about programs were anticipated by both (Turing 1949) and (von Neumann 1947). They stressed that programming has both a static and a dynamic aspect. The static text of the program itself is essentially an expression in some formal system of logic: a syntactic structure whose properties can be analyzed by logical methods alone. The dynamic process of running the program is part of the semantic meaning of this static text.

6 AUTOMATIC PROGRAMMING

Turing's friend Christopher Strachey was an early advocate, in the early 1950s, of using the computer itself to translate automatically from high-level 'mathematical' descriptions into low-level 'machine-language' prescriptions. His idea was to try to liberate the programmer from concern with 'how' to compute so as to be able to concentrate on 'what' to compute: in short, to think and write programs in a more natural and human idiom. Turing himself was not much interested in this idea, which he had already in 1947 pointed out as an 'obvious' one. In fact, he seems to have had a hacker's pride in his fluent machine-language virtuosity. He was able to think directly and easily in terms of bare bit patterns and of the unorthodox number representations such as the Manchester computer's reverse (i.e., low-order digits first) base-32 notation for integers. In this attitude, he was only the first among many who have stayed somewhat aloof from higher-level programming languages and higher-level machine architectures, on the grounds that a real professional must be aware of and work closer to the actual realities of the machine. One senses this attitude, for example, throughout Donald Knuth's monumental treatise on the art of computer programming.

It was not until the late 1950s (when FORTRAN and LISP were introduced) that the precise sequential details of how arithmetical and logical expressions are scanned, parsed, and evaluated could routinely be ignored by most programmers and left to the computer to work out. This advance brought an immense simplification of the programming task and a large increase in programmer productivity. There soon followed more ambitious language design projects such as the international ALGOL project, and the theory and practice of programming language design, together with the supporting software technology of interpreters and compilers, quickly became a major topic in computer science. The formal grammar used to define the syntax of ALGOL was not initially accompanied by an equally formal specification of its semantics; but this soon followed. Christopher Strachey and Dana Scott developed a formal 'denotational semantics' for programs, based on a rigorous mathematical interpretation of the previously uninterpreted, purely syntactical, lambda calculus of Church. It was, incidentally, a former student of Church, John Kemeny, who devised the enormously popular 'best-selling' programming language, BASIC.

7 DESCRIPTIVE AND IMPERATIVE ASPECTS

There are two sharply-contrasting approaches to programming and programming languages: the descriptive approach and the imperative approach. The descriptive approach to programming focusses on the static aspect of a computing plan, namely on the denotative semantics of program expressions. It tries to see the entire program as a timeless mathematical specification which gives the program's output as an explicit function of its input (whence arises the term 'functional' programming). This approach requires the computer to do the work of constructing the described output automatically from the given input according to the given specifications, without any explicit direction from the programmer as to how to do it.

The imperative approach focusses on the dynamic aspect of the computing plan, namely on its operational semantics. An imperative program specifies, step by step, what the computer is to do, what its 'flow of control' is to be. In extreme cases, the nature of the outputs of an imperative program might be totally obscure. In such cases one must (virtually or actually) run the program in order to find out what it does, and try to guess the missing functional description of the output in terms of the input. Indeed it is necessary in general to 'flag' a control-flow program with comments and assertions, supplying this missing information in order to make it possible to make sense of what the program is doing when it is running.

Although a purely static, functional program is relatively easy to understand and to prove correct, in general one may have little or no idea of the cost of running it, since that dynamic process is deliberately kept out of sight. On the other hand, although an operational program is relatively difficult to understand and prove correct, its more direct depiction of the actual process of computation makes an assessment of its efficincy relatively straightforward. In practice, most commonly-used high-level programming languages — even LISP and PROLOG – have both functional and operational features. Good programming technique requires an understanding of both. Programs written in such languages are often neither wholly descriptive nor wholly imperative. Most programming experts, however, recommend caution and parsimony in the use of imperative constructs. Some even recommend complete abstention. Dijkstra's now-classic Letter to the Editor (of the Communications of the ACM), entitled 'GOTO considered harmful' is one of the earliest and best-known such injunctions.

These two kinds of programming were each represented in pure form from the beginning: Gödel's purely descriptive recursive function formalism and Turing's purely imperative notation for the state-transition programs of his machines.

8 LOGIC AND PROGRAMMING LANGUAGES

In the late 1950s at MIT John McCarthy and his group began to program their IBM 704 using symbolic logic directly (McCarthy 1960). Their system, LISP, is the first major example of a logic programming language intended for actual use on a computer. It is essentially Church's lambda calculus (Church 1941), augmented by a simple recursive data structure (ordered pairs), the conditional expression, and an imperative 'sequential construct' for specifying a series of consecutive actions. In the early 1970s Robert Kowalski in Edinburgh (Kowalski 1974) and Alain Colmerauer in Marseille (Colmerauer 1973) showed how to program with another, only slightly augmented, system of symbolic logic, namely the Horn-clause-resolution form of the predicate calculus. PROLOG is essentially this system of logic, augmented by a sequentializing notion for lists of goals and lists of clauses, a flow-of-control notion consisting of a systematic depth-first, back-tracking enumeration of all deductions permitted by the logic, and a few imperative commands (such as the 'cut'). PROLOG is implemented with great elegance and efficiency using ingenious techniques originated by David H. D. Warren. The principal virtue of logic programming in either LISP or PROLOG lies in the ease of writing programs, their intelligibility, and their amenability to metalinguistic reasoning. LISP and PROLOG are usually taken as paradigms of two distinct logic programming styles (functional programming and relational programming), which on closer examination turn out to be only two examples of a single style (deductive programming). The general idea of purely descriptive deductive programming is to construe computation as the systematic reduction of expressions to a normal form. In the case of pure LISP, this means essentially the persistent application of reduction rules for processing function calls (Church's beta-reduction rule), the conditional expression, and the data-structuring operations for ordered pairs. In the case of pure PROLOG, it means essentially the persistent application of the beta-reduction rule, the rule for distributing AND through OR, the rule for eliminating existential quantifiers from conjunctions of equations, and the rules for simplifying expressions denoting sets. By merging these two formalisms one obtains a unified logical system in which both flavors of programming are available both separately and in combination with each other.

LISP, PROLOG, and their cousins have thus demonstrated the possibility, indeed the practicality, of using systems of logic directly to program computers. Logic programming is more like the formulation of knowledge in a suitable form to be used as the axioms of automatic deductions by which the computer infers its answers to the user's queries. In this sense this style of programming is a bridge linking computation in general to AI systems in particular. Knowledge is kept deliberately apart (in a 'knowledge base') from the mechanisms which invoke and apply it. Robert Kowalski's well-known equational aphorism '*algorithm = logic + control*' neatly sums up the necessity to pay attention to both descriptive and imperative aspects of a program, while keeping them quite separate from each other so that each aspect can be modified as necessary in an intelligible and disciplined way.

The classic split between procedural and declarative knowledge again shows up here: some of the variants of PROLOG (the stream-parallel, committed-choice nondeterministic languages such as ICOT's GHC) are openly concerned more with the control of events, sequences, and concurrencies than on the management of the deduction of answers to queries. The uneasiness caused by this split will remain until some way is found of smoothly blending procedural with declarative within a unified theory of computation.

Nevertheless, with the advent of logic programming in the wide sense, computer science has outgrown the idea that programs can only be the kind of action-plans required by Turing–von Neumann symbol-manipulating robots and their modern descendants. The emphasis is (for the programmer, but not yet for the machine designer) now no longer entirely on controlling the dynamic sequence of such a machine's actions, but increasingly on the static syntax and semantics of logical expressions, and on the corresponding mathematical structure of the data and other objects which are the denotations of the expressions.

It is interesting to speculate how different the history of computing might have been if in 1936 Turing had proposed a purely descriptive abstract universal machine rather than the purely imperative one that he actually did propose; or if, for example, Church had done so. We might well now have been talking of 'Church machines' instead of Turing machines.

Let us call a *Church machine* an automaton whose states are the expressions of some formal logic. Each of these expressions denotes some entity, and there is an obvious semantic notion of *equivalence* among the expressions: 'A and B are equivalent' means 'A and B denote the same entity'. For example, the expressions

$$(23 + 4)/(13 - 4), \quad 1.3 + 1.7, \quad (\lambda\ z.\ (2z + 1)^{1/2})(4)$$

are equivalent, because they all denote the number three. A Church machine computation is a sequence of its states, starting with some given state and then continuing according to the transition rules of the machine. If the sequence of states eventually reaches a terminal state, and (therefore) the computation stops, then that terminal state (expression) is the output of the machine for the initial state (expression) as input. In general the machine computes, for a given expression, another expression *which is equivalent to it* but which is *as simple as possible*. For example, the expression '3' is as simple as possible, and is equivalent to each of the above expressions, and so it would be the output of a computation starting with any of the expressions above. These simple-as-possible expressions are said to be in 'normal form'. The 'program' which determines the transitions of a Church machine through its successive states is a set of 'rewriting' rules, together with a criterion for applying some one of them to any expression. A rewriting rule is given by two expressions, called the 'redex' and the 'contractum' of the rule, and applying the rule to an expression changes (rewrites) it to another expression. The new expression is a copy of the old one, except that the new expression contains an occurrence of the contractum in place of one of the occurrences of the redex.

For example: if the initial state is $(23 + 4)/(13 - 4)$ then the transitions are:

$$\begin{array}{lll}
(23 + 4)/(13 - 4) & \text{becomes} & 27/(13 - 4), \\
27/(13 - 4) & \text{becomes} & 27/9, \\
27/9 & \text{becomes} & 3.
\end{array}$$

but if the initial state is $\lambda \mathrm{z}.(2\mathrm{z} + 1)^{1/2}$ (4), then the transitions are:

$$\begin{array}{lll}
(\lambda \mathrm{z}.(2\mathrm{z} + 1)1/2)(4) & \text{becomes} & ((2\times 4) + 1)^{1/2}, \\
((2 \times 4) + 1)^{1/2} & \text{becomes} & (8 + 1)^{1/2}, \\
(8 + 1)^{1/2} & \text{becomes} & 9^{1/2}, \\
9^{1/2} & \text{becomes} & 3.
\end{array}$$

Most of us are trained in early life to act like a simple purely arithmetical Church machine. We all learn some form of numerical rewriting rules in elementary school, and use them throughout our lives (but of course Church's lambda notation is not taught in elementary school, or indeed at any time except when people later specialize in logic or mathematics; but it ought to be). Since we cannot literally store in our heads all of the infinitely many redex-contractum pairs <23 + 4, 27>, <2+2, 4>, etc., infinite sets of these pairs are logically coded into simple finite algorithms. Each such algorithm (for addition, subtraction, and so on) yields the contractum for any given redex of its particular type. We hinted earlier that an expression is in normal form if it is as simple as possible. To be sure, that is a common way to think of normal forms, and in many cases it fits the facts. Actually, to be in normal form is not necessarily to be in as simple a form as possible. What counts as a normal form will depend on what the rewriting rules are. Normal form is a relative notion: given a set of rewriting rules, an expression in normal form is *one which contains no redex.*

In designing a Church machine care must be taken that no expression is the redex of more than one rule. The machine must also be given a criterion for deciding which rule to apply to an expression which contains distinct redexes, and also for deciding which occurrence of that rule's redexes to replace, in case there are two or more of them. A simple criterion is always to replace the leftmost redex occurring in the expression.

A Church machine, then, is a machine whose possible states are all the different expressions of some formal logic and which, when started in some state (i.e., when given some expression of that logic) will 'try' to compute its normal form. The computation may or may not terminate: this will depend on the rules, on the initial expression, and on the criterion used to select the redex to be replaced. Some of the expressions for some Church machines may have no normal form. Since for all interesting formal logics there are infinitely many expressions, a Church machine is not a finite-state automaton; so in practice the same provision must be made as in the case of the Turing machines for adjoining as much external memory as needed during a computation.

Church machines can also serve as a simple model for parallel computation and parallel architectures. One has only to provide a criterion for replacing more than one redex at the same time. In Church's lambda calculus one of the rewriting rules ('beta-reduction') is the logical version of executing a function call in a high-level programming language. Logic programming languages based on Horn-clause-resolution can also be implemented as Church machines, at least as far as their static aspects are concerned.

In the early 1960s Peter Landin, then Christopher Strachey's research assistant, undertook to convince computer scientists that not merely LISP, but also ALGOL, and indeed all past, present, and future programming languages are essentially the abstract lambda calculus in one or another concrete manifestation. One need add only an abstract version of the 'state' of the computation process and the concept of 'jump' or change of state. Landin's abstract logical model combines declarative programming with procedural programming in an insightful and natural way (Landin 1966).

Landin's thesis also had a computer-design aspect, in the form of his elegant abstract logic machine (the SECD machine) for executing lambda calculus programs. The SECD machine language is the lambda calculus itself: there is no question of 'compiling' programs into a lower-level language (but more recently Peter Henderson has described just such a lower-level SECD machine which executes compiled LISP expressions). Landin's SECD machine is a sophisticated Church machine which uses stacks to keep track of the syntactic structure of the expressions and of the location of the leftmost redex (Landin 1963).

We must conclude that the descriptive and imperative views of computation are not incompatible with each other. Certainly both are necessary. There is no need for their mutual antipathy. It arises only because enthusiastic extremists on both sides sometimes claim that computing and programming are 'nothing but' the one or the other. The appropriate view is that in all computations we can expect to find both aspects, although in some cases one or the other aspect will dominate and the other may be present in only a minimal way. Even a pure functional program can be viewed as an implicit 'evaluate this expression and display the result' imperative (as in LISP's classic read–eval–print cycle).

9 LOGIC AND ARTIFICIAL INTELLIGENCE

In AI a controversy sprang up in the late 1960s over essentially this same issue. There was a spirited and enlightening debate over whether knowledge should be represented in procedural or declarative form. The procedural view was mainly associated with Marvin Minsky and his MIT group, represented by Hewitt's PLANNER system (Hewitt 1969) and Winograd's application of it (Winograd 1972) to support a rudimentary natural language capability in his simple simulated robot SHRDLU. The declarative view was associated with Stanford's John McCarthy, and was represented by Green's QA3 system (Green 1969) and by Kowalski's eloquent advocacy of my resolution logic (Robinson 1965a, 1965b), suitably restricted to Horn clauses and applied parsimoniously according to restrictions which he and others had devised in the late 1960s and early 1970s, as a deductive programming language. Kowalski was able to make the strong case that he did because of Colmerauer's development of PROLOG as a practical *logic programming* language (Kowalski 1974). Eventually Kowalski found an elegant way to end the debate, by pointing out a procedural interpretation for the ostensibly purely declarative Horn clause sentences in logic programs[9].

There is a big epistemological and psychological difference between simply *describing* a thing and *giving explicit instructions for constructing* it, which corresponds to the difference between descriptive and imperative programming. Of course, some descriptions are already explicit recipes for construction, as for example

the square root of the sum of three cubed and three squared

But one cannot always so immediately see how to construct the denotation of a descriptive expression *efficiently*. For example, the meaning of the descriptive expression

the smallest integer which is the sum of two cubes in two different ways.

seems quite clear. We certainly understand the expression, but those who don't already (probably from reading of Hardy's famous visit to Ramanujan in hospital) know that it denotes 1729 will have to do some work, including some searching, to figure it out for themselves. It is easier to see that 1729 is the sum of two cubes in two different ways if one is given as a hint the two equations

$$1729 = 13 + 123 \qquad\qquad 1729 = 103 + 93$$

[9]Donald Michie [personal communication, 16 May 1992] writes: "The first practical demonstration of Kowalski's fundamental idea was Maarten [van Emden]'s successful implementation and execution of Quicksort using Horn-clause Logic with an SL resolution theorem prover ...It was I who pushed Maarten into the Quicksort attempt." This makes the point that at Edinburgh in the early 1970s a theorem prover based on Kowalski and Kuehner's *linear resolution with selection function*, when implemented with the *structure-sharing techniques* pioneered by Boyer and Moore, and when limited to Horn-clause problems, already amounted to the *logic programming* which was soon to be closely associated with PROLOG. One should also mention even earlier anticipations of resolution-based logic programming: the 1969 QA3 question–answering system (Green 1969), and the 1969 ABSYS system (Foster and Elcock 1969). See (Elcock 1992) and also (Kowalski 1988).

but it needs at least a little work to find these equations oneself. Then to see that 1729 is the *smallest* integer with this property, one has to see somehow that all smaller integers lack it, and this means checking each one, either literally, or by some clever shortcut. To *find* 1729, then, as the denotation of the expression, one has to carry out the all of this work, in some form or another. There are of course many different ways to organize the task, some of which are much more efficient than others, some of which are less efficient, but more intelligible, than others. So to write a general computer program which would automatically and efficiently reduce the expression

the smallest integer which is the sum of two cubes in two different ways

to the expression '1729' and equally well handle other similar expressions, is not at all a trivial task.

10 AI AND PROGRAMMING

Automatic programming has never really been that. It is no more than the automatic translation of one program into another. So there must be some kind of program (written by a human, presumably) which starts off the chain of translations. An assembler and a compiler both do the same kind of thing: each accepts as input a program written in one programming language and delivers as output a program written in another programming language, with the assurance that the two programs are equivalent in a suitable sense. The advantage of this technique is of course that the source program is usually more intelligible and easier to write than the target program, and the target program is usually more efficient than the source program because it is typically written in a lower-level language, closer to the realities of the machine which will do the ultimate work. The advent of such automatic translations opened up the design of programming languages to express 'big' ideas in a style 'more like mathematics' (as Christopher Strachey put it). These big ideas are then translated into smaller ideas more appropriate for machine languages. Let us hope that one day we can look back at all the paraphernalia of this program-translation technology, which is so large a part of today's computer science, and see that it was only an interim technology. There is no law of nature which says that machines and machine languages are condemned, intrinsically, to being low-level. Surely we must strive towards machines whose topmost levels match our own.

Turing and von Neumann both made important contributions to the beginnings of AI, although Turing's contribution is the better known. His essay *Computing Machinery and Intelligence* is surely the most quoted single item in the entire literature of AI, if only because it is the original source of the famous so-called Turing Test (Turing 1950). The recent revival of interest in artificial neural models for AI applications recalls von Neumann's deep interest in computational neuroscience, a field he richly developed in his later years and which was absorbing all his prodigious intellectual energy during his final illness. When he died in early 1957 he left behind an uncompleted manuscript which was posthumously published (von Neumann 1958) as the book *The Computer and the Brain.*

11 LOGIC AND PSYCHOLOGY IN AI

I do not mean to say that there is anything wrong with logic; I only object to the assumption that ordinary reasoning is largely based on it. *M. L. Minsky, The Society of Mind*

AI has from the beginning been the arena for an uneasy coexistence between logic and psychology as its leading themes, as epitomized in the contrasting approaches to AI of John McCarthy and Marvin Minsky. McCarthy has maintained since 1957 that AI will come only when we learn how to write programs (as he put it) which have common sense and which can take advice. His putative AI system is a (presumably) very large knowledge base made up of declarative sentences written in some suitable logic (until quite recently he has taken this to be the first order predicate calculus), equipped with an inference engine which can automatically deduce logical consequences of this knowledge (McCarthy 1958). Many well-known AI problems and ideas have arisen in pursuing this approach: the Frame Problem, Nonmonotonic Reasoning, the Combinatorial Explosion, and so on.

This approach demands a lot of work to be done on the epistemological problem of declaratively representing knowledge and on the logical problem of designing suitable inference engines. Today the latter field is one of the flourishing special subfields of AI. Mechanical theorem-proving and automated deduction have always been a source of interesting and hard problems. After over three decades of trying, we now have well-understood methods of systematic deduction which are of considerable use in practical applications.

Minsky maintains that humans rarely use logic in their actual thinking and problem solving, but adds that logic is not a good basis even for artificial problem solving — that computer programs based solely on McCarthy's logical deductive knowledge-base paradigm will fail to display intelligence because of their inevitable computational inefficiencies; that the predicate calculus is not adequate for the heuristically effective representation of most knowledge; and that the exponential complexity of predicate calculus proof procedures will always severely limit what inferences are possible.

Because it claims little or nothing, the view can hardly be refuted that humans undoubtedly are in some sense (biological) machines whose design, though largely hidden from us at present and obviously exceedingly complicated, calls for some finite arrangement of material components all built ultimately out of 'mere' atoms and molecules and obeying the laws of physics and chemistry. So there is an abstract design which, when physically implemented, produces (in ourselves, and the animals) intelligence. Intelligent machines can, then, be built. Indeed, they can, and do routinely, build and repair themselves, given a suitable environment in which to do so. Nature has already achieved NI — natural intelligence. Its many manifestations serve the AI research community as existence proofs that intelligence can occur in physical systems. Nature has already solved all the AI problems, by sophisticated schemes only a very few of which have yet been understood.

12 THE STRONG AI THESIS

Turing believed, indeed was the first to propound, the Strong AI thesis that artificial intelligence can be achieved simply by appropriate programming of his universal computer. Turing's Test is simply a detection device, waiting for intelligence to occur in machines: if a machine is one day programmed to carry on fluent and intelligent-seeming conversations, will we not, argued Turing, have to agree that this intelligence, or at least this apparent intelligence, is a property of the program? What is the difference between apparent intelligence, and intelligence itself? The Strong AI thesis is also implicit in McCarthy's long-pursued project to reconstruct artificially something like human intelligence by implementing a suitable formal system. Thus the Turing Test might (on McCarthy's view) eventually be passed by a deductive knowledge base, containing a suitable repertory of linguistic and other everyday human knowledge, and an efficient and sophisticated inference engine. The system would certainly have to have a mastery of (both speaking and understanding) natural language. Also it would have to exhibit to a sufficent degree the phenomenon of 'learning' so as to be capable of augmenting and improving its knowledge base to keep it up-to-date both in the small (for example in dialog management) and in the large (for example in keeping up with the news and staying abreast of advances in scientific knowledge). In a recent vigorous defense of the Strong AI thesis, Lenat and Feigenbaum argued that if enough knowledge of the right kind is encoded in the system it will be able to 'take off' and autonomously acquire more through reading books and newspapers, watching TV, taking courses, and talking to people (Lenat and Feigenbaum 1991).

It is not the least of the attractions of the Strong AI thesis that it is empirically testable. We will know if someone succeeds in building a system of this kind: that indeed is what Turing's Test is for.

13 EXPERT SYSTEMS

Expert systems are limited-scale attempted practical applications of McCarthy's idea. Some of them, such as the Digital

Equipment Corporation's XCON system for configuring VAX computing systems (Bachant and McDermott 1984), and the highly specialized medical diagnosis systems, such as KARDIO (Bratko *et al.* 1989), have been quite useful in limited contexts, but there have not been as many of them as the more enthusiastic proponents of the idea might have wished. The well-known book by Feigenbaum and McCorduck on the Fifth Generation Project was a spirited attempt to stir up enthusiasm for Expert Systems and Knowledge Engineering in the United States by portraying ICOT's mission as a Japanese bid for leadership in this field (Feigenbaum and McCorduck 1983).

There has indeed been much activity in devising specialized systems of applied logic whose axioms collectively represent a body of expert knowledge for some field (such as certain diseases, their symptoms, and treatments) and whose deductions represent the process of solving problems posed about that field (such as the problem of diagnosing the probable cause of given observed symptoms in a patient). This, and other, attempts to apply logical methods to problems which call for inference-making, have led to an extensive campaign of reassessment of the basic classicial logics as suitable tools for such a purpose. New, nonclassical logics have been proposed (fuzzy logic, probabilistic logic, temporal logic, various modal logics, logics of belief, logics for causal relationships, and so on) along with systematic methodologies for deploying them (truth maintenance, circumscription, nonmonotonic reasoning, and so on). In the process, the notion of what is a logic has been stretched and modified in many different ways, and the current picture is one of busy experimentation with new ideas.

14 LOGIC AND NEUROCOMPUTATION

Von Neumann's view of AI was a 'logico-neural' version of the Strong AI thesis, and he acted on it with typical vigor and scientific virtuosity. He sought to formalize, in an abstract model, aspects of the actual structure and function of the brain and nervous system. In this he was consciously extending and improving the pioneering work of McCullogh and Pitts, who had described their model as 'a logical calculus immanent in nervous activity'. Here again, it was logic which served as at least an approximate model for a serious attack on an ostensibly nonlogical problem. A particularly interesting example of his foray into the practical engineering of artificial neural networks can be found in his demonstration of how reliability (fault tolerance) can be obtained by proper design methods even when the elementary neuron devices are prone to error (von Neumann 1956).

Von Neumann's logical study of self-reproduction as an abstract computational phenomenon was not so much an AI investigation as an essay in quasi-biological information processing. It was certainly a triumph of abstract logical formalization of an undeniably computational process. The self-reproduction method evolved by Nature, using the double-helix structure of paired complementary coding sequences found in the DNA molecule, is a marvellous solution of the formal problem of self-reproduction. Von Neumann was not aware of the details of Nature's solution when he worked out his own logical, abstract version of it as a purely theoretical construction, shortly before Crick and Watson unravelled the structure of the DNA molecule in 1953. Turing, too, was working at the time of his death on another, closely-related problem of theoretical biology — morphogenesis — in which one must try to account theoretically for the unfolding of complex living structural organizations under the control of the programs coded in the genes. This is not exactly an AI problem. One cannot help wondering whether Turing may have been disappointed, at the end of his life, with his lack of progress towards realizing AI. If one excludes some necessary philosophical clarifications and preliminary methodological discussions, nothing had been achieved beyond his invention of the computer itself.

The empirical goal of finding out how the human mind actually works, and the theoretical goal of reproducing its essential features in a machine, are not much closer in the early 1990s than they were in the early 1950s. After forty years of hard work we have 'merely' produced some splendid tools and thoroughly explored plenty of blind alleys. We should not be surprised, or even disappointed. The problem is a very hard one.

The same thing can be said about the search for controlled thermonuclear fusion, or for a cancer cure. Our present picture of the human mind is summed up in Minsky's recent book *The Society of Mind*, which offers a plausible general view of the mind's architecture, based on clues from the physiology of the human brain and nervous system, the computational patterns found useful for the organization of complex semantic information-processing systems, and the sort of insightful interpretation of observed human adult- and child-behavior which Freud and Piaget pioneered. Logic is given little or no role to play in Minsky's view of the mind (Minsky 1985).

Minsky rightly emphasizes (as logicians have long insisted) that the proper role of logic is in the context of justification rather than in the context of discovery. Newell, Shaw, and Simon's well-known propositional calculus theorem-proving program, the Logic Theorist, illustrates this distinction admirably (Newell *et al.* 1957). The Logic Theorist is a discovery simulator. The goal of their experiment was to make their program discover a proof (of a given propositional formula) by 'heuristic' means, reminiscent (they supposed) of the way a human would attack the same problem. As an algorithmic theorem-prover (one whose goal is to show formally, by any means, and presumably as efficiently as possible, that a given propositional formula is a theorem) their program performed nothing like as well as the best nonheuristic algorithms. The logician Hao Wang soon rather sharply pointed this out (Wang 1960), but it seems that the psychological motivation of their investigation had eluded him (as indeed it has many others). They themselves very much muddled the issue by contrasting their heuristic theorem-proving method with a ridiculously inefficient, purely fictional, 'logical' one which consisted of enumerating all possible proofs in lexicographical order and waiting for the first one to turn up with the desired proposition as its conclusion. This Swiftian parody may have been no more than a rhetorical flourish which got out of control, but it strongly suggested that they believed it really is more efficient to seek proofs heuristically, as in their program, than algorithmically with a guarantee of success. Indeed in the exuberance of their comparison they also provocatively coined the wicked but amusing phrase 'British Museum algorithm' for this lexicogaphic-enumeration-of-all-proofs method — the intended sting in the epithet being that just as, *given enough time*, a systematic lexicographical enumeration of all possible texts will of course eventually succeed in listing any given text in the vast British Museum Library, so a logician, *given enough time*, must eventually succeed in proving any given provable proposition by proceeding along similar lines. Their implicit thesis was that *a proof-finding algorithm which is guaranteed to succeed for any provable input is necessarily unintelligent.* This may well be so: but that is not at all the same as saying that it is necessarily inefficient.

Interestingly enough, something like this thesis was anticipated in (Turing 1947):

> ...if a machine is expected to be infallible, it cannot also be intelligent. There are several mathematical theorems which say almost exactly that.

15 CONCLUSION

Logic's abstract conceptual gift of the universal computer has needed to be changed remarkably little since 1936. Until very recently, all universal computers have been realizations of the same abstraction. Minor modifications and improvements have of course been made, perhaps the most striking one being internal memories organized into addressable cells, designed to be randomly accessible, rather than merely sequentially searchable (although external memories remain essentially sequential, requiring search). Other improvements have consisted largely of building into the finite hardware some of the functions which would otherwise have to be carried out by software (although in the recent RISC architectures this trend has actually been reversed). For over fifty years, successive models of the basic machine have been 'merely' faster, cheaper, physically smaller versions *of the same device.* In the past, then, computer science has pursued what is in fact an essentially *logical* quest: exploration of the literally unbounded possibilities of von Neumann's version of the universal Turing machine. The technological challenge, of continuing to improve its physical realizations, has been largely left to the electrical engineers and computer architects, who have performed miracles and show every sign of continuing to do so.

In the future, we must hope that the logician and the engineer will find it possible and natural to work even more closely together to devise new kinds of *higher-level* computing machines which, by making programming easier and more natural, will help to bring artificial intelligence closer. That future has been under way for at least the past decade. Today we are already beginning to explore the possibilities of, for example, the Connection Machine (Hillis 1985), various kinds of neural network machines, and massively parallel machines for logical knowledge-processing.

It is this future that the bold and imaginative Fifth Generation Project has been all about. Japan's ten-year-long ICOT-based effort has stimulated (and indeed challenged) many other technologically advanced countries to undertake ambitious logic-based research projects in computer science. As a result of ICOT's international leadership and example, the computing world has been reminded not only of how central the role of logic has been in the past, as generation has followed generation in the modern history of computing, but also of how important a part it will surely play in the generations yet to come.

REFERENCES

Aspray, W., and Burks, A. W. (1987). Papers of John von Neumann on *Computing and Computer Science*. MIT Press.

Aspray, W. (1990). *John von Neumann and the Origins of Modern Computing*. MIT Press.

Bachant, J., and McDermott, J. (1984). *R1 revisited: four years in the trenches.* The AI Magazine, *Volume 5*, Number 3, Fall 1984.

Bratko, I., Mozetic, I., and Lavrac, N. (1989). *KARDIO: a study in Deep and Qualitative Knowledge for Expert Systems*. MIT Press.

Carpenter, B. E. and Doran, R. W. (editors) (1946). *A. M. Turing's ACE Report of 1946 and other Papers*. MIT Press.

Church, Alonzo (1936). *An unsolvable problem of elementary number theory.* American Journal of Mathematics, *58*, pp. 345 – 363.

Church, Alonzo (1936). *A note on the Entscheidungsproblem.* Journal of Symbolic Logic, *Volume 1* (1936), pp. 40 – 41. Correction, ibid., pp. 101 – 102.

Church, Alonzo (1941). *The calculi of lambda-conversion.* Annals of Mathematics Studies *Number 6.* Princeton University Press.

Colmerauer, A., Kanoui, H., Pasero, R., and Roussel, P. (1973). *Un Systeme de communication homme-machine en Francais.* Groupe d'Intelligence Artificielle, Universite d'Aix Marseille II, Luminy, France.

Davis, Martin (1965). *The Undecidable.* Evergreen Press.

Feigenbaum, Edward A., and McCorduck, Pamela (1983). *The Fifth Generation: Artifical Intelligence and Japan's Computer Challenge to the World.* Addison-Wesley.

Foster, J. M., and Elcock, E. W. (1969). *Absys-1: an incremental compiler for assertions Q– an introduction.* In Machine Intelligence, 4, edited by D. Michie, pp. 423 – 429.

Gödel, Kurt (1931). *Uber formal unentscheidbare Satze der Principia Mathematica und verwandte Systeme I* . Monatshefte fØr Mathematik und Physik, *38*, pp. 173 – 198. English translations in (Davis 1965), (van Heijenoort 1971).

Green, C. C. (1969). *Application of Theorem-proving to Problem-solving.* Proceedings of IJCAI 1 (Walker, D. E., and Norton, L. M., editors), Washington D. C., 1969, pp. 219 – 240.

Henderson, Peter (1980). *Functional Programming.* Prentice-Hall.

Hewitt, C. (1969). *PLANNER: a Language for proving Theorems in Robots.* Proceedings of IJCAI 1 (Walker, D. E., and Norton, L. M., editors), Washington D. C., pp. 295 – 301.

Hillis, W. D. (1985). *The Connection Machine.* MIT Press.

Hodges, Andrew (1983). *Alan Turing: The Enigma.* Simon and Schuster.

Kowalski, R. A. (1984). *Predicate calculus as a programming language.* IFIP 1974 Proceedings, North Holland, pp. 569 – 574.

Kowalski, R. A. (1988). *The early years of Logic Programming.* Communications of the Association for Computing Machinery, *31*, pp. 38 – 42.

Landin, P. J. (1963). *The Mechanical Evaluation of Expressions.*

Computer Journal, *6*, pp. 308 – 320.

Landin, P. J. (1966). *The Next 700 Programming Languages*. Communications of the Association for Computing Machinery, *9*, pp. 157 – 166.

Lenat, Douglas B., and Feigenbaum, Edward A. (1991). *On the Thresholds of Knowledge*. Artificial Intelligence, *47*, pp. 185 – 250.

Macrae, Norman (1992). *John von Neumann*. Pantheon Books.

McCarthy, John (1958). *Programs with Common Sense*. Proceedings of a Symposium on the Mechanisation of Thought Processes, *Volume 1*, pp. 77 – 84. National Physical Laboratory, London.

McCarthy, John (1960). *Recursive Functions of Symbolic Expressions and their Computation by Machine*. Communications of the Association for Computing Machinery, *3*, pp. 184 – 195.

McCullogh, W. S., and Pitts, W. (1943). *A Logical Calculus of Ideas Immanent in Nervous Activity*. Bulletin of Mathematical Biophysics, *5*, pp. 115 – 133.

Minsky, Marvin (1985). *The Society of Mind*. Simon and Schuster.

Newell, A., Shaw, J. C., and Simon, H. A. (1957). *Empirical explorations with the logic theory machine: a case study in heuristics*. Proceedings of the Western Joint Computer Conference, pp. 218 – 239.

Randell, Brian (1972). *On Alan Turing and the Origins of Digital Computers*. Machine Intelligence, *7*, pp. 3 – 20.

Randell, Brian (1982) (editor). *The Origins of Digital Computers*. Third Edition, Springer Verlag.

Robinson, J. A. (1965) *A machine-oriented logic based on the resolution principle*. Journal of the Association for Computing Machinery, *12*, pp. 23 – 41.

Robinson, J. A. (1965). *Automatic deduction with hyper-resolution*. International Journal of Computer Mathematics, *1*, pp. 227 – 234.

Shannon, Claude E., and McCarthy, John (1956) (editors). *Automata Studies*. Annals of Mathematics Studies *Number 34*. Princeton University Press.

Stern, Nancy (1981). *From ENIAC to UNIVAC. An Appraisal of the Eckert–Mauchly Computers*. Digital Press.

Taub, A. H. (1960) (editor). *John von Neumann: Collected Works*.

6 volumes. Macmillan, 1960 – 1963.

Turing, Alan Mathison (1936). *On Computable Numbers, with an application to the Entscheidungsproblem.* Proceedings of the London Mathematical Society, series 2, *Volume 42* (1936–37), pp. 230 –265. A Correction, ibid., *Volume 43* (1937), pp. 544 – 546.

Turing, Alan Mathison (1945). *Proposal for Development in the Mathematics Division of an Automatic Computing Engine (ACE).* Completed by the end of 1945[10]. Published in (Carpenter and Doran 1986), pp. 20 – 105.

Turing, Alan Mathison (1947). *Lecture to the London Mathematical Society on 20 February 1947.* Published in (Carpenter and Doran 1986), pp. 106 – 124.

Turing, Alan Mathison (1947). *Intelligent Machinery.* Written in September 1947. First published in Machine Intelligence *5*, edited by Bernard Meltzer and Donald Michie, Edinburgh University Press, 1969.

Turing, Alan Mathison (1949). *Checking a Large Routine.* In The Collected Works of A. M. Turing, *Volume 1*, Mechanical Intelligence, 1992, pp. 129 – 131.

Turing, Alan Mathison (1950). *Computing Machinery and Intelligence* . Mind, *Volume 59*, Number 236.

van Heijenoort, Jean (1971) (editor). *From Frege to Gödel: A Source Book in Mathematical Logic*, 1879 – 1931. Harvard University Press, 1967.

von Neumann, John (1945). *First Draft of a Report on the EDVAC* Written in June 1945. Moore School of Electrical Engineering, University of Pennsylvania, Contract No. W-670-ORD-4926, June 1945. Excerpts in (Randell 1982). First full publication in (Stern 1981). Also in (Aspray–Burks 1987), pp. 17 – 82.

Burks, Arthur W., Goldstine, Hermann H., and von Neumann, John (1946). *Preliminary Discussion of the Logical Design of an Electronic Computing Instrument.* Part I, *Volume 1* of a Report prepared for U.S. Army Ordnance Department under contract W-36-034-ORD-7481. (Taub 1960), *Volume 5*, pp. 34 – 79. Also in (Aspray–Burks 1987), pp. 97 – 142.

Goldstine, Hermann H., and von Neumann, John (1947). *Planning*

[10]Cf. (Hodges 1983) note 6.5, p. 557.

and Coding of Problems for an Electronic Computing Instrument. Part II, *Volume 1* of a Report prepared for U.S. Army Ordnance Department under contract W-36-034-ORD-7481. (Taub 1960), Volume 5, pp. 80 – 151. Also in (Aspray–Burks 1987), pp. 553 – 602.

von Neumann, John (1956). *Probabilisitic Logics and the Synthesis of Reliable Organisms from Unreliable Components.* In (Shannon and McCarthy 1956), pp. 43 – 98. Also (Taub 1960), *Volume 5*, pp. 329 – 378, and (Aspray – Burks 1987), pp. 553 – 602.

von Neumann, John (1958). *The Computer and the Brain.* Yale University Press, 1958.

Wang, Hao (1960). *Towards Mechanical Mathematics.* IBM Journal of Research and Development, *4*, pp. 2 – 22.

Winograd, T. (1972). *Understanding Natural Language.* Academic Press.

Woodger, M. (1958). *The History and Present Use of Digital Computers at the National Physical Laboratory.* Process Control and Automation, November 1958, pp. 437 – 442. Reprinted in (Carpenter and Doran 1986).

2

Logic and learning: Turing's legacy

S. Muggleton

Oxford University Computing Laboratory,
11 Keble Road,
Oxford, OX1 3QD,
UK.

Abstract

Turing's best known work is concerned with whether universal machines can decide the truth value of arbitrary logic formulae. However, in this paper it is shown that there is a direct evolution in Turing's ideas from his earlier investigations of computability to his later interests in machine intelligence and machine learning. Turing realised that machines which could learn would be able to avoid some of the consequences of Gödel's and his results on incompleteness and undecidability. Machines which learned could continuously add new axioms to their repertoire. Inspired by a radio talk given by Turing in 1951, Christopher Strachey went on to implement the world's first machine learning program. This particular first is usually attributed to A.L. Samuel. Strachey's program, which did rote learning in the game of Nim, preceded Samuel's checker playing program by four years. Neither Strachey's nor Samuel's system took up Turing's suggestion of learning logical formulae. Developments in this area were delayed until Gordon Plotkin's work in the early 1970's. Computer-based learning of logical formulae is the central theme of the research area of Inductive Logic Programming, which grew directly out of the earlier work of Plotkin and Shapiro. In the present paper the author describes the state of this new field and discusses areas for future development.

1 ALAN TURING AND THE HISTORY OF LOGIC AND LEARNING

1.1 The Hilbert program

At the 1928 International Mathematical Congress David Hilbert, one of the greatest mathematicians of the previous thirty years, set out three central questions for logic and mathematics. Was mathematics

1. *complete* in the sense that every mathematical statement could either be proved or disproved,
2. *consistent* in the sense that false statements could never be derived by a sequence of valid steps and
3. *decidable* in the sense that there existed a definite method which could decide the truth or falsity of any mathematical assertion?

Hilbert expected a positive answer to all three questions. Within three years Kurt Gödel (1931) had shown that not even arithmetic could be both complete and consistent. Within a decade both Alonzo Church (1936) and Alan Turing (1936) had shown the undecidability of certain mathematical assertions.

Turing's solution to this problem was based on defining a machine which emulated an ideal human computer who calculated with a pen and a one-dimensional roll of paper. A universal machine was one which could, when loaded with the appropriate definitions, simulate any other computing machine. Turing had purposefully devised a machine which had abilities equivalent to a human computer. However Gödel's result had been proved by showing that certain statements which were evidently true to a human could not be proved within a limited logical system. It seems that Turing noticed the clash for in his PhD. thesis (1939) on ordinal logics Turing attempted to circumvent Gödel's result. The idea was to introduce a set of 'oracles', each capable of deciding the truth of unprovable statements. The use of such oracles would allow for a complete logic. However they required an element of non-mechanical *intuition*. This approach failed to reconcile Turing's belief in purely mechanical intelligence with Gödel's incompleteness result.

1.2 The war period

It would appear that an alternative solution to the incompleteness problem presented itself to Turing through his war-time work as a cryptographer. Later (Turing, 1948) he was to say that

> There is a remarkably close parallel between the problems of the physicist and those of the cryptographer. The system on which a message is enciphered corresponds to the laws of the universe, the intercepted messages to the evidence available, the keys for a day or a message to important constants which have to be determined.

Clearly, the physicist, like the cryptographer is continuously completing his theories. By learning from experience intelligent machinery should be capable of avoiding many of the problems of Gödel's incompleteness results. The area of scientific discovery alluded to by Turing in the above quote is the theme of this present Machine Intelligence volume. The paper in this volume by Sternberg et al. (1993) describes recent progress in applying machine learning techniques to discovery of new scientific knowledge.

At Bletchley Park, in 1943, in numerous out-of-hours contexts Turing discussed the problem of machine intelligence with both Donald Michie and Jack Good. According to Andrew Hodges (1985), Turing's biographer,

> These meetings were an opportunity for Alan to develop the ideas for chess-playing machines that had begun in his 1941 discussion with Jack Good. They often talked about mechanisation of thought processes, bringing in the theory of probability and weight of evidence, with which Donald Michie was by now familiar. ... He (Turing) was not so much concerned with the building of machines designed to carry out this or that complicated task. He was now fascinated with the idea of a machine that could *learn*.

According to Michie, Turing put more effort into thinking about learning than anyone else in the group. He circulated for comment an unpublished typescript covering his ideas. Unfortunately it appears that this manuscript has not survived, though

Michie believes it to have been a conceptual predecessor of Turing's N.P.L. report (1948) (see section 1.5).

1.3 The Pilot ACE

At the end of the war Alan Turing joined in the race to design and implement the first general-purpose stored-program computing machine. In 1945, when working at the National Physical Laboratory, Turing authored the ACE report (Turing, 1946), describing a proposed 'large scale electronic digital computing machine'. The machine and its implications were described by Turing (1947) to an audience at the London Mathematical Society on 20th February 1947. The end of the lecture was dedicated to the problem of machine learning.

> Let us suppose we have set up a machine with certain initial instruction tables, so constructed that these tables might on occasion, if good reason arose, modify those tables. One can imagine that after the machine had been operating for some time, the instructions would have altered out of all recognition, but nevertheless still be such that one would have to admit that the machine was still doing very worthwhile calculations. Possibly it might still be getting results of the type desired when the machine was first set up but in a much more efficient manner.

In present day jargon this is known as speed-up learning. The machine does not increase its set of provable statements, but increases the efficiency of making the proofs. Turing goes on as follows

> In such a case one would have to admit that the progress of the machine had not been foreseen when its original instructions were put in. It would be like a pupil who had learnt much from his master, but had added much more by his own work. When this happens I feel that one is obliged to regard the machine as showing intelligence. As soon as one can provide a reasonably large memory capacity it should be possible to begin to experiment on these lines. .. One might reasonably hope to be able to make some real progress with a few million digits, especially if one confined one's investigation to some rather limited field such

> as the game of chess. It would probably be quite easy to find instruction tables which would enable the ACE to win against an average player. Indeed Shannon of Bell Telephone laboratories tells me he has won games playing by rule of thumb: the skill of his opponents is not stated. But I would not consider such a victory very significant. What we want is a machine that can learn from experience. The possibility of letting the machine alter its own instructions provides the mechanism for this, but this of course does not get us very far.

According to Turing learning is an indicator of intelligence. He realised that most experimentation with machine learning would have to wait until machines had a few million bytes of memory. Such a size of core-memory was not common until the 1970's. Turing also foresaw that chess would be an ideal domain for testing machine learning programs. Games, especially chess, have been extensively used for testing machine learning.Turing follows this passage by resolving Gödel's incompleteness problem in a discussion of interactive inductive machine learning.

> It has for instance been shown that with certain logical systems there can be no machine which will distinguish provable formulae of the system from unprovable, .. Thus if a machine is made for this purpose it must in some cases fail to give an answer. On the other hand if a mathematician is confronted with such a problem he would search around and find new methods of proof, .. fair play must be given to the machine. Instead of it sometimes giving no answer we could arrange that it gives occasional wrong answers. .. if a machine is expected to be infallible, it cannot also be intelligent. There are several mathematical theorems which say almost exactly that. But these theorems say nothing about how much intelligence may be displayed if a machine makes no pretence at infallibility. .. No man adds very much to the body of knowledge, why should we expect more of a machine? .. the machine must be allowed to have contact with human beings in order that it may adapt itself to their standards.

Interestingly, Shapiro's Model Inference System (Shapiro, 1983), developed in the early 1980's, works just like this. A Prolog program, which is a set of logical assertions, is interactively debugged by checking inferences against a human oracle. Shapiro divides all possible bugs in the program into three categories: incompleteness, incorrectness and non-termination. The correspondence with Hilbert's three questions about mathematics is striking, though this is not commented on by Shapiro.

1.4 Altering the program

Around this time Turing entered into an early discussion on the relative merits of neural-net learning and logic-based learning. The question is whether the structure of neural hardware is a prerequisite for machines to learn. In a letter dated 20th November 1946 (Hodges, 1985) to W. Ross Ashby, Turing says

> It would be quite possible for the machine to try out variations of behaviour and accept or reject them in the manner you describe and I have been hoping to make the machine do this. This is possible because, without altering the design of the machine itself, it can, in theory at any rate, be used as a model of any other machine, by making it remember a suitable set of instructions .. at worst at the expense of operating slightly slower than a machine specially designed for the purpose in question. Thus, although the brain may in fact operate by changing its neuron circuits by the growth of axons and dendrites, we could nevertheless make a model, within the ACE, in which the possibility was allowed for, .., instead of building a special machine.

Turing clearly sees the necessity of making use of an interpreter in machine learning. This allows the learning program to be distinguished from the program being learned and is nowadays standard practice. The idea of using an interpreter within learning may seem obvious now but it solved a paradox going back over a hundred years. In 1842, Ada Lovelace (Bowden, 1953), describing Babbage's planned Analytical Engine noted that

> The Analytical Engine has no pretensions whatever to *originate* anything. It can do *whatever we know how to order it* to perform.

In the following passage from his Mind article (Turing, 1950) Turing describes how secondary (interpreted) rules can be originated.

> How can the rules of a machine change? They should describe completely how the machine will react whatever its history might be, whatever changes it might undergo. The rules are thus quite time invariant .. The explanation of the paradox is that the rules which get changed in the learning process are of a rather less pretentious kind, claiming only an ephemeral validity. The reader may draw a parallel with the Constitution of the United States.

1.5 The 1948 NPL report

Prior to his resignation from the National Physical Laboratory, Turing submitted a report (Turing, 1948) devoted almost entirely to his ideas on machine learning. Michael Woodger says that Turing's report caused a furore at N.P.L with his prognostications of intelligent machinery. 'Turing is going to infest the countryside' some declared 'with a robot which will live on twigs and scrap iron'. They were reacting to the following section in Turing's report.

> In order that the machine should have a chance of finding things out for itself it should be allowed to roam the countryside, and the danger to the ordinary citizen would be serious. .. although this method is probably the 'sure' way of producing a thinking machine it seems to be altogether too slow and impractical.

Though machines learning from interacting with the real world must have sounded wild at the time, it is similar in spirit to the experiments in learning to fly a plane and balancing a pole described in this volume by Michie and Camacho (1993) and Sammut (Sammut, 1993). The present work grew out of investigations by Michie and Chambers (1968). Turing cautions about the danger to citizens and the slowness and impracticality of on-line learning from real-world data. The research reported in this volume avoids this problem by carrying out all learning within the safe confines of a simulator.

Much of what Turing wrote in the 1948 report is still relevant to present research in machine learning. Returning once more to the connection between learning and Gödel's incompleteness theorem Turing writes as follows.

> The argument from Gödel's and other theorems rests essentially on the condition that the machine must not make mistakes .. Gauss was asked at school to do the addition 15+18+21+...+54 .. and immediately wrote down 483, presumably calculating it as (15+54)(54-12)2.3. .. imagine a situation where the children were given a number of additions to do, of which the first 5 were all arithmetic progressions, but the 6th was say 23+34+45+ ... +100+112+122+ ... 199. Gauss might have given the answer to this as if it were an arithmetic progression, not having noticed that the 9th term was 112 instead of 111.

This is a neat exposition of the problem of noise in learning. The question is whether it is better to learn a simple rule which correctly covers 5 out of the 6 progressions or to make an exception of the last incorrectly coded progression. This topic is addressed in the papers by Tsukimoto and Morita (1993) and Srinivasan et al. (1993) in this volume.

In the report Turing describes the development of the human infant cortex as that of transforming an unorganised machine into a universal one. He then goes on to describe an experiment in which, using a hand-simulated program, he managed to train an unorganised state-transition machine to become a universal Turing machine. The result is a machine that *learns how to accept and interpret instructions.* This goes beyond the aspirations of almost all modern machine learning algorithms. Commenting on this experiment Turing notes that

> One particular kind of phenomenon I had been hoping to find in connection with the P-type machines. This was the incorporation of old routines into new. One might have 'taught' (i.e. modified or organise) a machine to add (say). Later one might teach it to multiply by small numbers by repeated addition and so arrange matters that the same set of situations which formed the addition routine, as originally taught, was also used in the additions involved in the multiplication.

Here Turing is discussing the problem of incremental learning with background knowledge. The discussion of learning plus and then multiply is very reminiscent of Sammut's Marvin program

(Sammut and Banerji, 1986) which was taught in exactly this way.

In the following passage Turing discusses child development with relationship to machine learning.

> The training of the human child depends largely on a system of rewards and punishments, and this suggests that it ought to be possible to carry through the organising with only two interfering inputs, one for 'pleasure' or 'reward' (R) and the other for 'pain' or 'punishment' (P).

This is equivalent to the use of 'positive' and 'negative' examples in supervised machine learning. Later in his Mind article (Turing, 1950) Turing commented on this system.

> The use of punishments and rewards can at best be a part of the teaching process. Roughly speaking, if the teacher has no other means of communicating to the pupil, the amount of information which can reach him does not exceed the total number of rewards and punishments applied. .. It is necessary therefore to have some other 'unemotional' channels of communication. If these are available it is possible to teach a machine by punishments and rewards to obey orders given in some language, e.g. symbolic language. .. The use of this language will diminish greatly the number of punishments and rewards required.

There is a clear understanding here of the information amplifying effect of symbolically encoded background knowledge. Without it one gets only one bit of information per example. The 1948 N.P.L. report makes clear the kind of symbolic language that Turing intended.

> Starting with a UPCM (Universal Practical Computing Machine) we first put a program into it which corresponds to building in a logical system (like Russell's *Principia Mathematica*). This would not determine the behaviour of the machine completely ..

The language intended for learning is predicate calculus. This is the basis of what is now called 'Inductive Logic Programming' (ILP). Rather than building in Russell's Principia, researchers in this area generally make use of a Prolog theorem-proving interpreter. This provides the power of first-order predicate calculus through its use of methods based on Alan Robinson's (1965) resolution theorem-proving. ILP is discussed in Section 2.

1.6 Christopher Strachey and the first machine learning program

On 15th May 1951 Turing gave a lecture entitled 'Can Digital Machines Think' on the BBC's Third Programme. According to Hodges (1985)

> This short talk did not include any details of how he proposed to program a machine to think, beyond the remark that 'it should bear a close relation to that of teaching.' This comment sparked off an immediate reaction in a listener: Christopher Strachey ..

In May 1951 Strachey was working on writing a draught's program on the recently[1] working Pilot ACE machine. On the evening of the broadcast Strachey wrote to Turing.

> ...The essential thing which would have to be done first, would be to get the machine to programme itself from very simple and general input data ... It would be a great convenience to say the least if the notation chosen were intelligible as mathematics ... once the suitable notation is decided, all that would be necessary would be to type more or less ordinary mathematics and a special routine called, say, 'Programme' would convert this into the necessary instructions to make the machine carry out the operations indicated. This may sound rather Utopian, but I think it, or something like it, should be possible, and I think it would open the way to making a simple learning programme. .. as soon as I have finished the Draughts programme I intend to have a shot at it.

Strachey decided to test his ideas on learning with the game of Nim. In this game three piles of matches are laid out, and two players take turns to remove as many matches as they want from any pile. A non-mathematical friend of Strachey's had

[1]According to Mike Woodger (1958) the first program ran on the Pilot ACE in May 1950. However, the Pilot ACE operated without component error for more than half an hour for the first time in September 1950. The first 'large' program, which solved 17 simultaneous linear equations, ran on 26th June 1951.

noticed that any player who could achieve the position (n,n,0) had won, since it was only necessary to copy the opponents moves to reduce the heaps to (0,0,0). According to Hodges (1985)

> He (Strachey) had worked out a program which could keep a record of winning positions, and so improve its play by experience, but it could only store them individually, as (1,1,0), (2,2,0) and so on. This limitation soon allowed his novice friend to beat the program.

Strachey wrote

> This shows very clearly, I think, that one of the most important features of thinking is the ability to spot new relationships when presented with unfamiliar material ...

Strachey's was a simple rote learning program. It appears to the author to be the first implemented machine learning program. Strachey's comments show that he had understood the importance of generalisation within machine learning. In machine learning the patterns of (1,1,0) and (2,2,0) can be generalised to (n,n,0) using Plotkin's (1969) least general generalisation operator. Interestingly, Strachey's Nim rule is almost isomorphic to the colinearity rule in the King-Rook-King illegality domain described in (Muggleton *et al.* 1989). Machine learning of the colinearity rule has been shown (Muggleton *et al.* 1989) to require relational learning (ILP).

Later A.L. Samuel, in the USA, was also to start work on a draughts (or checkers) playing program. According to Donald Michie, Samuel inherited Strachey's draughts playing program. Samuel published a technical report in 1955 in which he had incorporated Strachey's ideas on rote learning into the checker player. In a later report (Samuel, 1959) the rote learning approach was extended by the use of parameter learning. Samuel (1967) extended this method even further to incorporate elements of simple logic learning.

2 INDUCTIVE LOGIC PROGRAMMING

In this volume Alan Robinson has the opening paper (Robinson, 1993) on Turing and the history of computation. It is a great honour to have been asked to write a related paper on Turing and machine learning. As every computer scientist knows Alan Robinson's (1965) paper on machine-oriented theorem proving has had and continues to have an enormous effect on computer science and artificial intelligence. The modern subject of Logic Programming is based on Robinson's theorem-proving techniques. Prolog is the language at the centre of Logic Programming.

For the following reasons pure Prolog is also an almost ideal target language for symbolic learning.

- Prolog has the expressiveness of a substantial subset of first-order predicate calculus. It is thus capable of expressing grammars, plans, mathematical and scientific theories as well as arbitrary computer programs.
- Prolog programs consist of conjunctions of clauses. Logical conjunction is both associative and commutative. Because of this clauses represent independent axioms which can be added to in any order.
- Pure Prolog programs, treated as sets of logical clauses, have a clear and simple semantics (LLoyd, 1984).
- Prolog can be efficiently interpreted using SLD resolution. The importance of an interpreter within a learning system was noted in Section 1.4.

Inductive Logic Programming (ILP) is a research area formed at the intersection of Machine Learning and Logic Programming. ILP systems develop predicate descriptions from examples and background knowledge. The examples, background knowledge and final descriptions are all described as logic programs. A unifying theory of Inductive Logic Programming is being built up (Muggleton, 1991) around lattice-based concepts such as refinement (Shapiro, 1983; Dzeroski and Lavrac, 1992) least general generalisation (Plotkin, 1971; Muggleton, 1991) inverse resolution (Muggleton and Buntine, 1988) and most specific corrections (Bain and Muggleton, 1991) In addition to a well established tradition of learning-in-the-limit convergence results (Plotkin, 1971; Shapiro, 1983; Deraedt and Bruynooghe, 1992) some results within Valiant's PAC-learning framework have been demonstrated for ILP systems (Page, 1992; Dzeroski, Muggleton and Russell, 1992)

2.1 Theory

In the general setting an ILP system S will be given a logic program B representing background knowledge and a set of positive and negative examples $\langle E^+, E^- \rangle$, typically represented as ground literals. In the case in which $B \not\models E^+$, S must construct a clausal hypothesis H such that

$$B \wedge H \models E^+$$

where B, H and E^- are satisfiable. In some approaches (Shapiro, 1983; Quinlan, 1990) H is found via a general-to-specific search through the lattice of clauses. This lattice is rooted at the top by the empty clause (representing falsity) and is partially ordered by θ-subsumption (H θ-subsumes H' with substitution θ whenever $H\theta \subseteq H'$). Two clauses are treated as equivalent when they both θ-subsume each other.

Following on from work by Plotkin (1971), Buntine (1988) demonstrated that the equivalence relation over clauses induced by θ-subsumption is generally very fine relative to the the equivalence relation induced by entailment between two alternative theories with common background knowledge. Thus when searching for the recursive clause for *member/2* (list-membership), infinitely many clauses containing the appropriate predicate and function symbols are θ-subsumed by the empty clause. Very few of these entail the appropriate examples relative to the base case for *member/2*.

Specific-to-general approaches based on Inverse Resolution (Muggleton and Buntine, 1988; Rouveirol, 1992; Sammut and Banerji, 1986) and relative least general generalisation (Buntine, 1988; Muggleton and Feng, 1992) maintain admissibility of the search while traversing the coarser partition induced by entailment. For instance Inverse Resolution is based on inverting the equations of Robinson's resolution operator to find candidate clauses which resolve with the background knowledge to give the examples. Inverse resolution (Muggleton and Buntine, 1988) can also be used to add new theoretical terms (predicates) to the learner's vocabulary. This process is known as *predicate invention.*

2.2 ILP applications

Many of the ILP applications to date have been developed using Muggleton and Feng's (1992) Golem. Bratko, Muggleton and Varsek (1992) showed that simple naive physics systems could be learned within an ILP setting. In this case, a qualitative model of water-filled U-tube was learned by Golem from 5 positive examples 6 negative examples and background knowledge representing Kuiper's (1986) QSIM theory.

Feng (1992) used Golem to construct a complete and correct set of diagnostic rules from a qualitative model of the power subsystem of a European Space Agency satellite.

Dolsak and Muggleton (1992) used Golem to construct design rules for finite element analysis used within CAD packages.

Golem has also had two major successes in discovering new scientific knowledge in the area of biomolecular modelling. Firstly (Muggleton *et al.* 1992) in the area of predicting protein secondary structure from primary amino acid sequences Golem produced reasonably intelligible rules with a higher accuracy than other approach tested on the same domain. Secondly (King *et al.* 1992) in a drug design domain Golem produced structure-activity prediction rules with an accuracy which is at least as good as the industry-standard Hansch regression technique. The advantage of the Golem approach lies in the fact that the rules are much easier for medicinal chemists to understand. These domains are discussed in detail in the paper within this volume by Sternberg et al. (1993).

3 FUTURE TRENDS AND DEVELOPMENTS

In his writings Alan Turing did not make the modern distinction between computer science and artificial intelligence. The universal Turing machine was in fact inspired by Turing's idea of an automatic mathematician. It would therefore not have surprised him in the least to see automatic theorem proving at the heart of Logic Programming. With this achievement firmly established it seems reasonable to take seriously Turing's other aspirations.

In Section 1 we showed that Turing saw machine learning as a central component in the future of computing. Present day machine learning is a specialised subject area of Artificial Intelligence. One way to increase the impact of machine learning might be to develop specialised conceptualising tools as Scientific Assistants. Such a tool would help scientists by suggesting interesting hypotheses from data and background knowledge, as has already been started in herbicide selection, cardiology (Bratko, Mozetic and Lavrac, 1989) and molecular chemistry (Muggleton *et al.* 1992; King *et al.* 1992; Sternberg *et al.* 1992). This kind of tool certainly seems like something worth aiming for.

However in Turing's vision of learning machines, the learning played a much more fundamental role. Every action involving communication between humans and computers, and even between one computer and another has the potential for triggering learning processes. It might be possible to achieve Turing's aims by making ILP an integral part of machine interfaces. Learning should eventually become to user-interfaces what theorem proving has become to program execution within a Logic Programming framework.

Shapiro's (1983) debugging system made a start in this direction. However his system was inefficient and was never incorporated into any widely-used program development system. Present ILP systems lack a standard model for their implementation. Although FOIL (Quinlan, 1990) and Golem (1992) are reasonably widely used and efficient, their approaches differ considerably.

If we were to take humans and animals as our model of computation then learning should be a part of every information processing task within computers. This seems like a very tall order. However, integrating learning sufficiently strongly into Logic Programming would in a sense achieve this end. The Japanese Fifth Generation project proved that Logic Programming could be used throughout an operating system. Although today's logic programming systems are likely to be obsolete within a decade, logic will maintain and increase its role within computing. It has been fifty years since Turing's initial investigations of logic and learning. Powerful logic-based learning systems will play a vital and increasingly central part within the next fifty years.

Acknowledgments

Thanks are due to Donald Michie for valuable information concerning Alan Turing's war-time discussions. His inspiration and support have helped generations of young scientists to share Alan Turing's vision.

REFERENCES

Bain, M. and Muggleton, S. (1991) Non-monotonic learning. In D. Michie, editor, *Machine Intelligence 12.* Oxford University Press, Oxford.

Bain, M. and Muggleton, S. (1993) Learning optimal chess endgame strategies. In K. Furukawa, D. Michie and S. Muggleton, editors, *Machine Intelligence 13.* Oxford University Press, Oxford.

Bowden, B.V. (1953). Lady Lovelace's memoir on the Analytical Engine. In B.V. Bowden, editor, *Faster than thought.* London. Republished document from 1842.

Bratko, I., Mozetic, I. and Lavrac N. (1989). *KARDIO: a study in deep and qualitative knowledge for expert systems.* MIT Press, Cambridge.

Bratko, I., Muggleton, S. and Varsek, A. (1992). Learning qualitative models of dynamic systems. In S. Muggleton, editor, *Inductive Logic Programming.* Academic Press, London.

Buntine, W. (1988). Generalised subsumption and its applications to induction and redundancy. *Artificial Intelligence*, 36(2):149–176.

Church, A. (1936). An unsolvable problem of elementary number theory. *American Journal of Mathematics*, 58:345–363.

de Raedt, L. and Bruynooghe, M (1992). An overview of the interactive Concept-Learner and Theory Revisor CLINT. In S. Muggleton, editor, *Inductive Logic Programming*, 163–192, London, 1992. Academic Press.

Dolsak, B. and Muggleton, S (1992). The application of inductive logic programming to finite element mesh design. In S. Muggleton, editor, *Inductive Logic Programming*, 453–472, London, 1992. Academic Press.

Dzeroski, S. and Lavrac, N. (1992). Refinement graphs for FOIL and LINUS. In S. Muggleton, editor, *Inductive Logic Programming*, 319–334. Academic Press, London.

Dzeroski, S., Muggleton, S. and Russell S. (1992). PAC-learnability of determinate logic programs. In *Proceedings of the International Conference on Learning Theory (COLT92)*, San Mateo, California. Kaufmann.

Feng, C. (1992). Inducing temporal fault dignostic rules from a qualitative model. In S. Muggleton, editor, *Inductive Logic Programming*, 473–494. Academic Press, London.

Gödel, K. (1931). Über formal unentscheidbare Sätze der Principia Mathematica und verwandter System I. *Monats. Math. Phys.*, 32:173–198.

Hodges, A. (1985). *The enigma of intelligence.* Unwin Paperbacks, Hemel Hempstead.

King, R., Muggleton, S., Lewis R. and Sternberg, M (1992). Drug design by machine learning: the use of inductive logic programming to model the structure-activity relationships of trimethoprim analogues binding to dihydrofolate reductase. *Proceedings of the National Academy of Sciences*, 89(23):11322-11326.

Kuipers, B. (1986). Qualitative simulation. *Artificial Intelligence*, 29:289–338.

Lloyd, J.W. (1984). *Foundations of Logic Programming.* Springer-Verlag, Berlin.

Michie, D. and Camacho R. (1993). Building symbolic represenatations of intuitive real-time skills from performance data. In K. Furukawa, D. Michie and S. Muggleton, editors, *Machine Intelligence 13.* Oxford University Press, Oxford.

Michie, D. and Chambers R.A. (1968). BOXES: An experiment in adaptive control. In E. Dale and D. Michie, editors, *Machine Intelligence 2*, 137–152. Oliver and Boyd, Edinburgh.

Muggleton, S. (1991). Inductive logic programming. *New Generation Computing*, 8(4):295–318.

Muggleton, S., King, R., and Sternberg, M. (1992). Protein secondary structure prediction using logic-based machine learning.

Protein Engineering, 5(7):647–657.

Muggleton, S., Bain, M., Hayes-Michie, J. and Michie, D. (1989). An experimental comparison of human and machine learning formalisms. In *Proceedings of the Sixth International Workshop on Machine Learning.* Kaufmann.

Muggleton, S. and Buntine, W. (1988). Machine invention of first-order predicates by inverting resolution. In *Proceedings of the Fifth International Conference on Machine Learning*, 339–352. Kaufmann.

Muggleton, S. and Feng, C. (1992). Efficient induction of logic programs. In S. Muggleton, editor, *Inductive Logic Programming*, 281–298. Academic Press, London.

Numao, M., Takashi, M. and Shimura, M. (1993). Inductive speed-up learning of logic programs. In K. Furukawa, D. Michie and S. Muggleton, editors, *Machine Intelligence 13.* Oxford University Press, Oxford.

Page, C. and Frisch, A. (1992). Generalisation and learnability: a study of constrained atoms. In S. Muggleton, editor, *Inductive Logic Programming*, 29–62. Academic Press, London.

Plotkin, G.D. (1969). A note on inductive generalisation. In B. Meltzer and D. Michie, editors, *Machine Intelligence 5*, 153–164. Edinburgh University Press, Edinburgh.

Plotkin, G.D. (1971). *Automatic Methods of Inductive Inference.* PhD thesis, Edinburgh University.

Quinlan, R. (1990). Learning logical definitions from relations. *Machine Learning*, 5:239–266, 1990.

Robinson, J.A. (1993). Turing, von Neumann and the universal machine. In K. Furukawa, D. Michie and S. Muggleton, editors, *Machine Intelligence 13.* Oxford University Press, Oxford.

Robinson, J.A. (1965). A machine-oriented logic based on the resolution principle. *Journal of the Association of Computing Machinery*, 12(1):23–41.

Rouveirol, C. (1992). Extensions of inversion of resolution applied to theory completion. In S. Muggleton, editor, *Inductive Logic Programming.* Academic Press, London.

Sammut, C. (1993). Recent progress with BOXES. In K. Furukawa, D. Michie and S. Muggleton, editors, *Machine Intelligence 13.* Oxford University Press, Oxford.

Sammut, C. and Banerji, R. (1986). Learning concepts by asking questions. In R. Michalski, J. Carbonnel, and T. Mitchell, editors, *Machine Learning: An Artificial Intelligence Approach. Vol. 2*, 167–192. Kaufmann, Los Altos, CA.

Samuel, A.L. (1959). Some studies in machine learning using the game of checkers. *IBM Journal of research and development*, 3:211–229.

Samuel, A.L. (1967). Some studies in machine learning using the game of checkers, 2 recent progress. *IBM Journal of research and development*, 11:601–617.

Shapiro, E.Y. (1983). *Algorithmic program debugging*. MIT Press.

Srinivasan, A., Muggleton, S., and Bain, M. (1993). The justification of logical theories. In K. Furukawa, D. Michie and S. Muggleton, editors, *Machine Intelligence 13*. Oxford University Press, Oxford.

Sternberg, M., King, R., and Muggleton, S. (1993). Machine learning and biomolecular modelling. In K. Furukawa, D. Michie and S. Muggleton, editors, *Machine Intelligence 13*. Oxford University Press, Oxford.

Tsukimoto, H. and Morita, C. (1993). The discovery of propositions in noisy data. In K. Furukawa, D. Michie and S. Muggleton, editors, *Machine Intelligence 13*. Oxford University Press, Oxford.

Turing, A. (1947). Lecture to the London Mathematical Society on 20 February 1947. Published in 'A.M. Turing's ACE Report of 1946 and other papers', MIT Press.

Turing, A. (1936). On computable numbers with an application to the Entscheidungsproblem. *Proceedings of the London Mathematical Society*, 42:230–265.

Turing, A. (1939). Systems of logic based on ordinals. *Proceedings of the London Mathematical Society*, pages 161–228.

Turing, A. (1946). Proposal for development in the mathematics division of an Automatic Computing Engine (ACE). Technical report, National Physical Laboratory, 1946. Published in 'A.M. Turing's ACE Report of 1946 and other papers', MIT Press.

Turing, A. (1948). Intelligent machinery. Technical report, National Physical Laboratory, 1948. First published in Machine Intelligence 5, Edinburgh University Press.

Turing, A. (1950). Computing machinery and intelligence. *Mind*,

1950. Reprinted in 'The Mind's I' ed. D. Hofstadter and D. Dennett, pub. by Basic Books, New York, 1981.

Woodger, M. (1958). The history and present use of digital computers at the National Physical Laboratory. *Process Control and Automation*, pages 437–442, November 1958. Reprinted in 'A.M. Turing's ACE Report of 1946 and other papers', MIT Press.

INDUCTIVE INFERENCE

3

A Generalization of the Least General Generalization

H. Arimura

T. Shinohara

S. Otsuki

Department of Artificial Intelligence

Kyushu Institute of Technology

H. Ishizaka

Fujitsu Laboratories, ISIS

Abstract

In this chapter, we present a polynomial time algorithm, called a k-minimal multiple generalization (k-*mmg*) algorithm, where $k \geq 1$, and its application to inductive learning problems. The algorithm is a natural extension of the least general generalization algorithm developed by Plotkin and Reynolds. Given a finite set of ground first-order terms, the k-*mmg* algorithm generalizes the examples by at most k first-order terms, while Plotkin's algorithm does so by a single first-order term. We apply the k-*mmg* algorithm to several learning problems in inductive logic programming, and knowledge discovery in databases.

1 INTRODUCTION

Inductive inference is a process to guess or identify an unknown general rule from its examples. An inference algorithm receives finite examples and produces a generalization of them as a hypothesis.

This paper is concerned with inference only from positive examples. For example, recently, a number of studies have been aimed at knowledge discovery in databases (Piatetsky-Shapiro and Frawley 1991). Usually, we think of a database as a collection of positive examples drawn from some rule that applies in the real world. To discover such a rule in the database, inductive inference from only positive examples rather than from both positive and negative examples is natural.

However, in general, any successful inference from positive data should avoid *overgeneralizations*. In other words, the key notion in inference from positive data is a *minimal* generalization. Consider the case where an unknown rule to be inferred is represented by a single first-order term. Then, an algorithm developed by Plotkin (1970) and Reynolds (1970) efficiently finds the minimal generalization of given examples. This generalization is called the *least general generalization* (*lgg*, for short). For this reason, the least general generalization algorithm plays an important role in model inference systems (Shapiro 1981; Ishizaka 1988) and inductive logic programming (Muggleton 1990).

On the other hand, when some rules in question are too complex to be represented in a single term, the least general generalization algorithm cannot be directly applied because it may produce an overgeneralization. For example, assume that the following examples are given:

$$\left\{ \begin{array}{c} app([],[],[]),\ app([b],[a],[b,a]),\ app([a],[],[a]), \\ app([],[a],[a]),\ app([a,b],[c,d],[a,b,c,d]) \end{array} \right\}$$

Clearly, the least general generalization of them becomes a most general term $app(x,y,z)$. However, if we generalize the examples by a set of several terms instead of a single term, we might get a less general generalization. For any positive integer k, we call a minimal generalization by at most k terms a *k-minimal multiple generalization* (*k-mmg*, for short). For example, the pair

$$\{\, app([],X,X), app([A|X],Y,[A|Z]) \,\}$$

is a 2-*mmg* of the examples above. Since the notion of 1-*mmg* coincides with the one of *lgg*, *k-mmg* is a generalization of the least general generalization.

Clearly, there exist several *k-mmg*s for a finite set of examples, while the *lgg* is unique. We say that a *k-mmg* is reduced with respect to a set of examples when it has no redundancy. We can show that the set of *all* the reduced *k-mmg*s for a set of examples is hard to compute. By contrast, *one* of the *k-mmg*s can be found in polynomial time under the assumption of compactness with respect to containment. This assumption ensures that containment relations are characterized by syntactic relations.

In Section 2, first we introduce *lgg* according to Plotkin (1970) and Reynolds (1970). Then we define *mmg* in Section 3 by generalizing *lgg* from a single first-order term to a set of first-order terms, and present a *k-mmg* algorithm that finds a *k-mmg* in polynomial time. The results in the section are obtained from our work (Arimura *et al.* 1991). In Section 4, we apply the *k-mmg* algorithm to problems in inductive logic programming. We introduce three subclasses of logic programs, unit clause programs, context-free transformations with a flat base and primitive Prologs, and show that these subclasses are polynomial time inferable from positive examples. The results in this section are obtained from previous works (Arimura *et al.* 1992b; Ishizaka *et al.* 1992). In Section 5, we show the method is also applicable to discovery of a set of rules that characterize a concept in a database.

2 PRELIMINARIES

For a finite set A, we denote by $\sharp A$ the number of elements in A. Let Σ be a finite set of *function symbols* and X be a countable set of *variables* disjoint from Σ, where a mapping *arity* from function symbols to positive numbers is associated with Σ. We call Σ an *alphabet.* A *first-order term* (or a *term*) is either a variable, a 0-ary function symbol, or a string $f(t_1, \ldots, t_n)$ that is recursively constructed from an n-ary function symbol f and terms $t_1, \ldots, t_n$. A term is *ground* if it contains no variable. We denote by $\mathcal{TP}$ the set of first-order terms and by $\mathcal{T}$ the set of ground first-order terms. The *size of term* p is the total number $|p|$ of occurrences of function symbols and variables in p, and *the size of set P of terms* is $||P|| = \sum_{p \in P} |p|$.

A *substitution* is a mapping θ from terms to themselves such that for any term $t = f(t_1, \ldots, t_n)$, $\theta(t) = f(\theta(t_1), \ldots, \theta(t_n))$. A set of replacement $\{x_1 := t_1, \ldots, x_n := t_n\}$ denotes the substitution that maps variable x_i to term t_i and any other variable to itself.

We define a binary relation $\leq'$ on $\mathcal{T}$ by $p \leq' q \Longleftrightarrow p = \theta(q)$ for some substitution θ. If $p \leq' q$ then we say p is an *instance* of q, p is a *generalization* of q, p is more *specific* than q, or q is more *general* than p. We also define $<'$ by $p \leq' q \Longleftrightarrow p \leq' q$ but $q \not\leq' p$. Let $\mathcal{TP}_{\equiv}$ be the set of representatives of the equivalence classes of $\mathcal{TP}$ modulo $\equiv$, where $p \equiv q \Longleftrightarrow p \leq' q$ and $q \leq' p$. In Section 2 and Section 3, we write $\mathcal{TP}$ for $\mathcal{TP}_{\equiv}$ without further notice.

Let P be a finite set of terms. If $P \subseteq L(p)$, then we say a term p is a *common generalization* of S. The *least general generalization* (*lgg*, for short) of P is a common generalization p of P such that $p \leq' q$ for any common generalization q of P. The *lgg* of P is computable in polynomial time in $||P||$ by the anti-unification algorithm (Plotkin 1970; Reynolds 1970).

The *language* defined by term p is the set $L(p) = \{w \in \mathcal{T} \mid w \leq' p\}$, that is, the set of ground instances of p. A set L of ground terms is a *tree pattern language* if $L = L(p)$ for some p. By definition, $p \leq' q \Longrightarrow L(p) \subseteq L(q)$. In this case, the converse also holds.

Lemma 3.1 (Reynolds 1970) If $\sharp\Sigma > 1$, then $L(p) \subseteq L(q) \Longleftrightarrow p \leq' q$.

By Lemma 3.1, we can define the *lgg* of S as a term that defines the minimum language containing S with respect to set inclusion $\subseteq$.

3 POLYNOMIAL TIME K-MMG ALGORITHM

In this section, we present a polynomial time algorithm to compute a k-minimal multiple generalization of a finite set of ground terms. Let k be a positive integer. A *k-multiple term* is a set P of at most k first-order terms, and the *language* defined by P is the set $L(P) = \bigcup_{p \in P} L(p)$. Two k-multiple generalizations

P and Q are *equivalent* if they define the same language. We denote by $\mathcal{TP}^k$ the class of k-multiple terms. If $S \subseteq L(P)$, we say P is a *k-multiple generalization* of S. A *k-minimal multiple generalization* (*k-mmg*, for short) of S is a k-multiple generalization P of S such that $L(Q) \not\subset L(P)$ for any k-multiple generalization Q of S. Note that there may be more than one *k-mmg* of S if $k > 1$, while the *lgg* of S is unique. For example, let $S = \{f(a,a), f(a,b), f(b,b)\}$ and $\Sigma = \{a, b, f(\cdot,\cdot)\}$. Then, $P_1 = \{f(x,x), f(a,b)\}$ and $P_2 = \{f(a,a), f(y,b)\}$ are both 2-*mmg* of S.

In this chapter, we are concerning with an efficient algorithm that, given S, finds one of the *k-mmg*s of S. Indeed, it is reasonable to cease trying to find all the answers because we can show by a reduction from an NP-complete problem that it is hard to compute all the *k-mmg*s of S in polynomial time.

A set P in $\mathcal{TP}^k$ is *reduced with respect to* S iff $S \subseteq L(P)$, but $S \not\subseteq L(Q)$ for any proper subset Q of P. Any *k-mmg* of S is equivalent to a member Q in $\mathcal{TP}^k$ that is reduced with respect to S, and there are only finitely many such Qs. Thus, by the definition of *mmg*, if the decision of $L(P) \subseteq L(Q)$? is computable (indeed, it is possible from Theorem 3.2 below), we can use an exhaustive search method to compute a *k-mmg* of S. However, this simple method may not efficiently work. Because even for a fixed $k > 0$, S may have exponentially many reduced *k-mmg*s. Furthermore, the observation on hardness of computing all solutions mentioned above also leads to the difficulty.

Recall that k-multiple generalizations are defined through the class of languages defined by sets in $\mathcal{TP}^k$. Consider the following property of the class: for any terms $p, q_1, \ldots, q_m$ $(1 \leq m \leq k)$,

$$L(p) \subseteq L(q_1) \cup \cdots \cup L(q_m) \implies L(p) \subseteq L(q_i) \text{ for some } 1 \leq i \leq m.$$

We call this property the *compactness with respect to containment* of $\mathcal{TP}^k$. If Σ is sufficiently large, $\mathcal{TP}^k$ has this property. Lassez and Marriott (1986) showed the property in the case where Σ is infinite. The next theorem improves their result.

Theorem 3.2 (Arimura *et al.* 1992a) Let $k > 0$ and Σ be an alphabet with $\sharp\Sigma > k$. Then the class $\mathcal{TP}^k$ has the compactness with respect to containment.

The condition on $\sharp\Sigma$ above is necessary. For example, let $k = 2$ and Σ consists of two symbols $a, f(\cdot,\cdot)$. For terms $p = x$, $q_1 = a$, and $q_2 = f(x, y)$, we see $L(p) \subseteq L(q_1) \cup L(q_2)$ but none of q_1 and q_2 is a generalization of p.

Hereafter, we assume $\sharp\Sigma > k$ for the compactness. Then we can use Theorem 3.2 to show the necessary and sufficient condition for a k-multiple generalization P consisting of exactly k terms to be a *k-mmg* of a finite set S. A k-multiple generalization P is said to be of *normal form with respect to* S iff p is the *lgg* of $S - L(P \setminus p)$ for any $p \in P$, where $P \setminus p$ is the set obtained by removing p from P. Assume that P is of normal form with respect to S, and also assume that there is some Q satisfying $S \subseteq L(Q) \subset L(P)$. If $\sharp Q < \sharp P$ then compactness shows that P is not reduced with respect to S: contradiction. On the other hand, if $\sharp Q = \sharp P$, then the compactness shows that P cannot be of normal form. Hence, the following holds.

Theorem 3.3 Let P be a multiple generalization of S that is reduced with respect to S and $\sharp P = k$. Then P is of normal form with respect to S iff P is *k-mmg* of S.

Now, we give a polynomial time algorithm $MMG(k, S)$ in Algorithm 1 that finds a *k-mmg* based on a greedy search. First, the MMG starts from any k-multiple generalization of S that is reduced with respect to S. The next theorem ensures that such a candidate can be efficiently found (we prove the theorem in the latter half of this section).

Theorem 3.4 (Arimura *et al.* 1991) For any $k > 0$ and any finite set S of ground terms, a set consisting of exactly k terms that is reduced with respect to S can be found in polynomial time with respect to $||S||$ if it exists.

Once a general candidate is obtained, the algorithm searches k-mmg in the direction from the general one to the most specific one. The algorithm tries to make P more specific by replacing

Algorithm 1: The algorithm $MMG(k, S)$

Input: A positive integer k and a finite set $S \subseteq \mathcal{T}$.
Output: A *k-mmg* of S.
Procedure:

```
1  if k = 1 then return lgg(S)
2  else
3      P(= {p_1, ..., p_k}) := REDUCED(k, S);
4      if P is found then
5          for each i = 1, ..., k do  /* tightening process */
6              replace p_i in P by lgg(S − L(P \ p_i));
7          return P;
8      else return MMG(k − 1, S);
```

each component p in P by the more specific term $lgg(S - L(P \setminus p))$. We call this process *tightening*. Any fixed point of this tightening process is of normal form with respect to S.

Lemma 3.5 In MMG in Algorithm 1, assume that a set of exactly k terms that is reduced with respect to S is found at Line 1. Then after executing the lines from Line 2 to Line 6, any P that MMG returns at Line 7 is of normal form with respect to S.

Hence, we show the main result of this section. Note that any finite set of ground terms has its *k-mmg* for any $k \geq 1$ (At worst, it is the lgg of S).

Theorem 3.6 (Arimura *et al.* 1991) Let k be a positive integer and Σ be an alphabet with $\sharp\Sigma > k$. Then, for any finite set S of ground terms, a *k-mmg* of S can be computed in polynomial time in $||S||$.

The key to an efficient *k-mmg* algorithm is to efficiently find a reduced k-multiple term which dominates the time complexity of total computation. In the rest of the section, we present a polynomial time algorithm $REDUCED$ (in Algorithm 2) to find a reduced ck-multiple term and prove Theorem 3.4. In particular, the algorithm $REDUCED$ searches k-multiple terms

Algorithm 2: The algorithm $REDUCED(k, S)$

where gci S is the greatest common instance of a set of terms.

Input: A positive integer k and a finite set S of ground terms.
Output: A set of exactly k terms reduced with respect to S.
Procedure:

```
for each V ⊆ S with ♯V = k do
    P_1 := ∅, ..., P_k := ∅;
    for each v ∈ V do
        let {v_1, ..., v_{k-1}} := V \ v;
        for each q_1 ∈ MAXT(v, v_1), ..., q_{k-1} ∈ MAXT(v, v_{k-1}) do
            P_v := P_v ∪ gci{q_1, ..., q_{k-1}};
    end for;
    for each p_1 ∈ P_1 , ..., p_k ∈ P_k do
        P := {p_1, ..., p_k};
        if S ⊆ L(P) then                /* check whether reduced */
            return P;
    end for;
end for;
```

that define a maximal language within reduced k-terms with respect to S instead of all reduced k-multiple terms.

A *k-pivot* of a k-multiple term P is a set V of just k ground terms in $L(P)$ such that there is some one-to-one correspondence f from V to P such that $v \in L(p_v)$ but for any $u \in V \setminus v$, $u \notin L(p_v)$, where p_v denotes the member $f(v)$ of P $(= \{ p_v \mid v \in V\})$. By definition, If P is a k-multiple generalization of S with $\sharp P = k$, P is reduced with respect to S iff P has a k-pivot drawn from S. Let $\mathcal{R}(V)$ be the class of k-multiple terms that has a k-pivot V, and $\max \mathcal{R}(V)$ be the subclass consisting of maximal members of $\mathcal{R}(V)$ with respect to set inclusion on their languages. Then, we have the following lemma.

Lemma 3.7 Assume that $\sharp P = k$ and P is a k-multiple term. If P defines a maximal language within reduced k-terms with respect to S, then P is a member of $\max \mathcal{R}(V)$ for some k-pivot V contained in S.

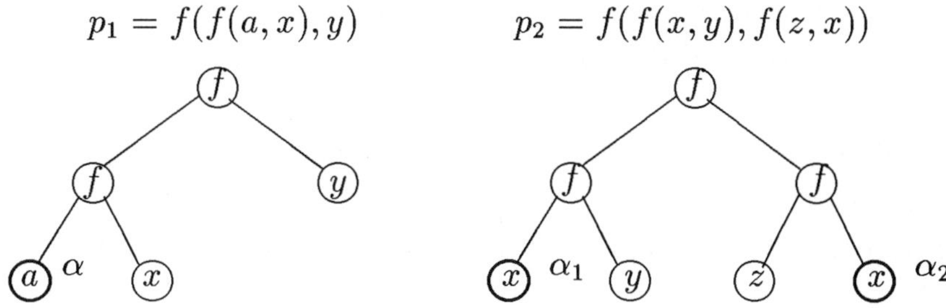

Figure 3.1. Two possible shapes p_1 and p_2 of max trees consistent with a positive term $f(f(a,a),f(b,a))$ and a negative term $f(f(b,a),f(a,a))$. See Lemma 3.9.

Next, we show how to compute members in $\max\mathcal{R}(V)$. We introduce the notion of consistent max trees. Let u_+, u_- be ground terms. Then, a term p is *consistent with a positive term* u_+ *and a negative term* u_- if $u_+ \in L(p)$ but $u_- \notin L(p)$. We say p is a *max tree* consistent with u_+ and u_- if

- p is consistent with u_+ and u_-, and
- for any consistent term q with u_+ and u_-, $q \not<' p$.

To characterize $\max\mathcal{R}(V)$, we use the set $MAXT(u_+, u_-)$, the set of max trees consistent with u_+ and u_-.

Lemma 3.8 Assume that $P = \{\, p_v \mid v \in V\}$ is k-multiple term in $\max\mathcal{R}(V)$. Then, each p_v is a maximally general term such that $v \in L(p_v)$ but for any $u \in V \setminus v$, $u \notin L(p_v)$. Moreover, each p_v is represented as the greatest common instance of some terms $\{\, q_u \mid u \in V \setminus v\}$ such that for every $u \in V \setminus v$, each q_u is a max tree consistent with positive v and negative u.

By Lemma 3.8 above, we can see that the algorithm $REDUCED$ shown in Algorithm 2 finds a k-multiple generalization of S that is reduced with respect to S.

Finally, we show that the set $MAXT(u_+, u_-)$ has at most polynomially many members, and it can be computed in polynomial time. A first-order term is identified with an ordered tree $tree(p)$ labelled by symbols in $\Sigma \cup X$ in the standard way. A node α *touches* a path $\beta_1, \ldots, \beta_n$ if for some $1 \le i \le n$ either

Algorithm 3: The algorithm $MAXT(t_+, t_-)$

where $w(\alpha)$ is the label of the node α of w, and w/α is the subtree of w whose root is α. See Lemma 3.9 for 1-$branch(t_+, \alpha)$ and 2-$branch(t_+, \alpha_1, \alpha_2)$.

Input: Ground terms t_+, t_-.
Output: The set of max trees consistent with t_+ and t_-.
Procedure:

```
T := ∅;
for each node α in t+ do
    if t+(α) ∈ Σ and t+(α) ≠ t−(α) th en
        T := T ∪ { 1-branch(t+, α) };
for each pair α1, α2 of nodes in t+ do
    if t+/α1 = t+/α2 but t−/α1 = t−/α2 then
        T := T ∪ { 2-branch(t+, α1, α2) };
for each pair p, q ∈ T do
    if L(p) ⊂ L(q) then T := T − {p};
return T;
```

$\alpha = \beta_i$ or α is a child of β_i. The following lemma tells us the possible shapes of max trees (See figure 3.1).

Lemma 3.9 If p is a max tree with t_+ and t_-, then p is a generalization of t_+ such that $tree(p)$ satisfies either 1 or 2:

1. There is a node α labelled by a function symbol such that every node touches the path from the root to α, and all leaves other than α are labelled by mutually distinct variables.
2. There are nodes α_1, α_2 labelled by the same variable, say x, such that every node touches a path from the root to α_1 or α_2, and all leaves other than α_1, α_2 are labelled by mutually distinct variables that are different from x.

In Lemma 3.9, the choice of α and the choice of α_1, α_2 in t_+ determine the unique generalizations p of t_+ satisfying 1 and p satisfying 2, respectively. We refer to these ps as 1-$branch(t_+, \alpha)$ and 2-$branch(t_+, \alpha_1, \alpha_2)$, respectively. By Lemma 3.9, we can compute the set $MAXT(t_+, t_-)$ by the algorithm $MAXT$ shown in Algorithm 3 in polynomial time with respect to the size $|t_+|$.

By the observation above, the algorithm *REDUCED* in Algorithm 2 runs in time $O(m^{2k^2+1}n^{k+1})$ with respect to m and n, where m is the maximum size of terms in S and n is the number of terms in S because the number of different choices of k-pivot V is bounded by n^k, $\sharp MAXT(t_+, t_-)$ is bounded by $2m^2$, and the greatest common instance of terms can be computed by the unification algorithm in linear time in m. Hence, we can compute a k-multiple generalization reduced with respect to S in polynomial time in $||S||$ by the algorithm *REDUCED*. Hence we have Theorem 3.4.

4 APPLICATIONS IN ILP

In this section, we describe several applications of the *mmg* algorithm in inductive logic programming (ILP, for short). In some inductive inference algorithms for logic programs such as GEMINI (Ishizaka 1988) or CIGOL (Muggleton and Buntine 1988), the least general generalization plays a very important role in inferring heads of clauses. However, in general, a program consists of several clauses. In order to infer several heads using *lgg*, the inference algorithm has to divide a given set of positive examples (a finite subset of the least Herbrand model of a target program) into several appropriate subsets at first, then it can get candidates for heads of clauses by computing the *lgg* of each subset. This process, that is, dividing a set of positive examples appropriately then generalizing each obtained subset of examples, exactly corresponds to the *mmg* calculation. Hence, we believe that *mmg* is more useful than *lgg* in ILP.

The results we introduce here were obtained from our previous works (Arimura *et al.* 1991; Arimura *et al.* 1992b; Ishizaka *et al.* 1992) on inductive inferability of several subclasses of logic programs from positive facts using the *mmg* algorithm. Shinohara (1991) showed that a class of *linear Prologs* with at most k clauses is inferable from only positive facts. However, his result concerns just inferability but not *efficiency*. We introduce three subclasses of linear Prologs, unit clause programs, context-free transformations with a flat base and primitive Prologs. Each class is *efficiently* inferable from only positive facts.

In this section, the reader is assumed to be familiar with the rudiments of logic programs (Lloyd 1984). Furthermore, we assume that a first-order language $\mathcal{L}$, that has finitely many predicate and function symbols (we regard a constant symbol as a 0-ary function symbol), is given. An atom, a term, a clause, a logic program (program, for short), and related notions are defined over $\mathcal{L}$. We denote the set of predicate symbols and function symbols of $\mathcal{L}$ by Π and Σ respectively. For a program P, $M(P)$ denotes the *least Herbrand model* of P.

The following definitions concerned with the notion of identification in the limit are based on (Gold 1967). An *inference algorithm* $\mathcal{A}$ is an algorithm that iterates a process input request $\to$ computation $\to$ output. Let $g_1, g_2, \ldots$ be a sequence of outputs of $\mathcal{A}$ for an input sequence $e_1, e_2, \ldots$ We say that $\mathcal{A}$ *converges* to g for the input sequence $e_1, e_2, \ldots$ iff there exists $n \geq 1$ such that $g_i = g$ for any $i \geq n$.

An *enumeration* of a model M is a sequence $e_1, e_2, \ldots$ of elements in M such that every atom in M occurs as e_i for some $i \geq 1$. We say that $\mathcal{A}$ *identifies* a model M *in the limit from positive facts* iff $\mathcal{A}$ converges to a program P such that $M(P) = M$ for any enumeration of M. We say that $\mathcal{A}$ *identifies* a class of programs $\mathcal{P}$ *in the limit from positive facts* iff, for any $P \in \mathcal{P}$, $\mathcal{A}$ identifies $M(P)$ in the limit from positive facts.

Let $P_1, P_2, \ldots$ be a sequence of outputs of $\mathcal{A}$ for an enumeration $e_1, e_2, \ldots$ of a model M and S_i be the set $\{e_1, \ldots, e_i\}$. An inference algorithm $\mathcal{A}$ is *consistent* iff $S_i \subseteq M(P_i)$ for any i. An inference algorithm $\mathcal{A}$ is *conservative* iff $P_i = P_{i-1}$ for any i such that $e_i \in M(P_{i-1})$. An inference algorithm $\mathcal{A}$ is a *polynomial update time inference algorithm* iff there exists some polynomial f such that, for any stage i, after $\mathcal{A}$ feeds the input e_i it produces the output P_i in $f(||S_i||)$ steps. Any exponential update time inference algorithm can be converted into a cunning polynomial update time one, even if either consistency or conservativeness lacking. Hence, both conditions are necessary for the validity of polynomial update time inference.

A class of programs $\mathcal{P}$ is said to be (*consistently, conservatively, polynomial update time*) *inferable* from positive facts iff there exists an (consistent, conservative, polynomial update time) inference algorithm that identifies $\mathcal{P}$ in the limit from positive facts.

4.1 Unit clause programs

First we consider a class of very simple logic programs that consist of only unit clauses. We denote the class by $\mathcal{UCP}$ and a class of logic programs that consist of at most k unit clauses by k-$\mathcal{UCP}$. For any $P \in \mathcal{UCP}$, since each clause $C \in P$ is unit, it holds that $M(P) = \bigcup_{C \in P} L(C)$ where $L(C)$ is a set of all ground instances of C. Thus, if $\sharp(\Pi \cup \Sigma) > k$, then we can directly apply the *k-mmg* algorithm to infer $\mathcal{UCP}$.

Algorithm 4: Inference algorithm for k-$\mathcal{UCP}$

Input: An enumeration of a model $M(P)$ where $P \in k$-$\mathcal{UCP}$.
Output: An infinite sequence of programs in k-$\mathcal{UCP}$.
Procedure:

$H := \emptyset$; $S := \emptyset$;
repeat
  read the next fact e; $S := S \cup \{e\}$;
  if $e \notin M(H)$ **then** $H := MMG(k, S)$;
  output H;
forever

Angluin (1980) showed that if a target class has the property called *finite thickness*, that is, it contains only finitely many concepts including a given examples, then an inference algorithm that, at any stage, outputs a *minimal hypothesis consistent with given positive examples* can identify the class. A minimal hypothesis is a hypothesis that defines a minimal concept such as a minimal language or a least Herbrand model; in our context, it corresponds to $MMG(k, S)$. Unfortunately, the class k-$\mathcal{UCP}$ does not have finite thickness when $k > 1$. However, Angluin's result can be extended to the class with the property called *finite elasticity* (Wright 1989a, 1989b). Shinohara (1990) showed that the class of linear Prologs with at most k clauses is inferable from positive facts by showing the class has finite elasticity. Since k-$\mathcal{UCP}$ is a subclass of linear Prologs, it also has finite elasticity. Thus Algorithm 3 identifies k-$\mathcal{UCP}$. Consistency and conservativeness of Algorithm 4 are trivial. From Theorem 3.6, it is also clear that Algorithm 4 produces each hypothesis in polynomial steps in $||S||$. Thus we have the following theorem.

Theorem 3.10 (Arimura *et al.* 1991) Suppose that $\sharp(\Pi \cup \Sigma) > k$. Then the class k-$\mathcal{UCP}$ is consistently and conservatively polynomial update time inferable.

4.2 Context-free transformations with a flat base

Next we consider a slightly complex subclass $\mathcal{CFT}^{uniq}_{FB}$ of context-free transformations ($\mathcal{CFT}$, for short). The class $\mathcal{CFT}$ was originally introduced by Shapiro in his study on MIS (Shapiro 1981) and includes a lot of non-trivial programs such as *append*, *plus*, and so on. We introduce a restricted subclass $\mathcal{CFT}^{uniq}_{FB}$ of $\mathcal{CFT}$.

A *context-free transformation with a flat base* (CFT_{FB}) is a program that consists of two clauses C_0 and C_1:

$$\begin{aligned} C_0 &= p(s_1, \ldots, s_m) \\ C_1 &= p(t1, \ldots, t_m) \leftarrow p(x_1, \ldots, x_m) \end{aligned}$$

that satisfy the following conditions (a)–(c).

(a) Every argument s_i $(1 \le i \le m)$ of the head of C_0 is either a function symbol of arity 0 or a variable symbol.

(b) All arguments $x_1, \ldots, x_m$ of the body of C_1 are mutually distinct variables.

(c) For every $1 \le i \le m$, every argument x_i of the body of C_1 occurs exactly once in the term t_i of the head. Moreover, x_i does not occur in any argument t_j $(i \ne j)$ of the head.

A program P is a CFT^{uniq}_{FB} iff there exists at most one 2-mmg of $M(P)$. We denote the class of all CFT^{uniq}_{FB} programs by $\mathcal{CFT}^{uniq}_{FB}$.

It seems the class $\mathcal{CFT}^{uniq}_{FB}$ is too restrictive. Furthermore, the class $\mathcal{CFT}^{uniq}_{FB}$ is defined according to the least Herbrand model of each element. Since the least Herbrand model of a program is an infinite set in general, the definition seems to be problematic. Fortunately, however, the class is decidable in polynomial time. That is, given a program P, we can decide whether P is in $\mathcal{CFT}_{FB}^{uniq}$ in polynomial time of $size(P)$, where $size(P)$ is the size of P as an expression. In fact, we can show that several non-trivial programs in $\mathcal{CFT}$ are still in $\mathcal{CFT}_{FB}^{uniq}$. For example, the following CFTs are in $\mathcal{CFT}_{FB}^{uniq}$.

$$append([], X, X)$$
$$append([A|X], Y, [A|Z]) \leftarrow append(X, Y, Z)$$

$$suffix(X, X)$$
$$suffix(X, [A|Y]) \leftarrow suffix(X, Y)$$

$$plus(X, 0, X)$$
$$plus(X, s(Y), s(Z)) \leftarrow plus(X, Y, Z)$$

$$lesseq(0, X)$$
$$lesseq(s(X), s(Y)) \leftarrow lesseq(X, Y)$$

Algorithm 5 is an inference algorithm for $\mathcal{CFT}_{FB}^{uniq} \cup 2\text{-}\mathcal{UCP}$. Algorithm 5 does not change the current hypothesis H, if it is consistent with a newly given positive fact e. Suppose that

$$S = \{\, app([], [], []), app([b], [a], [b, a]), app([a], [], [a]), app([], [a], [a]), \\ app([a, b], [c, d], [a, b, c, d])\}$$

is the set of positive facts given so far and the current hypothesis cannot imply the last fact. First, the algorithm finds a pair of atoms

$$\{app([], X, X), app([A|X], Y, [A|Z])\}$$

by $MMG(2, S)$. Next, it tries to find a hypothesis consistent with S. This is done by enumerating every CFT_{FB}^{uniq} with $\{app([], X, X), app([A|X], Y, [A|Z])\}$ as its heads. Actually, the search is done by enumerating every possible instance of CFT_{FB}^{uniq} with pair of atoms obtained by the 2-*mmg* algorithm as its heads, because the candidate heads are possibly less general than the heads of a target program. If such a program P^* containing a clause with non-empty body is found, the algorithm outputs it. Otherwise the algorithm simply outputs $MMG(2, S)$

as an approximation of the target model. Since P^* is an instance of a CFT_{FB}^{uniq}, Algorithm 5 may need to transform P^* into a CFT_{FB}^{uniq} that has the same least Herbrand model with P^*. The transformation φ performs such model preserving generalization.

Algorithm 5: Inference algorithm for $\mathcal{CFT}_{FB}^{uniq} \cup 2\text{-}\mathcal{UCP}$

Input: Enumeration of model $M(P)$; $P \in \mathcal{CFT}_{FB}^{uniq} \cup 2\text{-}\mathcal{UCP}$.
Output: Infinite sequence of programs in $\mathcal{CFT}_{FB}^{uniq} \cup 2\text{-}\mathcal{UCP}$.
Procedure:

```
H := ∅; S := ∅;
repeat
    read the next fact e; S := S ∪ {e};
    if e ∉ M(H) then
        {h0, h1}:= MMG(2, S);
        find a hypothesis P* consistent with S
            whose heads are {h0, h1};
        if found then H := φ(P*);
        else H := {h0, h1};
    output H;
forever
```

Since the class $\mathcal{CFT}_{FB}^{uniq}$ is also a subclass of linear Prologs, it has finite elasticity. Hence, if Algorithm 5 outputs a minimal hypothesis consistent with S at any stage, it is ensured that the algorithm identifies $\mathcal{CFT}_{FB}^{uniq} \cup 2\text{-}\mathcal{UCP}$. As described in the next subsection, in general, there exist several 2-*mmg*s for an entire model $M(P)$ of a program P that contains a clause with non-empty body. If there exist several 2-*mmg*s of $M(P)$, then it becomes difficult to decide which 2-*mmg* is appropriate for the heads of a target program. The difficulty is directly concerned with the difficulty of finding a consistent minimal hypothesis as described in the next subsection. However, from the uniqueness of 2-*mmg* for the model of CFT_{FB}^{uniq}, it is possible to ensure that a consistent minimal hypothesis can be found by a very simple search as mentioned above. Consistency and conservativeness of Algorithm 5 is trivial. From the syntactical restriction on CFT_{FB}, the search for a consistent hypothesis P^* and the transformation P^* into a CFT_{FB}^{uniq} $\varphi(P^*)$ can be finished in polynomial time in $||S||$. Hence we can obtain the following theorem.

Theorem 3.11 (Arimura *et al.* 1992b) Suppose that $\sharp\Sigma > 2$. Then the class $\mathcal{CFT}_{FB}^{uniq} \cup 2\text{-}\mathcal{UCP}$ is consistently and conservatively polynomial update time inferable.

4.3 Primitive Prologs

Finally, we consider a class of programs called primitive Prologs. A *primitive Prolog* P is a program that satisfies the following conditions (a)–(d):

(a) Only one unary predicate symbol appears in P.

(b) P consists of at most two clauses.

(c) If P consists of two clauses, then both heads of clauses have no common instance.

(d) Atoms appearing in the body of a clause are most general atoms as $p(x)$.

(e) Variables appearing in the body of a clause are mutually distinct and also appear in the head of the clause.

In a word, a primitive Prolog is a program of the form:

$$p(t[x_1, \ldots, x_m]) \leftarrow p(x_1), \ldots, p(x_m)$$
$$p(s),$$

where $x_1, \ldots, x_m$ are mutually distinct variables, $t[x_1, \ldots, x_m]$ is any term containing the variables $x_1, \ldots, x_m$, s is any term, and $L(t[x_1, \ldots, x_m]) \cap L(s) = \emptyset$. We denote a class of primitive Prologs by $\mathcal{PP}$.

Although the class $\mathcal{PP}$ is so restricted, there exists a primitive Prolog P that has several 2-*mmg*s of its model $M(P)$. For example, consider the following primitive Prolog P:

$$p([a, b, a])$$
$$p([b|X]) \leftarrow p(X).$$

For the least Herbrand model

$$M(P) = \{p([a, b, a]), p([b, a, b, a]), p([b, b, a, b, a]), \ldots\},$$

there exist two kinds of 2-*mmg* of $M(P)$:

$$\{p([a,b,a]), p([b,X,Y,Z|W])\} \text{ and } \{p([b,a,b,a]), p([X,b,Y|Z])\}.$$

Actually the former is an instance of the heads of the program P. Hence, if an inference algorithm selects the former pair as the heads of a hypothesis, it will be able to identify the target program. However, if the inference algorithm selects the latter pair, then it may produce an overgeneralized hypothesis and fail to identify the target program consistently and conservatively.

Let P_1 be an instance

$$\begin{array}{l} p([a,b,a]) \\ p([b,X,Y,Z|W]) \leftarrow p([X,Y,Z|W]) \end{array}$$

of P with the former 2-*mmg* as its heads, and P_2 be a program consisting of the atoms in the latter 2-*mmg*. We know that P_1 is a correct hypothesis but P_2 is an overgeneralized one. However the algorithm is given only positive facts and both P_1 and P_2 are consistent with every fact. Hence, in order to achieve conservative inference, the algorithm has to decide which program has a smaller model. That is, to construct a consistent and conservative polynomial update time inference algorithm for the class $\mathcal{PP}$, a model containment problem for primitive Prologs (P_1 can be easily transformed into the original primitive Prolog) should be solved efficiently. Unfortunately the problem is still open.

Although the problem of consistent and conservative polynomial update time inferability of $\mathcal{PP}$ is also still open, we have shown an interesting polynomial update time algorithm that identifies $\mathcal{PP}$ consistently but not conservatively (Ishizaka *et al.* 1992). The algorithm concentrates its attention on a minimal size fact given so far to find a unit clause in a target program. This simple idea works well. Whenever a target primitive Prolog consists of a unit clause and a recursive clause with non-empty body as the previous example:

$$\begin{array}{l} p([a,b,a]) \\ p([b|X]) \leftarrow p(X), \end{array}$$

a fact $p([a,b,a])$ of minimal size in the model $M(P)$ should be an instance of the unit clause. Using this property, the algorithm can identify the correct heads in several 2-*mmg*s in the limit working unconservatively. Lange and Wiehagen (1991) showed an interesting inference algorithm that inconsistently runs in polynomial update time and infers pattern languages from positive examples. Our idea is similar to theirs.

5 APPLICATION IN KNOWLEDGE DISCOVERY IN DATABASES

In this section, we describe an application of the k-*mmg* algorithm to the problem of discovering knowledge in databases.

5.1 Attribute-oriented induction

There are different two approaches in the discovery of knowledge in databases (Piatetsky-Shapiro and Frawley 1991). One is, given positive examples and negative examples, to find *classification rules*, that is, rules separating positive examples from negative examples. Another is, given positive examples, to find *characteristic rules*, that is, rules characterizing the concept represented by the positive examples.

The latter one is closely related to inductive inference from positive examples and can be thought of as computing a generalization of databases. Since a database usually contains only positive examples of a relation in the real world, we believe discovery algorithms that learn concepts only from positive examples are useful. Thus, we deal with the discovery of characteristic rules from positive examples in this section.

Cai *et al.* (1991) proposed a heuristics algorithm to discover characteristic rules. Their algorithm LCHR generalizes a database by using a given conceptual hierarchy. In Table 3.1, we show an example of databases extracted from (Cai *et al.* 1991), which consists of tuples concerning information about graduate students. In Figure 3.2, we show the corresponding conceptual hierarchy for the attribute Birth_Place by a tree-like structure. The tree-like structure shows that a concept placed near the root is more general than one placed near the leaves.

Given a small positive integer k, called a *threshold*, a database, and the corresponding conceptual hierarchy, the algorithm LCHR iterates the following process. It non-deterministically selects a tuple from the database and replaces some of attribute values

Table 3.1. A university database (Cai *et al.* 1991)

Name	Category	Major	Birth_Place	GPA
Anderson	M.A.	Physics	Vancouver	3.5
Fraser	M.S.	Physics	Ottawa	3.9
Gupta	Ph.D	Math	Bombay	3.3
Liu	Ph.D.	Biology	Shanghai	3.4
Monk	Ph.D.	Computing	Victoria	3.8
Wang	M.S.	Statistics	Nanjing	3.2

GPA = grade point average

Table 3.2. A generalized database obtained by the algorithm LCHR

Major	Birth_Place	GPA
science	Canada	excellent
science	foreign	good

by more general values with respect to the conceptual hierarchy. In this process, attribute values to be substituted should be carefully selected from tuples to avoid overgeneralization. Successive applications of this process eventually decrease the number of different tuples in the database. If the number of tuples becomes less than or equal to the threshold k, the algorithm terminates and outputs the resulting generalized database as a characteristic rule.

Table 3.2 is a generalized database obtained from the database in Table 3.1 by LCHR, where two columns for Name and Category that have no useful information have been removed (Cai *et al.* 1991). The generalized database computed by LCHR characterizes the given database by a kind of clustering. The paper (Cai *et al.* 1991) reported that the algorithm LCHR efficiently computes rules characterizing the database.

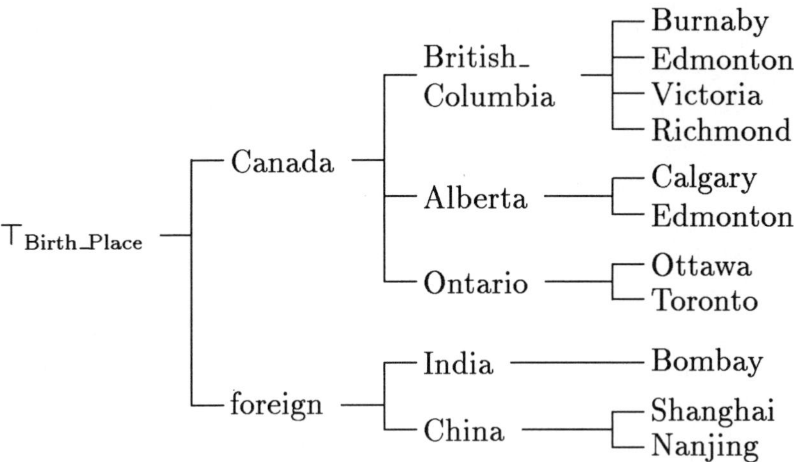

Figure 3.2. A concept hierarchy $\mathcal{D}$ for attribute Birth_Place (Cai *et al.* 1991).

5.2 Applying the *k-mmg* to attribute-oriented induction

As seen before, the algorithm LCHR finds a generalized database as specific as possible by careful applications of substituting concept values. In fact, we can see that what LCHR tries to find is a minimal multiple generalization in the sense of databases with conceptual hierarchy. In the remaining part, the method of application of *k-mmg* to attribute-oriented induction by the previous example.

First, we formalize the notion of generalized databases with conceptual hierarchy. A *concept hierarchy* or *domain* is a pair $(\mathcal{D}, \leq)$ of a set $\mathcal{D}$ of objects and a partial order $\leq$ on D. We assume that $(\mathcal{D}, \leq)$ is a tree with the greatest element $\top$. The relation $\leq$ is called the *generalization relation*. Hereafter, we write $\mathcal{D}$ for $(\mathcal{D}, \leq)$ if it is clear from context. A ground object is a minimal element of $\mathcal{D}$. Assume countably many attributes $a_1, a_2, \ldots$ and the corresponding domains $\mathcal{D}_1, \mathcal{D}_2, \ldots$. Then, a *scheme* is an m-tuple $\mathcal{R} = (a_1, \ldots, a_m)$ of attributes. For the corresponding tuple $(\mathcal{D}_1, \ldots, \mathcal{D}_m)$ of domains, a *gen-*

eralized tuple on $\mathcal{R}$ is an m-tuple $p = (o_1, \ldots, o_m)$ such that $o_1 \in \mathcal{D}_1, \ldots, o_m \in \mathcal{D}_m$. The *value* of a_i in p is o_i $(1 \leq i \leq m)$. A *general ized database* on $\mathcal{R}$ is a finite set R of generalized tuples on $\mathcal{R}$, equivalently, a subset of $\mathcal{D}_1 \times \cdots \times \mathcal{D}_m$. If R consists only of ground objects, then we say R is a *ground* database. Usual relational databases are ground databases. We denote by $\mathcal{DB}_{\mathcal{R}}$ the class of generalized databases on R.

Then we introduce a generalization relation between generalized databases by extending $\leq$ on concept hierarchies. A tuple p is *more general than* q if for every attribute a_i $(1 \leq i \leq m)$, the value of a_i in p is more general than that of a_i in q. A generalized database P is *more general than* Q, denoted by $Q \sqsubseteq P$, if for every q in Q, there is some p in P that is more general than q. The careful reader may find that the relation $(\mathcal{DB}_{\mathcal{R}}, \sqsubseteq)$ is the generalization relation used in the framework of Cai *et al.* (1991), and what their heuristics algorithm searches for is k-minimal multiple generalizations, where k is the threshold.

Let E be a ground database that the learning algorithm is given, and $k > 0$ be the predetermined threshold value. Our method consists of the following three stages:

1. Transform ground database Q on the scheme R into a set S of ground first-order terms that represent tuples in Q.
2. Apply the *k-mmg* algorithm to S. It finds a k-minimal multiple generalization P of S in polynomial time in the size of $||S||$.
3. Transform P to the corresponding generalized database R on $\mathcal{R}$. R is a minimally generalized database of Q.

To apply the *k-mmg* algorithm directly to the problem, we represent a concept hierarchy $(\mathcal{D}, \leq)$ by a specific class $\mathcal{TP}_{\mathcal{D}}$ of first-order terms with generalized relation $\leq'$. For each object o in $\mathcal{D}$, the alphabet $\Sigma_{\mathcal{D}}$ of $\mathcal{TP}_{\mathcal{D}}$ contains a function symbol o, where o is of arity 0 if it is a ground element in $\mathcal{D}$, and of arity 1 otherwise.

Recall that we identified a first-order term p on $\Sigma_{\mathcal{D}}$ with an ordered tree $tree(p)$ labelled by symbols in $\Sigma_{\mathcal{D}}$ in Section 3. The mapping $tree$ maps a term on $\Sigma_{\mathcal{D}}$ to a tree consisting of a single path from the root to the leaf. We use this correspondence to define a mapping τ from $\mathcal{D}$ to $\mathcal{TP}_{\mathcal{D}}$. Hereafter, we consider $\mathcal{D}$ as an ordered tree labelled by objects in $\mathcal{D}$. Let o be an object in $\mathcal{D}$ and π be a branch from the root $\top$ of $\mathcal{D}$ to o. If o is not a ground object, we add a node labelled by a variable, say x, to π as the child of o.

Then, the single path π labelled by function symbols (and possibly a variable) determines the unique first-order term p such that $tree(p)$ is the ordered tree π. Intuitively, a path $\pi = o_1, o_2, \ldots, o_n$ is mapped to a term $o_1(o_2(\ldots(o_n)\ldots))$. Consider the concept hierarchy shown in figure 3.2. Objects China and Shanghai have the following correspondence with first-order terms:

$$\begin{array}{lcl} o = \text{China} & & \tau(o) = \top(\text{foreign}(\text{China}(x))) \\ & \Longleftrightarrow & \\ o' = \text{Shanghai} & & \tau(o') = \top(\text{foreign}(\text{China}(\mathit{Shanghai}))) \end{array}$$

For a generalized tuple $p = (o_1, \ldots, o_m)$, we define the corresponding first-order term $\tau(p)$ by $\tau(p) = db(\tau(o_1), \ldots, \tau(o_m))$ using a special m-ary function symbol db, and the set $\tau(R)$ of first-order terms by $\tau(R) = \{\tau(p) \,|\, p \in R\}$. Let $\mathcal{TP}_D = \{\tau(o) \,|\, o \in \mathcal{D}\}$ and $\mathcal{TP}^*_{\mathcal{R}} = \{\tau(R) \,|\, R \in \mathcal{DB}_{\mathcal{R}}\}$.

Then, $(\mathcal{D}, \leq)$ and $(\mathcal{TP}_{\mathcal{D}}, \leq')$ are order isomorphic. Moreover, $(\mathcal{TP}_{\mathcal{D}}, \leq')$ is closed under generalizations. Thus, if we take a relation $\sqsubseteq'$ defined by $Q \sqsubseteq P$ if for every q in Q, there is some p in P such that $q \leq' p$, $\mathcal{DB}_{\mathcal{R}}$ and $\mathcal{TP}^*_{\mathcal{R}}$ become isomorphic. By careful observation of the result in Section 3, the reader can see that k-mmg algorithm computes a minimal k-multiple term P containing a given finite set S of ground terms with respect to $\sqsubseteq'$.

Hence, we can efficiently find a minimally general database R in $\mathcal{DB}_{\mathcal{R}}$ containing a ground database E under threshold k as follows. First, we transform E to a set of first-order terms $S = \tau(E)$, then compute a k-mmg P of S, and again transform P to a generalized database $R = \tau^{-1}(P)$. We now show an example of computing a minimally generalized database from a given ground database E on the scheme $\mathcal{R} = (\text{Major}, \text{Birth\_Place}, \text{GPA})$ by the k-mmg algorithm.

Step 1. Transform a ground database E to the set S of ground first-order terms by mapping τ (Figure 3.3).

Major	Birth_Place	GPA
⊤(science(Phy))	⊤(Can(B_C(Van)))	⊤(excellent(3.5))
⊤(science(Phy))	⊤(Can(Ont(Ott)))	⊤(excellent(3.9))
⊤(science(Math))	⊤(foreign(Ind(Bom)))	⊤(good(3.3))
⊤(science(Bio))	⊤(foreign(Chi(Shan)))	⊤(good(3.4))
⊤(science(Comp))	⊤(Can(B_C(Vic)))	⊤(excellent(3.8))
⊤(science(Stat))	⊤(foreign(Chi(Nan)))	⊤(good(3.2))

Figure 3.3. A set S of ground first-order terms

Step 2. Apply the *k-mmg* algorithm to S.

1. It first searches for k-multiple terms reduced with respect to S (Figure 3.4).
2. Then it computes k-minimal multiple generalization P by tightening Q with respect to S (Figure 3.5).

Step 3. Transform P to the corresponding generalized database R on $\mathcal{R}$ by mapping τ^{-1}. Then, R is a minimally general generalized database obtained from E with respect to the order $(\mathcal{DB}_{\mathcal{R}}, \sqsubseteq)$ (Figure 3.6).

Major	Birth_Place	GPA
$\top(x_1)$	$\top(\text{Can}(x_2))$	$\top(x_3)$
$\top(\text{science}(y_1))$	$\top(y_2)$	$\top(y_3)$

Figure 3.4. A k-multiple term Q reduced with respect to S

We can easily see that the generalized database computed by our method (figure 3.3) corresponds to what the LCHR algorithm finds (figure 3.2). In the work in (Cai *et al.* 1991), the correctness and the time complexity of the algorithm LCHR

Major	Birth_Place	GPA
$\top$(science(x_1))	$\top$(Can(x_2))	$\top$(excellent(x_3))
$\top$(science(y_1))	$\top$(foreign(y_2))	$\top$(good(y_3))

Figure 3.5. A 2-mimimal multiple generalization P of S

Major	Birth_Place	GPA
science	Can	excellent
science	foreign	good

Figure 3.6. A transformed generalized database R

are not clear. On the other hand, we can estimate those of our method from the correctness and the time complexity of the *k-mmg* algorithm.

Unfortunately, the *k-mmg* algorithm may not find all of the answers. This means that our method may lose some of characteristic rules that Cai et al. method can find. Thus, the study of a *k-mmg* algorithm that can find any *k-mmg*s nondeterministically seems to be interesting from the viewpoint of knowledge discovery in databases.

REFERENCES

Angluin, D. (1980) Inductive inference of formal languages from positive data. *Information and Control*, 45:117–135.

Arimura, H., Shinohara, T. and Otsuki, S. (1991) A polynomial time algorithm for finding finite unions of tree pattern languages. In *Proc. of the 2nd International Workshop on Nonmonotonic and Inductive Logic.* LNAI 659, pp. 118–131. Springer, 1993.

Arimura, H., Shinohara, T. and Otsuki, S. (1992a) Polynomial time inference of unions of two tree pattern languages. *IEICE trans. Inf. and Syst.*, E75-D(7):426–434.

Arimura, H., Ishizaka, H. and Shinohara, T. (1992b) Polynomial time inference of a subclass of context-free transformations. In *Proceedings of 5th Annual ACM Workshop on Computational Learning Theory*, pp.136–143.

Cai, Y., Cercone, N. and Han, J. (1991) Attribute-oriented induction in relational databases. In G. Piatetsky-Shapiro and W. J. Frawley, editors, *Knowledge Discovery in Databases*, pp. 213–228. AAAI Press/The MIT Press.

Gold, E. M. (1967) Language Identification in the Limit. *Information and Control*, 10:447–474.

Ishizaka, H. (1988) Model inference incorporating generalization. *Journal of Information Processing*, 11(3):206–211.

Ishizaka, H., Arimura, H. and Sinohara, T. (1992) Efficient inductive inference of primitive prologs from positive data. In S. Doshita, K. Furukawa and T. Nishida, editors, *Proc. ALT '92*, pp. 135–146.

Lange, S. and Wiehagen, R. (1991) Polynomial-time inference of arbitrary pattern languages, *New Generation Computing*, 8(4):361–370.

Lassez, J-L. and Marriott, K. (1986) Explicit representation of terms defined by counter examples. *Journal of Automated Reasoning*, 3:301–317.

Lloyd, J. W. (1984) *Foundations of Logic Programming.* Springer-Verlag.

Muggleton, S. (1990) Inductive logic programming. In S. Arikawa, S. Goto, S. Ohsuga, and T. Yokomori, editors, *Proc. ALT '90*, pp. 42–62. Ohmsha.

Muggleton, S. and Buntine, W. (1988) Machine invention of first-order predicates by inverting resolution. In *Proc. 5th International Conference on Machine Learning*, pp. 339–352.

Piatetsky-Shapiro, G. and Frawley, W. J. editors. (1991) *Knowledge Discovery in Databases.* AAAI Press/The MIT Press.

Pitt, L. and Valiant, L. G. (1988) Computational limitations on learning from examples. *JACM*, 35(4):965–984.

Plotkin, G. D. (1970) A note on inductive generalization. In B. Meltzer and D. Michie, editors, *Machine Intelligence 5*, pp. 153–163. Edinburgh University Press.

Reynolds, J. C. (1970) Transformational systems and the algebraic structure of atomic formulas. In B. Meltzer and D. Michie, editors, *Machine Intelligence 5*, pp. 135–151. Edinburgh University Press.

Shapiro, E. Y. (1981) Inductive inference of theories from facts. Technical Report 192, Yale University Computer Science Dept.

Shinohara, T. (1991) Inductive inference of monotonic formal systems from positive data. *New Generation Computing*, 8:371–384.

Wright, K. (1989a) Identification of unions of languages drawn from an identifiable class. In *Proceedings of 2nd Annual Workshop on Computational Learning Theory*, pp. 328–333. Morgan Kaufmann.

Wright, K. (1989b) *Inductive Inference of Pattern Languages.* PhD thesis, University of Pittsburgh.

4

The Justification of Logical Theories based on Data Compression

A. Srinivasan

S. Muggleton

Oxford University Computing Laboratory,
11 Keble Road, Oxford, UK

M. Bain

The Turing Institute,
36 North Hanover Street,
Glasgow, UK.

Abstract

Non-demonstrative or inductive reasoning is a crucial component in the skills of a learner. A leading candidate for this form of reasoning involves the automatic formation of hypotheses. Initial successes in the construction of propositional theories have now been followed by algorithms that attempt to generalize sentences in the predicate calculus. An important defect in these new-generation systems is the lack of a clear model for theory justification. In this chapter we describe a method of evaluating the significance of a hypothesis based on the degree to which it allows compression of the observed data with respect to prior knowledge. This can be measured by comparing the lengths of the input and output tapes of a reference Turing machine which will generate the examples from the hypothesis and a set of derivational proofs. The model extends an earlier approach of Muggleton by allowing for noise. The truth values of noisy instances are switched by making use of correction codes. The utility of compression as a significance measure is evaluated empirically in three independent domains. In particular, the results show that the existence of compression distinguishes a larger number of significant clauses than other significance tests. The method also appears to distinguish noise as incompressible data.

1 INTRODUCTION

The ability to form inductive hypotheses is a fundamental requirement for a learner. This is evident even in the most rudimentary robots that plan actions based on continuity assumptions about the state of the world around them. Literature on the automated formation of hypotheses has largely concentrated on their construction. Initial successes in the induction of propositional theories (Michalski 1983; Quinlan 1986; Clark and Niblett 1989) have been followed by algorithms that attempt to generalize sentences in the predicate calculus (DeRaedt and Bruynooghe 1992; Muggleton and Feng 1990; Quinlan 1990; Rouveirol 1991). However, while the need to justify inductive theories has long been recognized (Carnap 1952; Popper 1972) and is an important topic in the work by Plotkin (1971), it remains largely unexplored within machine learning. In particular, although propositional theories have been justified using statistical measures of significance (see, for example, Clark and Niblett 1989), probability estimates (Cestnik 1991), or simplicity (Quinlan and Rivest 1989), there appears to be no clear method for evaluating first-order theories.

In this chapter we describe a model that addresses the issue of hypothesis justification within the framework of learning first-order theories. There are two principal features in our approach. First, given a set of alternative first–order hypotheses for some data, the one chosen is that which is least likely to have explained the data by chance. Second, the model appears to be unique in that it accounts for the use of relevant prior knowledge by the hypotheses. Significance measures used in propositional learners are typically concerned only with the accuracy of a theory and the number of examples it explains. Simply adopting such a measure in a first-order learner would mean giving up the ability to judge if the explanation was based on chance coincidences in the prior knowledge.

2 INCREMENTAL HYPOTHESIS CONSTRUCTION

We adopt the logical framework described by Shapiro (1983). In this, the task of the learning process is to infer some (unknown) target logic program P. There is an intended interpretation (model) for this program M_P. We assume the presence of some predicates whose interpretations are fixed and do not change in M_P. These predicates constitute the background knowledge B. At any given stage i of incremental hypothesis construction, a learner has accumulated examples known to be included in M_P (denoted E_i^+) and those known to be excluded from it (denoted E_i^-). Current learning systems attempt to construct hypotheses H_i under the following constraints:

$$E_i = E_i^+ \cup E_i^-$$

$$E_i^+ \cap E_i^- = \emptyset$$

$$B \wedge H_i \vdash E_i^+$$

$$B \wedge H_i \wedge E_i^- \not\vdash \Box$$

There are in general, infinitely many consistent hypotheses that satisfy these requirements. A snapshot of the construction of one such hypothesis by a learning (induction) machine is shown in Figure 4.1. Despite satisfying the constraints listed

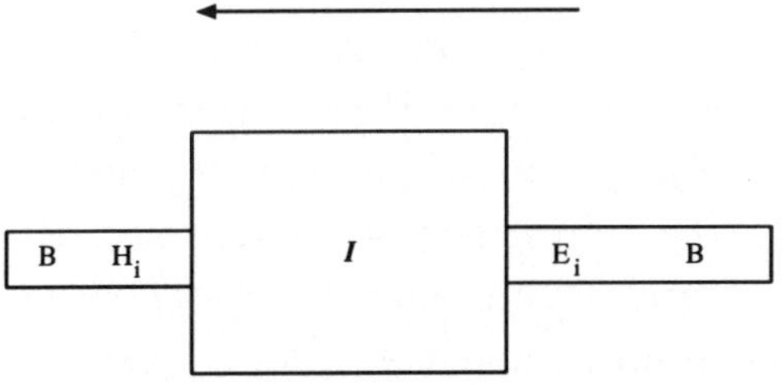

Figure 4.1. A stage in incremental hypothesis construction

above, a hypothesis H_i may be incorrect or incomplete with respect to M_P. This is remedied at subsequent stages when training examples exposing the problem become available to the learner.

There are some practical difficulties with this model of learning:

- Examples may arise from a noisy data source and at a given stage, we may have a theory that appears inconsistent. There is no mechanism for accepting such a theory.
- Even with noise-free data, there may be several consistent hypotheses to explain the data. There is no method of evaluating the explanatory power of these hypotheses.
- There is no direction as to how the examples for training are to be selected.

This chapter is concerned with the first two issues. We note in passing that the normal approach adopted to address the third problem involves either random data-sampling or asking queries of an oracle. The former approach may require a large number of examples before converging on the target program. Recent efforts (Bain 1991; Morales 1991) attempt to actively guide the generation of new examples.

3 INCREMENTAL HYPOTHESIS EVALUATION

In this section we consider evaluating competing hypotheses constructed at a single stage in the incremental learning process described earlier. Although in the 1950s Carnap (1952) and others suggested 'confirmation theories' aimed at providing statistical significance tests for logic-based inductive inference, various difficulties and paradoxes encountered with these approaches meant that they were never applied within machine learning programs (Mortimer 1988). Instead machine learning researchers have for the most part made use of Occam's razor with various *ad hoc* definitions of complexity. However the lack of a clear model underlying such approaches makes it difficult to associate any independent meaning to the simplicity of a hypothesis.

We concentrate on an approach described by Muggleton in (Muggleton 1988). This addresses the theory evaluation question using ideas from algorithmic information theory (Chaitin 1987; Kolmogorov 1965; Solomonoff 1964). In this approach, the utility of a Horn clause program is measured on a reference Turing machine. Unlike the machine in Figure 4.1 this reference machine behaves like a deduction machine. The input to this machine is a program that has two distinct parts: a Horn clause theory and a proof specification. The machine uses the latter to output examples derivable by the theory (Figure 4.2).

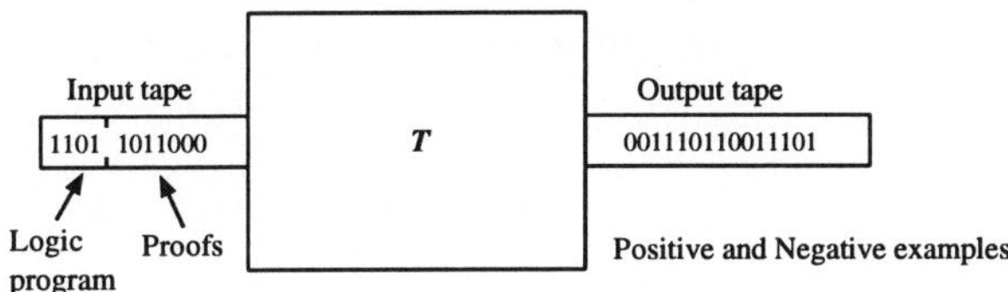

Figure 4.2. A Turing machine model for evaluating logic programs

Following algorithmic information theory, the Horn clause theory is said to be *compressive* if the length of an encoding of the input tape (in bits) is shorter than that of the output tape. The use of a reference machine (as opposed to a universal one) is motivated by demonstrating that the probability of obtaining a compressive theory by chance decreases exponentially with the amount of compression for any machine. This gives a clear meaning to the notion of compression. For completeness, we reproduce the proof of this result here.

Theorem 4.1 Let Σ_n be the set of all binary strings of length n, T be an arbitrarily chosen reference Turing machine and the k-bit-compressible strings of length n, $K_{n,k}$, be defined as $\{y : y \in \Sigma_n, x \in \Sigma_{n-k}, T(x) = y\}$. The set $K_{n,k}$ has at most 2^{n-k} *elements.*

Proof Since Turing machines are deterministic T either induces a partial one-to-one or many-to-one mapping from the elements of Σ_{n-k} to the elements of $K_{n,k}$. Thus $|K_{n,k}| \leq |\Sigma_{n-k}| = 2^{n-k}$. □

Corollary 4.2 The probability of a binary string generated by tossing an unbiased coin being compressible by k bits using any Turing machine T as a decoding mechanism is at most 2^{-k}.

These results provide the theoretical justification for the material in this chapter. Hypotheses are evaluated on the basis of the compression they produce. Theories with higher compression are preferred as they are less likely to explain the data by chance. This deductive evaluation of hypotheses nicely complements the induction process depicted in Figure 4.1. Evaluating the hypothesis constructed at any stage requires us to check the compression produced by the hypothesis at that stage. As seen in Figure 4.3, this is, in some sense, like *reversing* the theory construction stage.

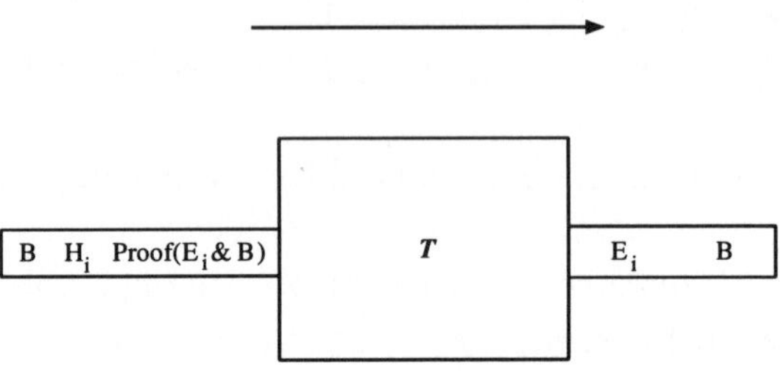

Figure 4.3. A Turing machine model for incremental hypothesis evaluation

In Figure 4.3, the machine T has the following behaviour:

$$T(\mathrm{I}(B \wedge H_i, \mathrm{Proof}(B \wedge E_i | B \wedge H_i))) = \mathrm{O}(B \wedge E_i)$$

where I, O, and $Proof$ are input, output, and proof encodings for T. The k-bit compression achieved by the theory on the input tape is then:

$$k \;=\; |\mathrm{O}(B \wedge E_i)| - |\mathrm{I}(B \wedge H_i, \mathrm{Proof}(B \wedge E_i | B \wedge H_i))|$$

To be of practical value, we also have to specify the encodings I, O, and $Proof$ (this aspect was left unexplored in Muggleton 1988). Clearly, we would like an encoding that is guaranteed to be optimal. However, results from algorithmic information theory have shown that finding an optimal encoding is equivalent to the halting problem. Consequently, we have to be satisfied with a sufficient test for significance based on the encoding scheme adopted. In the next section we describe an efficient coding scheme that can be used to evaluate Horn clause theories for the setting described by Figure 4.3.

4 AN ENCODING SCHEME FOR THE EVALUATION OF LOGICAL THEORIES

4.1 Input tape encoding

The basic premise of the compression model is that efficient (ideally optimal) encodings are found for the input and output tapes of a machine. The components of the input tape for the machine in Figure 4.3 are shown in Figure 4.4. In this diagram, the first three sections (up to the encoding of B) constitute header information.

Size-of-B	No-of-clauses	Symbol-Descrip	B	H_i	Proofs

Figure 4.4. Sections of the input tape for machine T

The reference machine T interprets the input tape as follows:

- The size of the background knowledge (number of atoms and/or clauses in B) allows the machine to distinguish between it and the hypothesis constructed. The number of clauses is used for two purposes. It states how many clauses to expect on the input tape and it is also used to construct a special clause separator symbol. Although the background knowledge can consist of clauses, it is common practice with current first-order learning systems to represent it by a ground model (Muggleton and Feng 1990; Quinlan 1990). The need to specify a symbol description in the header is elaborated shortly.
- A header is generated for the output tape using B and the examples specified by the proof encoding. This header has the same form as that on the input tape. The difference is that the clause count and symbol description refer to the output tape.
- Each example on the output tape is generated by its proof encoding. The machine acts as a logic program interpreter. For each example, the proof encoding specifies the clauses in the hypothesis and background knowledge that are used to derive the example.
- The machine outputs the atoms and/or clauses in B with-

out interpreting them on to the output tape.

4.1.1 *Background knowledge and hypothesis encoding*

A logic program can be viewed as a sequence of symbols. A near optimal choice for encoding these symbols involves the use of prefix codes. We assume a vocabulary S of symbols where each symbol $s \in S$ appears with relative frequency p_s. A prefix code is a function

$$Prefix : S \rightarrow \{0,1\}^*$$

which has the property that no code is a prefix of any other code. This property ensures that codes are self-delimiting. Information Theory (Shannon and Weaver 1963) tells us that the optimal code length for symbol s is $-log_2 p_s$ bits. Huffman coding (Gallager 1968) is a prefix coding which achieves approximately this code length for each symbol.

In order for the machine to 'understand' the encoding of symbols in B and the hypothesis, it is necessary to define the frequencies of the different symbols used. This can then be used to construct a code-book for the message on the input tape. The components of this symbol description header are shown in Figure 4.5.

PSyms	Zero	FSyms	Zero	Vars	Zero	P-arity	F-arity

Figure 4.5. Sections of the symbol description header

Predicate, function and variable symbols have different codes. A prefix table, such as that of the predicate symbols, consists of the individual symbol counts in order of their appearance. This sequence of natural numbers is sufficient for a unique reconstruction of the codes used in the theory. The clause separator symbol (constructed using the clause count) is treated as though it were a predicate symbol. The arities for predicate and function symbols are also number sequences whose orders correspond to those in the prefix tables. Clearly the clause separator 'predicate' symbol has no arity. 'Zero' is defined to be the encoding of the natural number 0 and acts as a separator for different sections of the header. Separators are not necessary to delimit the arities since their number is determined by the predicate and function symbol counts. In order to avoid infinite regress we must find a universal coding for the natural numbers that appear in the header. Natural numbers can be encoded using prefix codes given an appropriate prior distribution. Rissanen (1982) shows that an optimal distribution can be defined for which the code length $L(n)$ is bounded as follows:

$$log_2 n < L(n) < log_2 n + r(n)$$

where $r(n)/log_2 n \to 0$ and $r(n) \to \infty$ as $n \to \infty$. For example, a universal code for numbers could be constructed using N 0s followed by a 1 followed by $2^N - 1$ binary encoded digits $N = 0, 1, 2....$ This gives the following codes for natural numbers:

$0 \to 1$
$1 \to 010$
$2 \to 011$
$3 \to 001000$
...
$11 \to 00010000000$

The approximate code-length for the natural number n using this code is $\lceil log_2(n+2) \rceil + \lceil log_2 \lceil log_2(n+2) \rceil \rceil$. The prior probability distribution assumed for these numbers is $p(n) = (n\log_2{}^2 n)^{-1}$. In general, the lengths of all such universal codes for natural numbers have $log_2 n$ as the dominant term and to first approximation can be taken as bounded by $log_2(n+2) + 2log_2 log_2(n+2)$.

We assume that logical theories are expressed as a set of Prolog clauses. The following grammar gives the syntax of our encoding of theories:

Theory ::= { Clause }$^{\text{No of clauses}}$
Clause ::= Atom Clause | Stop
Atom ::= PredSym [Negated] {Term}$^{\text{Arity(PredSym)}}$
Term ::= ['0'] FuncSym {Term}$^{\text{Arity(FuncSym)}}$ | ['1'] VarSym
Negated::= '0' | '1'

Example 4.3 Consider encoding a theory that consists of the following clauses

normal(Year) :- year(Year), not leap4(Year).
leap4(Year) :- mod(Year,4,0).

The predicate symbols in order of appearance are normal/1, year/1, leap4/1 and mod/3. In addition, there is a clause separator 'predicate'. The function symbols in order of appearance are 4/0 and 0/0. There is a single variable symbol that occurs five times. The encodings for the symbol counts and arities follow (**P**, **D**, and **B** stand for Parse, Decimal and Binary). For the example, we shall use the natural number coding described earlier:

P	P-count					F-count			V-count	
D	1	1	2	1	0	1	1	0	5	0
B	010	010	011	010	1	010	010	1	001010	1
P	P-arity					F-arity				
D	1	1	1	3		0	0			
B	010	010	010	001000		1	1			

This is a total of 44 bits. Huffman codes for the clause separator and the predicate symbols are 00, 10, 110, 01, and 111 respectively. The function symbols have the codes 0 and 1. Since there is only one variable it does not need a code. Thus the coding for the theory itself is as follows (predicate names are abbreviated for typesetting reasons only):

P	norm(...)	yr(...)	not l4(...)	Sep	l4(...)	mod(...)	Sep
P	Atom	...					
B	10 1	110 0 1	01 1 1	00	01 1	111 1 0	00

This is a further 21 bits. The encoding illustrates several points. First, since we are only concerned with Prolog clauses, the head of the clause does not have to be flagged for negation. Further, codes are not needed once the identity of the symbol is determined from the count and arity information. This is evident in the case of encoding the variable: all that is needed is to indicate where Year occurs (using the flag '1'). It is further illustrated in the encoding for mod/3. From the header, it is clear that the predicate has three arguments. From the symbol counts, these have to be single occurences of Year, 4/0 and 0/0. Thus once the position of the variable is known (from the '1' flag), there is no need to flag the remaining arguments as being function symbols. From the encoding of the second argument (by '0'), the third is determined and does not have to be stated.

Although for longer theories we would expect that the header information would be considerably shorter than the statement of the theory, prefix coding may not be very efficient for small theories. Clearly, each symbol type (predicate, function, variable) can be coded differently with bits at the front of the input tape indicating the type of coding adopted for each symbol. This will change the contents of the header. Within our implementation, we can select the most efficient amongst three different coding schemes for a symbol: universal natural number code, a fixed-length code, or a prefix code (listed in order of increasing header information). For each scheme, we use the non-integral code-length as an optimal estimate. The assumption here is that this value can be reached when sufficiently long messages are encoded.

4.1.2 *Proof encoding*

The reference machine T takes the theory and a proof encoding and generates the examples. Derivational proofs are represented as sequences of choices to be taken by a Prolog interpreter. We illustrate this with a simple example.

Example 4.4 Consider a hypothesis that consists of the following clause:

p(X,Y,Z) :- q(X,Y), r(Y,Z).

Background knowledge consists of the following ground atoms:

q(1,2).   r(2,3).
q(2,3).   r(2,4).

```
q(3,4).    r(3,4).
q(4,5).    r(3,6).
```

Consider deriving p(1,2,3) using this hypothesis and background knowledge. In deriving the atom, the interpreter has to first choose which clause of *p/1* to execute. In this case, there is only one such clause and thus, no choice. The first atom in the body of the chosen clause q(X,Y) can be matched against any one of the set of four ground unit clauses in the background. Specifying the choice for the proof (that is q(1,2)) requires 2 bits. This choice determines the choice of the variable Y and constrains the interpreter to one of two possible choices for r(Y,Z). Specifying the atom r(2,3) requires 1 bit. The complete choice-point encoding for the atom p(1,2,3) is specified by the string 011.

Note that the choice-point encoding is procedural: hypotheses that are more efficient to execute have lower choice complexity. This provides a natural bias towards learning efficient clauses. We want to be able to encode a sequence of proofs; one for each example on the output tape. This can be achieved by preceding the series of proofs by an encoding of the number of examples. This encoding of proofs is sufficient for examples which are derivable from range-restricted (generative) theories. However, it has to be extended to accomodate for the following:

1. For non-generative clauses, substitutions have to be provided for variables that do not occur in the body of the clause (since these will never be bound by any choice specification). The function codes for any substitutions needed appear after the choice specifications.
2. Incorrect or incomplete theories can still be used for compressing data to a certain degree. The theory in Example 4.3 is an example of a useful, though incorrect, theory for distinguishing leap years from normal ones.

To address the second issue we distinguish three categories of results obtained from the theory:

1. Correct. In this case the truth-value of the derived fact agrees with the intended interpretation.

2. Error of commission. The truth-value of the derived fact is the opposite of the intended interpretation.
3. Error of omission. The fact cannot be derived.

Each choice-point encoding is preceded by a prefix code indicating its category. The prefix codes for the categories are constructed using three numbers indicating the counts in each category. These numbers are coded using the universal coding scheme and precede the proof encoding on the input tape. Clearly, the total number of examples no longer has to be specified. The encoding of the errors of omission is the same as their explicit encoding.

4.2 Output tape encoding

The output tape encoding is almost the same as that of the input tape (see Section 4.1). The difference is that instead of the hypothesis and proofs, examples are explicitly encoded as atoms.

4.3 A comparison with FOIL's encoding scheme

The selection criterion used by the model in previous sections is one of minimizing description length. Within a first-order framework, the learning system FOIL (Quinlan 1990) uses an encoding length criterion motivated by the Minimum Description Length principle described in (Rissanen 1978). Viewed as a Turing machine model, the 'output' tape length is computed using the following function:

$$log_2(|E|) + log_2(\begin{pmatrix} |E| \\ |E^+| \end{pmatrix})$$

where $|E|$ is the number of examples in the training set, and $|E^+|$ is the number of positive instances covered by the hypothesis. The 'input' tape encoding is simply the encoding of the hypothesised clause. This is given by

$$\sum(1 + log_2(|PredSyms|) + log_2(|Args|)) - log_2(n!)$$

where n is the number of literals in the clause, $|PredSyms|$ is the number of predicate symbols in the background knowledge,

$|Args|$ is the number of possible arguments. The number of bits for the input tape must be less than the number of bits for the output tape. Comparing this to our encoding method we note the following:

1. There is no notion of proof encoding and consequently no apparent instructions on how to reproduce the examples on the output tape.
2. The hypothesis length grows logarithmically with the number of predicate symbols in the background knowledge. A particular clause with given coverage can be invalidated by adding predicates to the background knowledge. There is thus no distinction between relevant and irrelevant background knowledge. In contrast, in our model, the effect of irrelevant background knowledge is equal on input and output tapes.
3. Negative examples are treated asymmetrically.
4. The factor of $log_2(n!)$ is used to correct for the ordering of literals in clauses. The assumption is that any literal ordering has equivalent information content. However in our model the literal ordering affects the choice complexity for proving examples, and thus the corresponding lengths of proofs.

5 COMPRESSION AS A SIGNIFICANCE MEASURE

In this section, we illustrate the utility of using compression as a measure of confidence in clauses learned for three different problems:

1. Prediction of protein secondary structure. The prediction of protein secondary structure from primary sequence is an important unsolved problem in molecular biology. Recently it has been shown that the use of relational learning algorithms (see King and Sternberg 1990; Muggleton *et al.* 1991) can lead to improved performance.
2. Modelling drug structure–activity relationships. The design of a pharmaceutical drug often requires an understanding of the relationship between its structure and chem-

ical activity. Rules learned to model this relationship have been recently been shown to perform better than existing numerical methods (King *et al.* 1992).

3. Learning rules of illegality for the KRK chess end-game. Despite its simplicity, the KRK problem remains the test-bed for ILP techniques. We evaluate the compression measure with different levels of noise. We use a simple noise model that randomly changes the sign of a fixed percentage of the data-set. This is similar to the Classification Noise Process introduced by Angluin and Laird in (Angluin and Laird 1988) [1] .Training and testing is done with 1000 examples in each set.

The error rate of a clause on training data (that is, the *reclassification* or *resubstitution* error rate) can give a highly misleading estimate of the accuracy of the clause on the whole domain. The protein folding domain provides a striking example of this. Figure 4.6 shows a correlation diagram of the accuracies of clauses on training and unseen (test) data for the protein structure prediction problem. Each point in this diagram represents a first-order clause constructed for predicting the position of an α-helix. [2] The rank correlation between the training and test set accuracies in the diagram is 0.3.

In this section we evaluate compression as a method of identifying 'stable' clauses in the following sense. In general, we find a difference between the accuracy of a clause over training and test data. Consider the variation of this difference over a set of clauses judged to be significant by some measure. Small variations indicate that the significance measure has largely avoided selecting clauses that over-fitted the training data.

[1]In this, a noise of η implies that (independently) for each example, the sign of the example is reversed with probability η. This is not the only random noise process possible. For example, a noise of η here corresponds to a class-value noise of 2η in that adopted by Quinlan (1986), and Donald Michie (private communication) advocates a process that preserves the underlying distribution of positive and negative examples.

[2]The clauses were learnt by the learning program Golem (Muggleton and Feng 1990). Note that in this domain the accuracy of a rule which predicted all positions to be part of an α-helix is 0.5.

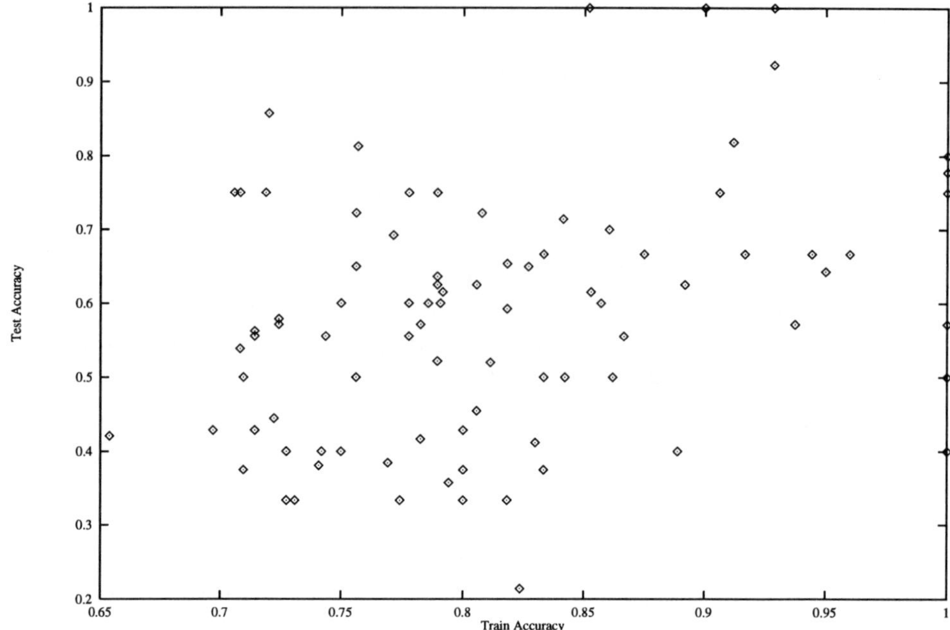

Figure 4.6. Training and test set accuracies of clauses predicting protein structure

We compare the compression measure against the following alternatives:

1. Training set accuracy. The motivation for this is that the observed sample accuracy is a consistent estimate of the domain accuracy of clauses. In the presence of noise, this measure is not very efficient (as seen from Figure 4.6).
2. Training set coverage. These are the numbers of examples that can be derived using a clause. The motivation for this comes from the expectation that the greater the cover, the less likely that sample accuracy will be far from domain accuracy. We consider the coverage of a clause as a percentage of the total training set.
3. Bayesian posterior variance. This measure is described in (Cussens 1992). It concerns the variance of the posterior distribution for the domain accuracy. This distribution is obtained by using Bayes' rule to update a uniform prior.

If a clause covers n examples, t_p of which are true positives then the value of this measure is:

$$VarianceMeasure = 1 - 12\frac{(t_p + 1)(n - t_p + 1)}{(n + 2)^2(n + 3)}$$

The comparison is done as follows.

- For each domain, we first find the compressive clauses. These are the clauses that the compression measure reports as having avoided over-fitting (that is, significant).
- The standard deviation of the difference in training and test accuracies s of compressive clauses is determined.
- For each of the other measures, we order clauses in increasing value of the measure. We want to select the clauses reported significant by the measure. Unlike the compression measure, there is no obvious threshold to determine this clause-set. Consequently, we chose a threshold that selects the largest clause-set with an s value guaranteed to be less than or equal to that of compressive clauses. This ensures that the set of significant clauses is at least as 'stable' as the compressive one.
- We compare the measures on the basis of the number of clauses reported significant by each measure.

Figure 4.7 shows the number of clauses selected as significant by each measure. The thresholds required to obtain these clauses are shown in Figure 4.8.

The diagrams illustrate the following points:

- Training accuracy appears to be clearly the worst judge of clauses. Compression usually distinguishes a larger number of significant clauses than the other measures. Although there is probably not enough data to claim that it does significantly better, we expect it to perform well in domains such as the one predicting protein structure. A characteristic of this domain is that the clauses are highly relational and the background knowledge is quite complex. In these cases, simple measures relying on coverage and accuracy may not be enough.

- In practice, the selection threshold for a significance measure is pre-determined. A compression-based choice is easy: all that is required is to select clauses that are compressive. On the other hand, the threshold for the other measures is not obvious.

Domain	Total clauses	Sd of sig. clauses	Number of significant clauses			
			Compr.	Acc.	% Cover	Var.
Proteins	89	0.09	16	0	7	9
Drugs	107	0.05	103	52	79	76
KRK(5%)	19	0.08	19	19	19	19
KRK(10%)	36	0.13	31	0	12	23
KRK(20%)	40	0.15	33	17	36	33

Figure 4.7. A comparison of significance measures

Domain	Selection thresholds for significant clauses			
	Compression	Accuracy	% Cover	Variance
Proteins	>0	>0.65	>2.9	>0.97
Drugs	>0	=1.0	>2.1	> 0.99
KRK(5%)	>0	>0.50	>0	>0.86
KRK(10%)	>0	>0.54	>11.1	>0.92
KRK(20%)	>0	>0.7	>0.59	>0.83

Figure 4.8. Selection thresholds for significance measures

6 COMPRESSION AS A NOISE-METER

The results in the previous section suggest that compressive hypotheses have good 'accuracy–stability'. We now consider extending this to estimate the amount of noise in the domain. Besides being an interesting quantity in itself, this estimate can be used to provide loose bounds on the expected accuracy of hypotheses on unseen (test) data. Since we are dealing with noisy domains, it is necessary to distinguish between the *real* sample accuracy (on noise-free data) of a hypothesis and its *observed* one (on noisy data).

Consider a noise-free training set classified by an oracle. Let p be the fraction of examples in this data whose classification by a hypothesis agrees with the oracle's classification (this is the real sample accuracy). Clearly, $p \leq 1$. Suppose this data is then subject to noise that changes the class values of a fraction $1-q$ of the examples. That is, a fraction q of the examples is not affected by the noise process. For noisy domains $q \leq 1$. Then, on unseen data from the same source the observed accuracy of the hypothesis can at best be $pq+(1-p)(1-q) = 1-p-q+2pq$. If the hypothesis does not fit any noise at all, then its observed accuracy on the training set will be q. Its corresponding real accuracy p can at best be 1. On the other hand, if the hypothesis fits all the noise, the observed accuracy on the training set will be 1. The corresponding real accuracy will be q. Therefore, a rough estimate is that the observed accuracy of any hypothesis on independent test data from the same distribution lies in the interval $[2q^2 - 2q + 1, q]$.

The problem with this analysis is that in general, the value of q is unknown. All that is known is that q will be the observed training accuracy of a hypothesis that exactly avoids fitting any noise. One possible method of identifying such a hypothesis is to exploit the feature of noisy data as being incompressible by a hypothesis. This suggests the following procedure for determining the amount of noise in the domain. Progressively consider hypotheses with higher training accuracy. Determine the compression of each hypothesis. Estimate q as the training accuracy of the most compressive hypothesis. Clearly, given that we can never prove that our input tape encoding is the shortest one possible, adopting this definition may result in incorrectly condemning some data as noise. Thus, the value of q obtained using this procedure will be an under-estimate (alternatively, the amount of noise will be over-estimated). We tested this on the KRK domain. The experiments involved introducing different levels of noise (using the classification noise model) into a

% Noise introduced	% Noise estimated
0	0
5	5.4
10	10.3
15	15.3
20	20.2
30	30.1
40	40.1

Figure 4.9. Noise estimation using compression

data set of 10000 examples. The noise level was then estimated using the 'most compressive' hypothesis. The results tabulated in Figure 4.9 clearly show the tendency to over-estimate the amount of noise. The relatively large size of the training set is also important to obtain reliable estimates using the compression model. This issue is explored further in the next section.

7 COMPRESSION-GUIDED LEARNING: A CASE STUDY

Based on the positive empirical results of the previous sections, we now describe a case study that involves incorporating the compression model in the learning process. We concentrate on the non-monotonic learning framework called closed-world specialization (Bain and Muggleton 1991). This method progressively corrects a first-order theory by inventing (and possibly generalizing) new abnormality predicates. The process has recently been used to construct a complete and correct solution for the standard KRK illegality problem (Bain 1991). However, a key issue remains to be addressed: there is no mechanism by which a non-monotonic learning strategy can reliably distinguish true exceptions from noise. For example, a strategy based on closed-world specialization would continue specialising until a correct theory is obtained. In noisy domains, this will necessarily result in fitting the noise. To address this, we incorporate the Turing machine compression model in Figure 4.3 within the non-monotonic learning framework. A simple search procedure is developed to find as compressive an explanation as possible for the data. The result of this compression-guided non-monotonic learning is evaluated on the KRK domain.

7.1 Closed-World Specialization

Figure 4.10 shows an algorithm that performs the alternate operations of specialization and generalisation characteristic of closed-world specialization.

It is worth noting here that:

1. As in (McCarthy 1986), there is an assumption that the exceptions to a rule are fewer than the examples that satisfy it.
2. The call to *generalize* results in an attempt to induce a (possibly over-general) rule by a learning algorithm.
3. All rules are added to the theory. Further, all negative examples covered by an over-general clause are taken to be exceptions and the clause is specialized with a (new) abnormality predicate.

7.2 Compression-based clause selection

Each correction performed by the CWS algorithm is an attempt to improve the accuracy of the theory, at the expense of increasing its size. Clearly, if the correction was worth while, the gain in accuracy should outweigh the penalty incurred in increasing the theory size. In encoding terms, each correction increases the theory encoding on the input tape and decreases the proof encoding. In the model in Figure 4.3, a net decrease in the length of the input tape occurs when the correction succeeds in identifying some pattern in the errors (that is, the errors are not noise). The new theory consequently compresses the data further by exploiting this pattern. Using this feature, we evaluate the utility of updating a theory by checking for an increase in compression. We note the following consequences of using the compression model within the non-monotonic framework adopted:

```
start:
  PosE = positive examples of target concept
  NegE = negative examples of target concept
  return learn(PosE,NegE)

learn(Pos,Neg):
  ClauseList = []
  repeat
    C = generalize(Pos,Neg)
    if C ≠ []
      PosC = positive examples covered by C
      NegC = negative examples covered by C
      Pos = Pos - PosC
      Neg = Neg - NegC
      ClauseList = ClauseList + (C,PosC,NegC)
  until C = []

  Theory = []
  foreach (Clause,PosC,NegC) in ClauseList
    if |NegC| ≠ 0
      Theory = Theory + specialise(Clause,PosC,NegC)
    else
      Theory = Theory + Clause
  return Theory

specialize(HornClause,Pos,Neg):
  hd(V1,...,Vn) = head of HornClause
  Body = body of HornClause
  ab = a new predicate symbol
  SpecializedClause = hd(V1,...,Vn) ← Body, not ab(V1,...,Vn)
  PosE = positive examples of ab formed from Neg
  NegE = negative examples of ab formed from Pos
  return SpecializedClause + learn(PosE,NegE)
```

Figure 4.10. Non-monotonic inductive inference using closed-world specialization (CWS)

1. We can be confident of not having fitted the noise only if the theory itself is compressive.
2. With the closed-world assumption, all examples are covered. Consequently, the output tape has to be encoded only once. Input tapes for alternative theories are compared against this encoding.
3. Consider an over-general clause in the current theory. The proof encoding described ensures all variables of the clause are bound to ground terms. Specializing this clause involves adding a negated literal to its body. By appending this literal to the body, we are guaranteed that it will be ground. This ensures safety of the standard Prolog computation rule used by the Turing machine.
4. Recall that the proof encoding for each example has two parts: a choice-point specification and a proof tag. Since the negative literal appended to a clause can never create bindings, the choice-point specification remains unaltered. The size-accuracy trade-off referred to earlier therefore reduces to a trade-off between increasing theory size and decreasing tag size. Not having to recalculate the choice-point encoding for each specialization is a major benefit as this is an extremely costly exercise.

While the aim is to obtain the most compressive subset of the clauses produced by the CWS algorithm, it is unnecessary to examine all subsets since clauses constructed as generalizations of an abnormality predicate cannot be considered independent of the parent over-general clause. For example, it makes no sense to consider the following set of clauses for explaining leap years:

normal(Year) :- year(Year), not leap4(Year).
leap400(Year) :- modulo(Year,400,0).

Despite this, there may still be an intractably large number of clause-sets to consider. Consequently, we adopt a greedy strategy of selecting clauses in order of those that give the most gain (in compression). This strategy has to confront two important issues: devising a reliable method of deciding on the 'best' clause to add to the theory and the fact that adding this clause may not produce an immediate increase in compression.

A simple way to address the first problem is to select the clause that corrects the most errors. Since decreasing errors is the only way to shorten the input tape, the gains are larger for theories that make fewer errors. This works well if all clauses are of approximately the same descriptional complexity. A better estimate would account for the complexity of individual clauses as well. This can be done using average estimates of the cost of encoding predicates, functions, and variables. In the experiments in the next section, this more sophisticated estimate has proved unnecessary. This is because the clauses fitting noisy data tend to correct fewer errors and, therefore, are considered later using the simpler estimate. For the other clauses, the gain from correcting errors dominates the loss from increased theory size.

To address the problem of local minima, it is clearly desirable to have a method of looking ahead to see if a (currently non-compressive) clause will be part of the final theory. To decide this, we calculate an estimate of the compression produced by the most accurate theory containing the clause. The clause is retained if this expected compression is better than the maximum achieved so far. Each time an actual increase in compression is produced, the theory is updated with all clauses that have been retained. Figure 4.11 shows how the estimate is calculated.
The estimated compression will usually be optimistic because it assumes that all errors can be compressed. Of course, one way to guarantee an optimistic estimate is to assume that there will be no increase in theory size (as opposed to the scaled estimate in Figure 4.11). However, this gives no heuristic power and usually only prolongs a futile search for a correct theory. Figure 4.12 summarises the main steps in the compression-based selection of clauses as described here. The following points deserve attention:

1. At any given stage, only some clauses produced by CWS are candidates to be added to the theory (recall the earlier statement that over-general clauses have to be considered

```
estimate(Theory):
  Correct = number of examples correctly classified by Theory
  Max = number of examples that learner can classify correctly
  Outbits = length of output tape (in bits)
  OldTheory = length of Theory (in bits)
  OldTags = current length of correction tags (in bits)
  Choices = length of choice-point encoding (in bits)

  NewTheory = OldTheory × Max / Correct
  NewTags = error tag length to classify correctly Max examples
  EstInbits = NewTheory + Choices + NewTags
  return (Outbits - EstInbits)
```

Figure 4.11. Estimating the compression from a theory

before their specializations).

2. The 'best' clause refers to the clause selected using the simple error-count measure, or the more sophisticated one that accounts for the estimated theory increase. To obtain the latter requires a knowledge of the number of predicate, function, and variable symbols in the clause.
3. Consider the situation where the estimated compression from adding the 'best' clause is no better than the compression already obtained. Figure 4.12 does not acknowledge the possibility that some of the other clauses can do better. It is possible to rectify this by progressively trying the 'next best' clause until all clauses have been tried.
4. The procedure in Figure 4.12 is reminiscent of post-pruning in decision-tree algorithms (the clauses are constructed first and then possibly discarded). A natural question that arises is whether it is possible to incorporate the compression measure within the specialization process The analogy to zero-order learning algorithms is whether tree pre-pruning is feasible. The answer is yes, and in practice it is the mechanism preferred as it avoids inducing all clauses. The procedure in Figure 4.12 however serves to bring out the main features succinctly.

```
start:
  ClauseList = clauses produced by CWS
  return select_clauses(ClauseList)

select_clauses(ClauseList):
  Theory = PartialTheory = []
  Compression = 0
  repeat
    Potential = clauses in ClauseList that can be added to theory
    C = 'best' clause in Potential
    if C ≠ []
      PartialTheory = PartialTheory + C
      NewCompression = compression of PartialTheory
      if NewCompression > Compression
        Theory = PartialTheory
        Compression = NewCompression
      else
        EstCompression = estimate(PartialTheory)
        if EstCompression ≤ Compression return Theory
  until C = []
  return Theory
```

Figure 4.12. Compression-based selection of clauses produced by CWS

7.3 Empirical evaluation

We evaluate the compression-guided closed-world specialization procedure on the KRK domain. However, contrary to normal practice, we chose to learn rules for KRK-legality (as opposed to KRK-illegality). This provides an extra level of exceptions for the specialization method. Given background knowledge of the predicates *lt/2* and *adj/2*, Figure 4.13 shows the target theory.

```
% legal(WK_file, WK_rank, WR_file, WR_rank,BK_file,BK_rank)
legal(A,B,C,D,E,F) :- not ab00(A,B,C,D,E,F).
ab00(A,B,C,D,C,E) :- not ab11(A,B,C,D,C,E).
ab00(A,B,C,D,E,D) :- not ab12(A,B,C,D,E,D).
ab00(A,B,C,D,E,F) :- adj(A,E), adj(B,F).
ab00(A,B,A,B,C,D).
ab12(A,B,C,B,D,B) :- lt(A,D), lt(C,A).
ab12(A,B,C,B,D,B) :- lt(A,C), lt(D,A).
ab11(A,B,A,C,A,D) :- lt(B,D), lt(C,B).
ab11(A,B,A,C,A,D) :- lt(B,C), lt(D,B).
```

Figure 4.13. A complete and correct theory for KRK-legality

It is possible to achieve an accuracy of about 99.6% without accounting for the second level of exceptions. In fact, the theory shown in Figure 4.14 is about 98% correct.

```
legal(A,B,C,D,E,F) :- not ab00(A,B,C,D,E,F).
ab00(A,B,C,D,C,E).
ab00(A,B,C,D,E,D).
ab00(A,B,C,D,E,F) :- adj(A,E), adj(B,F).
```

Figure 4.14. An 'approximately correct' theory for KRK-legality

We experiment with learning the legality theory with artificially introduced noise on different sized training sets. For the experiments, we again adopt the classification noise model referred to earlier. Finally, although the procedure described in Figure 4.12 is not dependent on any particular induction algorithm, the results quoted here use Golem (Muggleton and Feng 1990).

Figure 4.15 tabulates the percentage accuracy of the most compressive theory for different noise levels. Here 'accuracy' refers to accuracy on an independent (noise-free) test set of 10000 examples. Since the compression model only guarantees reliability for compressive theories, nothing can be said about those for which compression is less than 0 (irrespective of their accuracy on the test set). In Figure 4.15, an entry of '_' denotes that the theory obtained is non-compressive on the training data and consequently, no claim is made regarding its accuracy on the test set. The results highlight some important points. Compressive theories do appear to avoid fitting the noise to a large extent. The price for this reliability is reflected in the amount of data required. In comparison, it is possible that other techniques may require fewer examples. However, they either require various parameters to be set (Dzeroski 1991), use *ad hoc* constraints (Quinlan 1990), or need an additional data set for pruning (Brunk and Pazzani 1991). Further, most of them are unable to offer any guarantee of reliability (the approach followed in Dzeroski 1991 can select clauses above a user-set significance threshold). In this respect, our empirical results mirror PAC (Valiant 1984) results for learning with noisy data in propositional domains (Angluin and Laird 1988): with increasing noise, more examples are needed to obtain a good theory. It is also worth noting that the conditions covered by the second level of exceptions (the cases in which the White King is in between the White Rook and Black King) occur less than 4 times in every 1000 examples. This is only picked up in the noise-free data set of 10000 examples (in which there were 38 examples where the rules applied).

Extending the PAC analogy further, Figure 4.16 shows the results from a different perspective. For different levels of noise, this figure shows the number of training examples required for the 'approximately correct' theory of Figure 4.14 to be compressive. For example, at least 170 examples are required to obtain a compressive theory that is 98% accurate on noise-free data. While these numbers are approximate (they are obtained by

Noise (%)	Training Set Size 100	250	500	1000	5000	10000
0	–	99.7	99.7	99.7	99.7	100
5	–	98.1	98.1	99.7	99.7	99.7
10	–	–	98.1	98.1	99.7	99.7
15	–	–	98.1	98.1	99.7	99.7
20	–	–	–	98.1	99.7	99.7
30	–	–	–	–	98.1	98.1
40	–	–	–	–	–	98.1

Figure 4.15. Test-set accuracy for the 'most compressive' theory

extrapolating the compression produced by the theory for the different training sets in Figure 4.15) they do indicate the general trend of requiring larger example sets for increasing noise levels.

8 CONCLUSION

In this chapter we have developed a general encoding scheme for deciding the significance of first-order hypotheses by refining the approach found in (Muggleton 1988). The requirement to encode both hypotheses and proofs results in some unique advantages:

1. The resulting compression measure appears to be the first significance measure that accounts for the relevance and utility of background knowledge. This issue has been avoided to date by relational learning systems.
2. The measure appears to reliably distinguish noisy data by finding them to be incompressible with the background knowledge.
3. The encoding incorporates notions of efficiency in the same units (bits) as the program description. Michie (1977) discusses in detail the time–space tradeoff for encoded knowl-

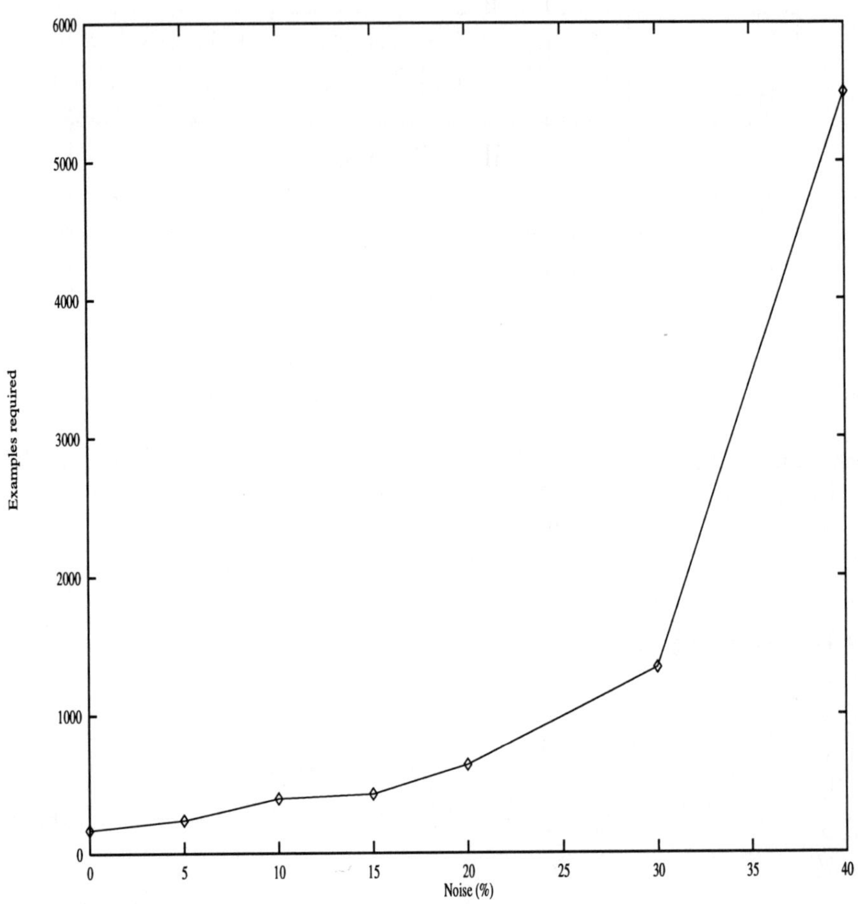

Figure 4.16. Examples required for a 98% correct and compressive theory

edge using the standard Machine Intelligence problem of perfect play in chess. Within Machine Learning, various methods including EBL (Mitchell *et al.* 1986) have been developed to learn the time-space tradeoff. EBL systems take a set of axioms and add redundant theorems in order to reduce proof lengths for specific types of problems. These systems do not contain any sound criterion of whether theorems being added are significant. A method of estimating the utility of adding these theorems is described in (Subramanian and Hunter 1992). The method does not appear to directly account for the space dimension. The compression model described in this chapter should be able to provide a criterion that accounts for this as well.

The question of how well hypotheses perform on unseen data was dealt with first by Gold (1967) and more recently within the PAC (Probably-Approximately-Correct) framework. The Gold and PAC frameworks describe the conditions under which a class of concepts can be said to be learnable. In the PAC framework it is explicitly assumed that the distribution of examples in the training and test sets are the same. Blumer *et al.* (1986), Board and Pitt (1989), and Li and Vitanyi (1989) have in various ways shown that a class of concepts is PAC-learnable if and only if it can be guaranteed that a learning algorithm is able to find a hypothesis which is smaller than the data. It remains to be shown that our concept of hypothesis size (that is, hypothesis and proofs) is equivalent to that adopted in these theoretical results.

The task of distinguishing between exceptions and noise is an issue that is typically ignored in the literature on non-monotonic reasoning. It is, however, of fundamental importance for a learning program that has to construct theories using real-world data. One way to approach the problem is to see if the exceptions to the current theory exhibit a pattern. The compression model we have used in this chapter does precisely this. Our empirical results suggest that by selecting the most compressive theory, it is possible (given enough data) to reliably avoid fitting most of the noise. Clearly, it would be desirable to confirm these results with controlled experiments in other domains. In practice, the method has found interesting rules on an independent problem of pharmaceutical drug design (King *et al.* 1992). Finally, the results also lend support to the link between compressive theories for first-order concepts and their PAC-learnability.

Acknowledgments

The authors would like to thank Donald Michie and the ILP group at the Turing Institute for their helpful discussions and advice. This work was carried out at Oxford University Computing Laboratory and the Turing Institute and was supported by the Esprit Basic Research Action project ILP (6020), the IED's Temporal Databases and Planning Project and the SERC Rule-Base Systems Project.

REFERENCES

Angluin, D. and Laird, P. (1988). Learning from noisy examples. *Machine Learning*, 2(4):343–370.

Bain, M. (1991). Experiments in non-monotonic learning. In *Proceedings of the Eighth International Workshop on Machine Learning*, pages 380–384, San Mateo, CA. Morgan Kaufmann.

Bain, M. and Muggleton, S. (1991). Non-monotonic learning. In Michie, D., editor, *Machine Intelligence, 12.* Oxford University Press.

Blumer, A., Ehrenfeucht, A., Haussler, D., and Warmuth, M. (1986). Classifying learnable geometric concepts with the Vapnik Chervonenkis dimension. In *Proceedings of the 18th ACM Symposium on Theory of Computing*, pages 273–282.

Board, R. and Pitt, L. (1989). On the necessity of occam algorithms. UIUCDCS-R-89-1544, University of Illinois at Urbana-Champaign.

Brunk, C. and Pazzani, M. (1991). An investigation of noise-tolerant relational concept learning algorithms. In Birnbaum, L. and Collins, G., editors, *Proceedings of the Eighth International Workshop on Machine Learning*, San Mateo. Morgan Kaufmann.

Carnap, R. (1952). *The Continuum of Inductive Methods.* Chicago University, Chicago.

Cestnik, B. (1991). Estimating probabilities: A crucial task in machine learning. In *Proceedings of the European Conference on Artificial Intelligence*, Stockholm, Sweden.

Chaitin, G. (1987). *Information, Randomness and Incompleteness – Papers on Algorithmic Information Theory.* World Scientific Press, Singapore.

Clark, P. and Niblett, T. (1989). The CN2 algorithm. *Machine Learning*, 3(4):261–283.

Cussens, J. (1992). Estimating rule accuracies from training data. In *ECAI Workshop on Logical Approaches to Learning.*

DeRaedt, L. and Bruynooghe, M. (1992). Interactive concept-learning and constructive induction by analogy. *Machine Learning*, 8(2):107–150.

Dzeroski, S. (1991). *Handling Noise in Inductive Logic Programming.* University of Ljubljana, (M.Sc. Thesis), Ljubljana.

Gallager, R. G. (1968). *Information Theory and Reliable Communication.* Wiley, New York.

Gold, E. (1967). Language identification in the limit. *Information and Control*, 10:447–474.

King, R., Muggleton, S., and Sternberg, M. (1992). Drug design by machine learning: The use of inductive logic programming to model the structure–activity relationships of trimethoprim analogues binding to dihydrofolate reductase. *Proc. of the National Academy of Sciences*, 89(23).

King, R. and Sternberg, M. (1990). A machine learning approach for the prediction of protein secondary structure. *Journal of Molecular Biology*, 216:441–457.

Kolmogorov, A. (1965). Three approaches to the quantitative definition of information. *Prob. Inf. Trans.*, 1:1–7.

Li, M. and Vitanyi, P. (1989). Inductive reasoning and Kolmogorov complexity. In *Proceedings of the Fourth Annual IEEE Structure in Complexity Theory Conference*, pages 165–185.

McCarthy, J. (1986). Applications of circumscription to formalizing common sense knowledge. *Artificial Intelligence*, 28:89–116.

Michalski, R. (1983). A theory and methodology of inductive learning. In Michalski, R., Carbonnel, J., and Mitchell, T., editors,

Machine Learning: An Artificial Intelligence Approach, pages 83–134. Tioga, Palo Alto, CA.

Michie, D. (1977). A theory of advice. In Elcock, E. and Michie, D., editors, *Machine Intelligence, 8*, pages 151–168. Horwood.

Mitchell, T., Keller, R., and Kedar-Cabelli, S. (1986). Explanation-based generalization: A unifying view. *Machine Learning*, 1(1):47–80.

Morales, E. (1991). Learning features by experimentation in chess. In Kodratoff, Y., editor, *EWSL '91*, pages 494–511, Berlin. Springer Verlag.

Mortimer, H. (1988). *The Logic of Induction.* Ellis Horwood, Chichester, England.

Muggleton, S. (1988). A strategy for constructing new predicates in first order logic. In *Proceedings of the Third European Working Session on Learning*, pages 123–130. Pitman.

Muggleton, S. and Feng, C. (1990). Efficient induction of logic programs. In *Proceedings of the First Conference on Algorithmic Learning Theory*, Tokyo. Ohmsha.

Muggleton, S., King, R., and Sternberg, M. (1991). Predicting protein secondary structure using inductive logic programming. *Protein Engineering*, 5:647–657.

Plotkin, G. (1971). *Automatic Methods of Inductive Inference.* PhD thesis, Edinburgh University.

Popper, K. (1972). *Conjectures and Refutations: The Growth of Scientific Knowledge.* Routledge and Kegan Paul, London.

Quinlan, J. (1986). Induction of decision trees. *Machine Learning*, 1:81–106.

Quinlan, J. (1990). Learning logical definitions from relations. *Machine Learning*, 5:239–266.

Quinlan, J. and Rivest, R. (1989). Inferring decision trees using the Minimum Description Length principle. *Information and Computation*, 80:227–248.

Rissanen, J. (1978). Modeling by shortest data description. *Automatica*, 14:465–471.

Rissanen, J. (1982). A universal prior for integers and estimation by Minimum Description Length. *Annals of Statistics*, 11:416–431.

Rouveirol, C. (1991). Itou: Induction of first-order theories. In *First International Workshop on Inductive Logic Programming*, Porto,

Portugal.

Shannon, C. and Weaver, W. (1963). *The Mathematical Theory of Communication.* University of Illinois Press, Urbana.

Shapiro, E. (1983). *Algorithmic Program Debugging.* MIT Press.

Solomonoff, R. (1964). A formal theory of inductive inference. *Information and Control*, 7:376–388.

Subramanian, D. and Hunter, S. (1992). Measuring utility and the design of provably good ebl algorithms. *Ninth International Machine Learning Conference.*

Valiant, L. (1984). A theory of the learnable. *Communications of the ACM*, 27:1134–1142.

5

Utilizing Structure Information in Concept Formation

K. Handa, M. Nishikimi and H. Matsubara

Electrotechnical Laboratory
Umezono 1-1-4, Tsukuba, JAPAN 305

Abstract
Inductive learning is a quickly growing area of machine learning, and the methodology of *concept formation* is to realize the task of inductive learning incrementally and without supervision. We have proposed a concept formation system CAFE, that creates a concept hierarchy from structured instances. The system uses a new classification algorithm *mutual induction* and an evaluation function known as *concept-predictability*. Our theory is that a system that deals with structures can not only expand possible domains of learning but also use structure information of instances as a domain theory, enabling accurate and fast learning. Such a system can even learn to neglect attributes that do not affect classification, provided appropriate structure information is given. We have evaluated CAFE in two domains (artificial and natural), and the results have proved our theory.

1 INTRODUCTION

An intelligent system, whether natural or artificial, has to adjust itself to its environment. It is unaware of its surroundings in the first instance, so it has to be able to acquire information while dealing with the world around it. Machine learning is, therefore, a growing area of artificial intelligence, where we pursue a system that can autonomously improve its performance.

Machine learning systems vary in their targets and techniques. Systems may learn parameters, operators, strategies, or concepts through reinforcement, deduction, induction, or analogy. Among these, inductive learning of concept hierarchy from examples is an exciting area, because we humans create and utilize concepts based on examples in our everyday intellectual activities.

Research into incremental and unsupervized learning of concept hierarchy has, until recently, ignored the possibility of using domain theory, or background knowledge. In this paradigm, a robust way of constructing a concept hierarchy that accurately predicts the values of attributes from a series of unclassified instances is sought.

Fisher's COBWEB (Fisher 1987), for example, assumes no knowledge of the domain, and thus instances are represented by a set of attribute–value pairs, where every attribute is treated equally. The system builds a concept hierarchy from scratch. This design presupposes that the system has no information about the relative importance of each attribute and thus all attributes have equal probability of contribution to the classification/discrimination of instances. It is a natural result of this presupposition that attributes that do not affect the classification retard COBWEB's learning and worsen its performance.

For example, let us consider the tasks necessary to predict the value of the Class attribute when instances are represented by a set of attributes [Class, Attr1, Attr2, Attr3, Attr4] but Attr3 and Attr4 have no correlation with Class. COBWEB requires many instances to learn to ignore Attr3 and Attr4. We call such attributes that do not contribute to classification *attribute noise.* Although great efforts have been made to cope with noise in attribute values and noise in classification (Iba *et al.* 1988; Carlson *et al.* 1990; Drastal 1991), *attribute noise* is left unchallenged in unsupervized learning. The *feature construction* technique in constructive induction (Schlimmer 1987; Matheus and Rendell 1989; Aha 1991; Drastal 1991) is an attempt to overcome this problem by reforming instances such that they do not contain *attribute noise*, but is still limited in supervized learning. In unsupervized learning, however, it is difficult to find a good guidance of *feature construction.*

Instances in natural domains, however, often have structure, or *part-of* relations, between themselves, which can be foreseen when we begin to classify them. This structure information affords a simple way of favoring some of the attributes. In the above example, if related attributes Attr1 and Attr2 are grouped together into one component (say Comp), the system classifies one instance as a body-part Body = [Class, Comp, Attr3, Attr4] and as its component Comp = [Attr1, Attr2]. In this situation, it is easier for the system to neglect unrelated attributes because they are not considered while classifying Comp.

Using part-of relations of instances as background knowledge, therefore, leads to a more flexible and realistic concept formation system. Among the few works on this topic is Thompson's LABYRINTH (Thompson and Langley 1991a), and which has had a great influence on our work. We provide a brief review of LABYRINTH in Section 2.

We previously proposed a concept formation system, CFIX, that can classify structured instances (Handa 1990). In this chapter we propose CAFE, an enhanced version of CFIX.

CAFE and LABYRINTH are directly influenced by COBWEB, and both systems share many features. But the fundamental differences are as follows:

- In the knowledge representation of CAFE, a component-part of an instance has a pointer to its body-part and can be further structured.
- CAFE classifies each instance (and all its components) twice; once to obtain temporary identifiers of the concepts to which its components belong, and then to revise the entire concept hierarchy (*Mutual Induction*).
- CAFE utilizes an evaluation function termed here as *concept-predictability* to classify components of instances into concepts of adequate specificity.

The first two characteristics enable CAFE to consider context while classifying components. Two otherwise identical components can form different concepts if they appear in bodies that are classified differently. The use of *concept-predictability* keeps information about each instance and, at the same time, avoids overly specified descriptions in the representation of a body.

This chapter describes the evaluation of CAFE and COBWEB through two experiments. The results show that CAFE learns faster and enables higher predictive accuracy in a domain with much *attribute noise.*

In the following section, we overview CAFE. We then report on the experimental results and finally discuss the related work and problems currently pending.

2 CAFE SYSTEM OVERVIEW

2.1 Knowledge Representation

A knowledge representation formalism used in CAFE resembles that of LABYRINTH, but is more relation oriented. Both systems accept an instance composed of components and have two types of attribute, one whose value is a primitive value (*primitive attribute*) and one that points to another concept in the hierarchy tree (*relational attribute*). In LABYRINTH, the relationships between components are all described in the body-part of instance. In CAFE, however, these relations are described in component-parts. In addition, CAFE not only permits components to be composed of lower-level components but also accepts instances related to each other. Therefore, we can apply CAFE to more relational domains, such as a network of relatives of people.

Figure 5.1 shows the difference between the representations of an object (learning instance) in CAFE and LABYRINTH using the same example as in (Thompson and Langley 1991a).

In the diagram, *relational attributes* of CAFE have subscripts s and d to indicate the source and destination parts of the relation, because all relations in CAFE are bi-directional. A relation may have an additional subscript to indicate how much the relation contributes to classification. The value of this factor defaults to 1 for a *primitive attribute* and to the number of *primitive attributes* of the pointed concept for a *relational attribute.*

Getting such instances one by one, CAFE gradually constructs a hierarchy tree of concepts as shown in Figure 5.2. The

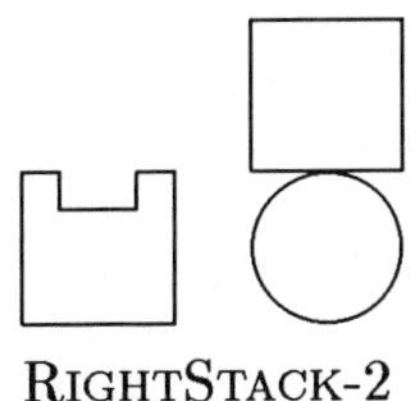

RIGHTSTACK-2

LABYRINTH:

```
(Rightstack-2 (component_1 (color blue) (shape odd))
              (component_2 (color red) (shape circular))
              (component_3 (color grey) (shape square))
              ((Left-of component_1 component_3))
              ((Left-of component_1 component_2))
              ((on component_3 component_2)))
```

CAFE:

```
(Rightstack-2 (left-part/s component_1)
              (right-part/s component_2))
(component_1 (left-part/d Rightstack-2)
             (left-of/s component_2)
             (color blue) (shape odd))
(component_2 (right-part/d Rightstack-2)
             (left-of/d component_1)
             (top-part/s component_21)
             (bottom-part/s component_22))
(component_21 (top-part/d component_2) (on/s component_22)
              (color grey) (shape square))
(component_22 (bottom-part/d component_2) (on/d component_21)
              (color red) (shape circular))
```

Figure 5.1. Representation of instance

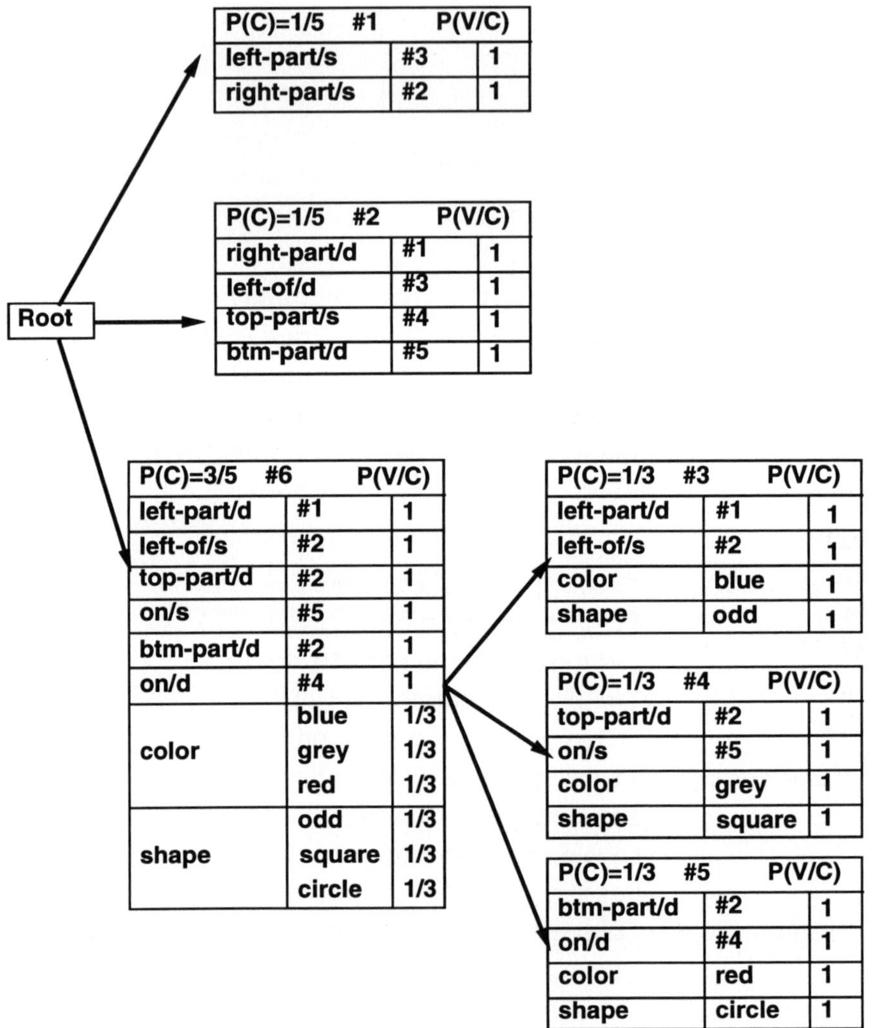

Figure 5.2. Hierarchy tree of concepts

diagram shows a tree constructed from the sole instance of Figure 5.1. In the diagram, each number preceded by # is a concept identifier. Each concept consists of a list of triplets. The first element of the triplet is an attribute name, the second is a list of values of the attribute, and the third is a list of frequencies of the attribute's taking the corresponding value. Again, the tree structure resembles that of LABYRINTH and hence that of COBWEB. The main difference is that while LABYRINTH holds a pointer to a generalized concept as a value of relational attributes, CAFE holds all pointers to non-generalized concepts because of the requirement of CAFE's evaluation function as described in the following section.

2.2 Classification and Learning

CAFE generates a concept hierarchy by using an *incremental hill-climbing* approach, in the same way as COBWEB and its descendants. Each time a new instance is given, CAFE classifies it and, at the same time, revises the concept hierarchy dynamically to a more plausible one. Since only the current concept is retained, backtracking is impossible but can be simulated by applying operators to the current hierarchy. Gennari *et al.* (1989) shows how COBWEB's two operators *merge* and *split* work for simulating backtracking. The two main characteristics of CAFE's learning that make it different from the COBWEB family are:

- **Mutual induction**:
  A mechanism for integrating structured objects into a concept hierarchy. CAFE classifies the body-parts of a structured object based on the classification information of component-parts, which is the same way as is done by LABYRINTH. In addition, in CAFE, the classification of component-parts is also influenced by the classification of body-parts. We named this mutual influence of classifications *Mutual Induction.*
- **Concept-predictability**:
  A function to decide how well the value of a relational attribute is predicted. Like COBWEB, CAFE also uses

category utility as an evaluation function to search for a plausible classification. In it, we must calculate the predictability of an attribute's value at each concept. Simply applying COBWEB's method to a relational attribute, however, leads to too low an estimation of predictability, because such values are pointers to other concepts and are different from each other. To overcome this problem, we introduced a *concept-predictability* function.

Table 5.1 shows the basic CAFE algorithm for classifying and learning with structured objects. Faced with a new object, CAFE tries to classify it twice (see the procedure `Cafe`): first to obtain temporary identifiers of concept for all *relational attributes*, next to get real identifiers based on the temporary ones. In the procedure `Cafe2`, component objects pointed to by the relational attributes of an object are first classified, after which the object itself is classified. After classification of the object, relational attributes of the component objects are updated to point to the identifier of the objects.

In the procedure `Cafe3`, `Cobweb´` is called as an internal function. `Cobweb´` is based on COBWEB, but with the addition of being able to use *concept-predictability* in the evaluation function, as described in the next section.

2.3 Evaluation Function

COBWEB tries to classify a new object by applying the divide-and-conquer strategy, classifying an object firstly at the top node of the hierarchy tree, going down to the lower subtrees. At each level of the tree, COBWEB selects the best subtree categorization of the tree by using an evaluation function based on Gluck's *category utility* (Gluck and Corter 1985). Candidate categorizations are:

- incorporating an object into one of the subtrees,
- merging two subtrees into one and incorporating the object into it,
- splitting a subtree into further subtrees,
- creating a new subtree for an object.

Table 5.1. The basic CAFE algorithm

```
Procedure Cafe(TABLE, ID, TREE, NEWTREE)
  /* Classify the object TABLE[ID] into TREE
     and return revised NEWTREE. */
  Update: TABLE;   /* An array of objects. */
  Input:  ID;      /* An identifier of object to be classified. */
          TREE;    /* A hierarchy tree of concepts. */
  Output: NEWTREE; /* An updated hierarchy tree. */
{
  Cafe2(TABLE, ID, TREE, _);
  Cafe2(TABLE, ID, TREE, NEWTREE);
}

Procedure Cafe2(TABLE, ID, TREE, NEWTREE)
  /* Classify TABLE[ID] into TREE and return revised NEWTREE. */
  Update: TABLE;
  Input:  ID, TREE;
  Output: NEWTREE;
{
  var OBJECT = TABLE[ID], TEMPTREE1 = TREE, TEMPTREE2;
  var CID, CID1, CID2; /* Concept identifiers. */
  For each component of OBJECT {
    Set CID to an identifier of the component;
    Cafe2(TABLE, CID, TEMPTREE1, TEMPTREE2);
    TEMPTREE1 = TEMPTREE2;
  }
  Cafe3(TABLE, TEMPTREE1, OBJECT, NEWTREE, CID1);
  Set CID1 to an identifier of OBJECT;
  For each CID2 in NEWTREE {
    If relational attributes of CID2 points to CID,
      Then update the pointer to CID1;
  }
  Update each element of TABLE with concerning NEWTREE;
}

Procedure Cafe3(TABLE, TREE, OBJECT, NEWTREE, CID);
  Input:  TABLE, TREE, OBJECT;
  Output: NEWTREE, CID;
{
  var ATTRIBUTE, VALUE;
  For each set of ATTRIBUTE and VALUE of OBJECT {
    If ATTRIBUTE is relational and TABLE[VALUE] has no identifier,
      Then set VALUE to nil;
  }
  Classify OBJECT into TREE by Cobweb′ and get updated NEWTREE;
  Set CID to an identifier of classified OBJECT;
}
```

The evaluation function takes the following form:

$$\frac{1}{K}\Big(\sum_{k=1}^{K} P(C_k)\sum_i\sum_j P(A_i = V_{ij}|C_k)^2 - \sum_i\sum_j P(A_i = V_{ij}|C)^2\Big), \qquad (5.1)$$

where k varies between subtrees, i between attributes, and j between the values of each attribute. C is the parent node and C_k is the kth subcategory (i.e. subtree) of C. This function evaluates the improvement in the predictability of attribute values as a result of this categorization. The term $P(A_i = V_{ij}|C_k)^2$ in the function is an indicator of the similarity of instances in C_k.

In CAFE, however, V_{ij} of *relational attribute* is a pointer to another concept in the hierarchy tree. There is no V_{ij} shared by more than two instances of C_k, therefore, the similarity of the instances is estimated too low. Hence, we propose new function named *concept-predictability* for such an attribute of C_k.

The function takes the form:

$$\frac{1}{L}\Big(\sum_{l=1}^{L}\sum_i\sum_j P(A_i = V_{lij}|C_k)^2\Big), \qquad (5.2)$$

where l varies with the level (or depth) of the hierarchy tree, i with the attributes, and j with the concepts at each level of the tree. V_{lij} is an identifier of the jth concept at level l to which a concept pointed to by the ith attribute can be generalized. $l = 0$ means the top node of the tree and L indicates the deepest level that contains multiple concepts that are generalizations of instances in C_k at the level. In a sense, Formula 5.2 averages the similarities of instances that are calculated with the assumption that all instances are generalized to each level. The summation stops at level L instead of at the leaf (deepest) level to avoid the value becoming too low while the tree is growing deeper and instances are, accordingly, being classified into a deeper leaf.

Figure 5.3 shows two examples of this calculation. In the upper example, concept `#1` has one relational attribute `rel` and the possible values are concepts `A`,`B`, and `E`. These are distributed in

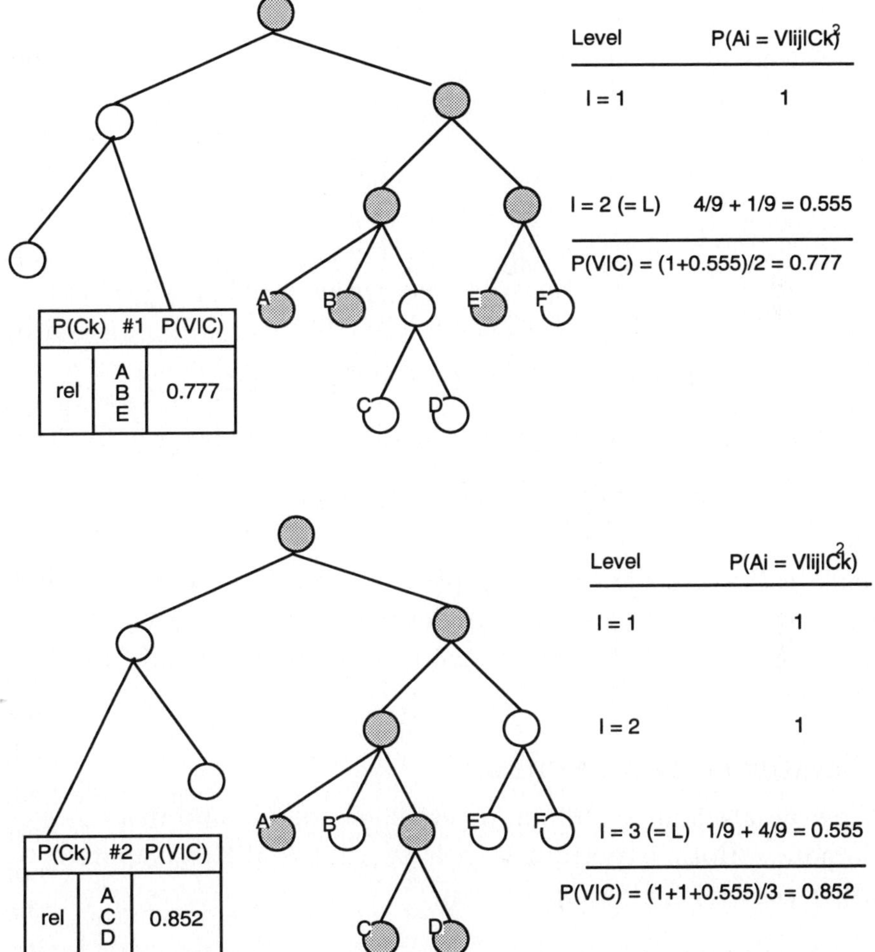

Figure 5.3. *Concept-predictability* function

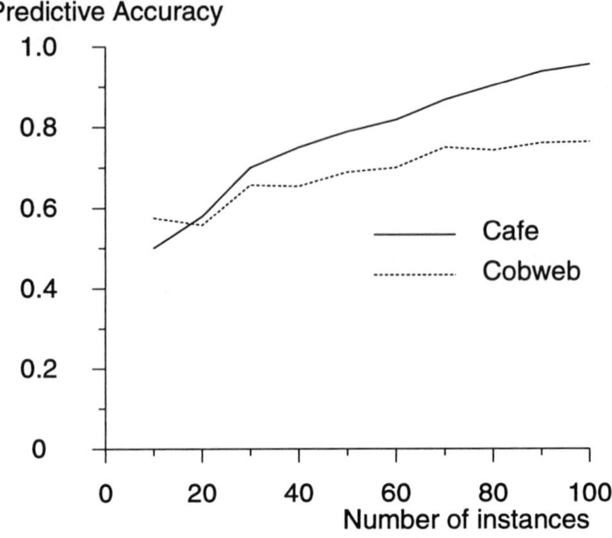

Figure 5.4. Learning curves in artificial domain (1)

the hierarchy tree as shown in the diagram. At level 1, formula

$$\sum_i \sum_j P(A_i = V_{lij} | C_k)^2$$

has a value of 1 since all concepts `A,B`, and `E` are generalized to the same node at this level. At level 2 ($= L$), the same formula has a value of 0.555. Hence, the concept-predictability of `#1` is calculated to be 0.777.

3 EVALUATION OF THE SYSTEM

We evaluated our system in an artificial domain and in a natural domain. Both experiments proved that CAFE is superior to other concept formation systems.

3.1 Artificial Domain

This experiment was done to test the stated theory:

- COBWEB's learning ability worsens as *attribute noise* increases.
- If the correlation between *contributive attributes* and *attribute noise* is cut off by structure information about instances, the learning ability of the system can be improved.

- Since CAFE utilizes such information on learning, its ability is relatively free of *attribute noise.*

Learning instances in our artificial domain have the following nine attributes:

```
locx, locy, locz, temperature, humidity,
sunlight-hour, air-pressure, soil-type,
water-quality
```

Each takes an integer number between 1 and 10 as its nominal value[1]. Only the first three attributes have meanings. The values of the remaining attributes are generated randomly (i.e. *attribute noise*). The values of `locx` and `locy` are also generated randomly but these values determine the value of `locz`, according to the following rule:

If `locx` ≤ 7 and `locy` ≤ 7, then `locz` is `high`, else `locz` is `low`.

Almost half the learning instances generated by this rule have a `high` value for `locz`. In our test, COBWEB was given only a plain list of attribute values because it can't handle any more complicated data, while CAFE was given structured data. The following is an example of ith instance given to CAFE(see Figure 5.1 for clarification of how to read it):

```
(instance_i
  (comp comp_i) (temperature 2) (humidity 5)
  (sunlight-hour 2) (air-pressure 10) (soil-type 3)
  (water-quality 8)
(comp_i
  (body instance_i) (locx 6) (locy 4) (locz low))
```

The performance task of the system is to predict the value of `locz` from test instances that are not given the value of `locz`.

We have run COBWEB and CAFE on 12 random orders of 100 training instances with another 20 test instances (10 `high` and 10 `low`). Test instances are given with the value of `locz` being set to nil (unknown). Figure 5.4 shows the learning curves for both systems, averaged over 14 trials, with different training and

[1] Attribute names have no meaning in themselves. Those are assigned simply to improve readability.

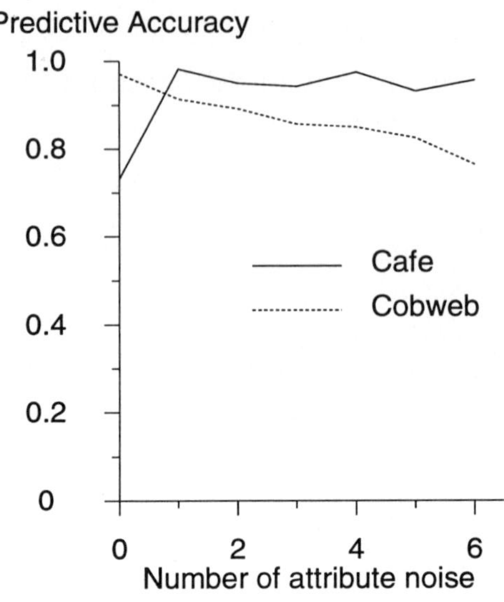

Figure 5.5. Learning curves in artificial domain (2)

and test instances. The *t-test* indicates the significant difference ($t > t_{0.05,26}$) between CAFE and COBWEB, every tenth step between the 60th and 100th training instances. Also, we have run each system in similar domains to determine how *attribute noises* affects learning performance. The form of instances in each domain varies only in the number of *attribute noises*. In our previous experiments, the number of *attribute noise* was 6. In this experiment we varied the number between 0 and 6. Predictive accuracy was recorded at the 100th training instance in each domain. Figure 5.5 shows the performance of both systems at each domain, averaged over 14 trials. This diagram clearly indicates the resilience of CAFE against *attribute noise*. The reason for the low value at an *attribute noise* of 0 was that CAFE cannot distinguish body part from components and consequently created an inappropriate concept hierarchy in an early stage. The number of training instances (100) was not sufficient for the system to recover from it.

3.2 Natural Domain

Our next experiment was to evaluate CAFE in a real world domain. Fortunately, we can use the data set for DNA promoters[2], already studied by several researchers in the field of machine learning (Towell *et al.* 1990; Ourston and Mooney 1990; Thompson *et al.* 1991b). This data set contains 53 positive (i.e.*promoter*) and 53 negative (i.e.*non-promoter*) instances, where features of each instance are 57 sequential DNA nucleotides (the values of which can be A, G, T, or C). There also exists a domain theory that may serve as background knowledge for the learning system. The theory can be summarized as follows:

In a *promoter*, the nucleotide sequence is composed of two contact regions and one conformation region. For each region, four typical patterns are known.

This gives us two types of background knowledge. One is the structure information of the data. We can utilize this knowledge in CAFE by forming an instance such that it is composed of three components, each corresponding to the three regions in the theory. The other is an initial bias on the distribution of values for the attributes in each region.

Our test was conducted in almost the same way as that described in (Thompson *et al.* 1991b), comparing the result with that shown in this chapter. The test differed in that we ignored the latter background knowledge to clarify the contribution of the structure information to learning task. We randomly selected 26 test instances (13 positive and 13 negative) from the data set and kept the remaining 80 as training instances. We included positive/negative information as additional attributes in the body part and component parts of training instances, but not in the test instances. Training instances were given one by one, and at every tenth instance, accuracy of predicting a class (positive or negative) from the test instances was recorded. Figure 5.6 shows the learning curves of COBWEB and CAFE, averaged over 40 trials, with different selections of test instances and

---

[2]We obtained this data set from FTP server at ics.uci.edu. This is a well-known source of data sets for machine learning, including the famous 'soybean', 'thyroid-disease', etc.

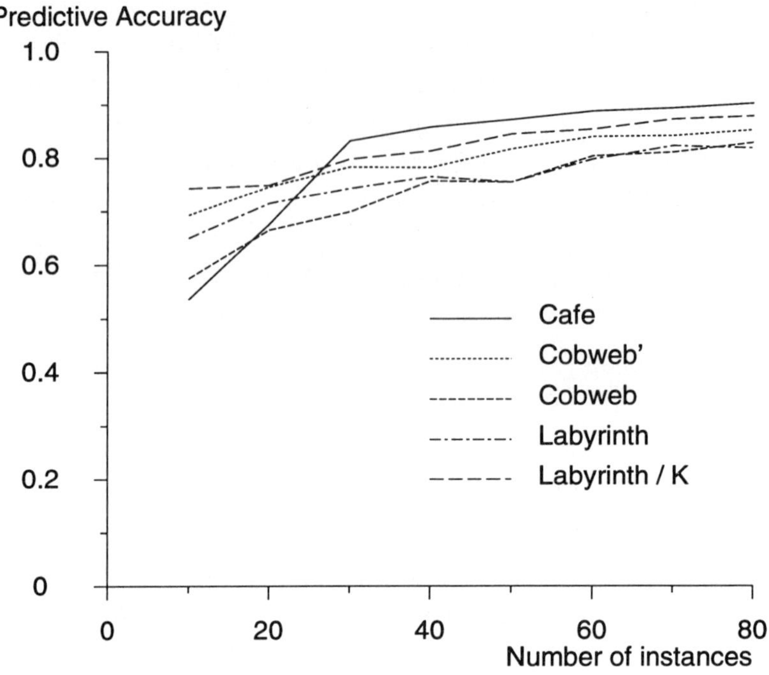

Figure 5.6. Learning curves in DNA sequence domain

different ordering of training instances. The *t-test* indicates that CAFE is superior to COBWEB with 99.5% certainty ($t > t_{0.005,78}$) at the 40th through to the 80th training instances.

The diagram also shows learning curves from Thompson *et al.* (1991b) [3]. LABYRINTH had been tested with (`Labyrinth/K` in the diagram) and without (`Labyrinth` in the diagram) the background knowledge of the second type, mentioned above. It is interesting to note that our version of COBWEB (`Cobweb´`) recorded much better performance than their's (`Cobweb`). Anyway, while we can't see any difference between the performance of `Labyrinth` and `Cobweb`, `Cafe` outperformed `Cobweb´` in our experiment. Thus, we can conclude that the difference in learning schemes between CAFE and LABYRINTH is suggestive. We also predict that the low performance of CAFE at fewer training

[3] These curves are not particularly accurate because data was manually picked up from the diagram in (Thompson *et al.* 1991b).

instances can also be improved by utilizing all of the background knowledge in the same way as `Labyrinth/K`.

4 DISCUSSION

4.1 Value-distribution of Relational Attributes

Let us assume that one attribute REL of an instance object OBJ points to another object Part, and that Part is incorporated in a concept hierarchy tree. When classifying OBJ, what value should attribute REL take? Or, which concept should be regarded being representative for Part? There are as many candidates as there are levels of the tree, from the root node that holds all instances to the leaf node whose only instance is Part itself.

LABYRINTH uses an operator named *attribute generalization* to avoid an overly-specified concept identifier. This operator replaces the values of relational attributes with their ancestors. LABYRINTH first uses the identifier of the leaf node as the value of the relational attribute, then decides if the operator should be applied while examining the information gained by applying it.

One problem of attribute generalization is that it demands an exhaustive search and requires much computation to find the best generalization. In the implementation of Labyrinth, Thompson and Langley (1991a) evaded this problem by using the first generalization found rather than the best. Moreover, LABYRINTH discards the original values of a relational attribute once they have been generalized. Therefore, when the concept hierarchy is altered through learning new instances, the values of relational attributes cannot reflect this change. With all these defects, attribute generalization can give a relatively rough measurement of similarity.

CAFE, in the meantime, sticks to the leaf node, and the evaluation function *concept-predictability* considers all the nodes Part passes to reach the top node. *Concept-predictability* needs less computation, and as it keeps every Part which is classified so far, CAFE can handle any changes in concept hierarchy. Compared to attribute generalization, it is a finer measurement of similarity, which, we believe, improves system performance.

4.2 Weight of an Attribute

Do all the attributes have potentially equal effects on classification? Assume an instance is represented by one primitive attribute A_p and one relational attribute A_r, and the relational attribute points to a part of the instance described by two primitive attributes. If these three primitive attributes have the same importance, attribute A_r naturally has a greater effect on the classification than attribute A_p. But how much more?

In CAFE, currently, we assign each attribute a *weight coefficient*. The weight coefficient for a primitive attribute is 1, and that for a relational attribute is the number of primitive attributes contained in what it points to. In the example above, A_r has a weight coefficient of 2. When category utilities are calculated, *within-class similarity* $\sum_j P(A = V_j|C)^2$ is weighted according to the coefficient and, thus, an attribute whose weight coefficient is 2 has twice as much influence as those with coefficient 1 upon revising the concept hierarchy.

This, however, is a rough approximation of the importance of each attribute. We plan to add a new routine to control the weight coefficient value while learning. In addition, dealing with a variable number of attributes is also an open problem.

Finally, the weight coefficient provides an easy way of changing the system's learning task from unsupervized to supervized; however, the usage of the word *supervized* is slightly different from what usually used as *supervized learning*. If we include class information of those instances as an additional attribute and supply it with a large weight coefficient, the system tries to classify instances by examining the attribute with high priority. This is almost equivalent to informing the system of the class into which each instance should be classified.

In our experiment in the DNA domain, we also included class information in attributes to make the conditions of experiment the same as (Thompson *et al.* 1991b) but without any special weight coefficient. Since the class information was treated in the same way as the other attributes, we classify the experiment as *unsupervized* learning.

5 CONCLUSION

We have described a concept formation system CAFE, that learns from structured instances. The system is equipped with a new classification algorithm *mutual induction* and an evaluation function *concept-predictability.* Experiments have shown good performance of CAFE both in an artificial domain and in a natural domain, which supports our theory that:

Structure information of instances serves as background knowledge for a learning system, and by utilizing this information, the system can improve its performance.

We have also examined the influence of *attribute noise* on learning tasks and shown the usefulness of structure information.

Although these results are encouraging to our research, many problems remain to be solved. One is dynamic evaluation of the *weight coefficient* of attributes. As mentioned in Section 1, *feature construction* is an attractive means of overcoming this problem. In unsupervized learning, however, we should construct a good estimation method for guiding it. Another problem is that of component matching. Generally, systems that handle structured objects must find the best match, or the bindings between the components in the objects and those in the concept. In CAFE, however, structured objects have unique attribute (relation) names for their components and thus the bindings are given *a priori*. Thompson and Langley (1991a) provide good guidance for this problem and we are pursuing research along the same lines. Evaluating CAFE in more a relational (and complex) domain is also within the range of our future task. Although the system already features a facility to represent flexible relation inter/intra instances, we need empirical study in a more complex domain than is covered by this chapter.

REFERENCES

Aha, D. W. (1991). Incremental Constructive Induction: An Instance-Based Approach. *Proc. of the 8th International Workshop on Machine Learning*, 554–8.

Carlson, B. and Weinberg, J. and Fisher, D. (1990). Search Control, Utility, and Concept Induction. *Proc. of the 7th International Workshop on Machine Learning*, 85–92.

Drastal, G. (1991). Informed Pruning in Constructive Induction. *Proc. of the 8th International Workshop on Machine Learning*, 132–6.

Fisher, D. (1987). Knowledge Acquisition via Incremental Conceptual Clustering. *Machine Learning*, *2*, 139–172.

Gennari, J. H. and Langley, P. and Fisher D. (1989). Models of Incremental Concept Formation. *Artificial Intelligence*, *40*, 11–62.

Gluck, M. and Corter, J. (1989). Information, Uncertainty and the Utility of Categories. *Proc. of the 7th Annual Conference of the Cognitive Science Society*, 283–7.

Handa, K. (1990). Concept Formation by Interaction of Related Objects *Proc. of the Pacific Rim Interaction Conference on Artificial Intelligence '90*, 613–8.

Iba, W. and Wogulis, J. and Langley, P. (1988). Trading Off Simplicity and Coverage in Incremental Concept Learning. *Proc. of the 5th International Workshop on Machine Learning*, 73–9.

Matheus, C. J. and Rendell, L. A. (1989). Constructive Induction on Decision Trees. *Proc. of the 11th IJCAI*, 645-650.

Ourston, D. and Mooney, R. J. (1990). Changing the Rules: A Comprehensive Approach to Theory Refinement. *Proc. of the 8th National Conference of the AAAI*, 815-820.

Schlimmer, J. C. (1987). Incremental Adjustment of Representations in Learning. *Proc. of the 4th International Workshop on Machine Learning*, 79–90.

Thompson, K. and Langley, P. (1991a). Concept Formation in Structured Domain. *Concept Formation: Knowledge and Experience in Unsupervised Learning*, 127–161. Morgan Kaufmann.

Thompson, K. and Langley, P and Iba, W. (1991b). Using Background Knowledge in Concept Formation. *Proc. of the 8th International Workshop on Machine Learning*, 117–121.

Towell, G. G. and Shavlik, J. W. and Noordewier, M. O. (1990). Refinement of Approximate Domain Theories by Knowledge-Based Neural Networks. *Proc. of the 8th National Conference of the AAAI*, 861–866.

6

The Discovery of Propositions in Noisy Data

Hiroshi Tsukimoto and Chie Morita

Systems & Software Engineering Laboratory,
Toshiba Corporation,
70 Yanagi-cho, Saiwai-ku, Kawasaki 210, Japan

Abstract

In science, the discovery of propositions in noisy data is important. This discovery is the unsupervised inductive learning with pruning. This chapter presents an efficient algorithm to induce an appropriate proposition from noisy data. A brief outline of the algorithm follows. The frequency data are transformed into a probability vector. The probability vector is transformed into a logical vector. This logical vector is then approximated by a classical logical vector. The classical logical vector is transformed into a logical proposition. This proposition is subsequently reduced to a minimum one. In this algorithm, the most original feature is the pruning method, which is executed at the approximation step. Such pruning is possible due to the transformation from a probability vector to a logical vector. The transformation is possible because (1) a proposition is represented as a vector in Euclidean space and (2) the probability vector can be transformed into a logical vector by a correspondence between the logical vector and the probability vector. Experimental results show that this algorithm performs well and also show that the propositions obtained by the algorithm basically satisfy MDL criteria. We apply this method to some problems to discover appropriate propositions or rules.

1 INTRODUCTION

The discoveries of laws and rules from a lot of data are important in science. The laws and rules are represented by quantitative formulas or qualitative propositions. Even when laws and rules are represented by quantitative formulas, such laws and rules are first represented as propositions.

This chapter deals with the discovery of propositions from noisy data. If the propositions are regarded as the structure of an object, the discovery of propositions is the acquisition of the structure of the object from noisy data. Since we can know the object only through the data, we regard the data as consisting of both structure and noise. The acquisition of propositions means the division of the data into the structure part and the noise part. For example, assume that the data are represented in a table consisting of events with the frequency of occurrence. (See Table 6.1.) Then, the division of the data into the structure part and the noise part means setting the threshold for the frequency of occurrence. The events whose frequency is less than the threshold are pruned. When the threshold is small, few data are pruned. This situation is called overfitting. Overfitting degrades prediction accuracy, so this degradation must be mitigated by pruning. A performance task for unsupervised learning is the prediction accuracy (Fisher 1987a,b) and the prediction accuracy depends on the threshold for the pruning. Therefore, setting the optimal threshold for pruning is an important problem.

Although some researchers (Quinlan 1986, Mingers 1989) deal with the pruning problem, they are concerned with supervised learning. A few researchers deal with the pruning problem in unsupervised learning. Nakakuki *et al.*(1990) present a mechanism to induce an optimal probabilistic model in MDL (Minimum Description Length) criteria from observed data. This mechanism corresponds to a pruning method, but the method is similar to a brute-force search. Fisher (1989) presents a pruning method which is similar to Quinlan's reduced error pruning (Quinlan 1987). Because these methods are a kind of search based on heuristics, they do not have a rigid theoretical background and they also consume much computation time.

This chapter presents an efficient algorithm to induce an appropriate proposition from noisy data. The most original feature in the algorithm is setting the optimal threshold for pruning. This pruning method does not use any heuristics, so it has a rigid theoretical basis for setting the optimal threshold for pruning. The pruning method is not a search, so it does not consume much computation time.

Table 6.1 shows a simple example about the weather. In this example, X stands for rain and Y stands for cloud. Using the algorithm presented in this chapter, an appropriate proposition can be induced from the given frequency data. In this example, the frequency data are (23, 2, 30, 45). This frequency data means that the number of days of rain and cloud are 23, and so on. The induced proposition is $\overline{X} \vee Y (= X \rightarrow Y)$, which means "Whenever it rains, it is cloudy." As probability $(\overline{X} \vee Y)$ is $0.95(= (23+30+45)/(23+2+30+45))$, this result is reasonable.

Event	Rain	Fine/Cloud	No. of days	Logical function
E1	Rain	Cloud	23	XY
E2	Rain	Fine	2	$X\overline{Y}$
E3	Not rain	Cloud	30	$\overline{X}Y$
E4	Not rain	Fine	45	$\overline{X}\overline{Y}$

Table 6.1. Events with frequency of occurrence

The outline of the algorithm is as follows:

1. The frequency data are transformed into a probability vector. (The probability vector is (0.23, 0.02, 0.3, 0.45) in the above case.)
2. The probability vector is transformed into a logical vector. (The logical vector will be explained intuitively in this section and formally in Section 2.)
3. The logical vector is approximated by a classical logical vector.
4. The classical logical vector is transformed into a logical proposition.

5. The proposition is reduced to the minimum one.

This algorithm also holds for multi-valued attributes. Incremental processing is also possible. The computational complexity does not depend on the number of events, but it does depend on the number of training examples.

The pruning is performed in the second and third steps. The most important step is step 2, that is, the transformation from a probability vector to a logical vector. This transformation is possible, because (1) a proposition is represented as a vector in Euclidean space, which is called a logical vector, and (2) a probability vector can be transformed into a logical vector by a correspondence between the logical vector and the probability vector.

We now explain why a proposition is represented as a logical vector. It is worth noticing that classical logic has properties similar to vector space. These properties are seen in Boolean algebra with atoms, which is a model for classical logic. Atoms in Boolean algebra have the following properties:

1. $a_i \cdot a_i = a_i$ (unitarity).
2. $a_i \cdot a_j = 0$ $(i \neq j)$ (orthogonality).
3. $\Sigma a_i = 1$ (completeness).

For example, the proposition $\overline{X} \vee Y$ is represented as $(1,0,1,1)$, where $XY = (1,0,0,0)$, $X\overline{Y} = (0,1,0,0)$, $\overline{X}Y = (0,0,1,0)$, $\overline{X}\overline{Y} = (0,0,0,1)$.

Figure 6.1 is the Hasse diagram of Boolean algebra of two variables. This diagram can be regarded as the projection of a four-dimensional hypercube to a two-dimensional space. The diagram shows that atoms in Boolean algebra correspond to unit vectors.

In other words, atoms in Boolean algebra are similar to the orthonormal functions in Hilbert space. Watanabe (1969) introduced 'logical spectra' to discuss the above properties; however it was just an analogy. This chapter shows that the space of logical functions actually becomes Euclidean space.

As Boolean algebra has properties similar to Hilbert space, constructing Euclidean space starts with Boolean algebra. The

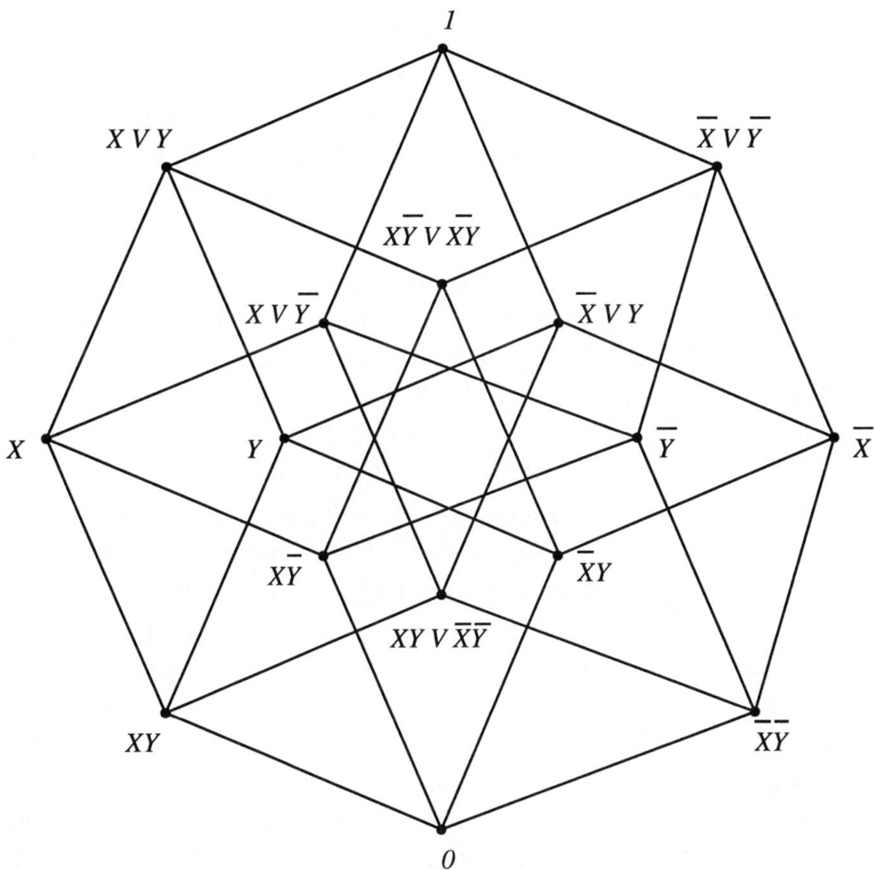

Figure 6.1. A Hasse diagram

process from Boolean algebra to Euclidean space is divided into four stages:

1. Represent Boolean algebra by elementary algebra. In other words, present an elementary algebra model for classical logic.
2. Extend the truth value ($\{0,1\} \to [0,1]$).
3. Eliminate the idempotent law partially to expand the model to the space where the idempotent law does not hold.
4. Introduce an inner product to the above space and construct a Euclidean space where logical functions are represented as vectors.

This theory can be regarded as the functional analysis of logical functions. A more detailed explanation will be found in Section 2.

We now explain why a probability vector can be transformed into a logical vector by a correspondence between the logical vector and the probability vector. In the Euclidean space, the information of logical functions will be defined. For example, the information of 'x' and 'xy' is one bit and two bits respectively. A principle will be presented about the correspondence between the logical vector and the probability vector. This principle is explained using a primitive case. $A \vee \overline{A}$ is tautology, therefore the information of $A \vee \overline{A}$ is 0 bit. Let $E1$ and $E2$ be events corresponding to A and $\overline{A}$ respectively. The probability distribution over $E1$ and $E2$, which corresponds to $A \vee \overline{A}$, should have no information, that is, the entropy of the probability distribution corresponding to tautology should be zero bit. The probability vector whose entropy is zero bit is (1/2, 1/2). The vector representation of $A \vee \overline{A}$ is (1,1), where the atoms are A and $\overline{A}$. Therefore logical vector (1,1) corresponds to probability vector (1/2, 1/2). This correspondence principle can be regarded as Baysian. The correspondence is possible due to the extension of logical model. The details will be discussed in Section 3.

In Section 4, we will explain the algorithm in detail using an example. We will briefly compare this algorithm with other methods to confirm that the algorithm is better than other methods. We will show experimental analyses, which examine correct prediction after training under the effect of noises. The experimental results show that this algorithm performs well. We also show that the propositions obtained by the algorithm basically satisfy MDL (Minimum Description Length) criteria (Rissanen 1978). We are at present applying this algorithm to several problems to discover propositions or rules.

2 LOGICAL VECTOR

We review the model for classical logic introduced in (Tsukimoto 1990). A more detailed explanation can be found in (Tsukimoto 1990,1991). Hereinafter, let $F, G, ...$ stand for propositions, f, g, ... stand for functions, $X, Y, ...$ stand for propositional variables and $x, y, ...$ stand for variables.

2.1 Some definitions

Definition 6.1 Let $f(x)$ be a real polynomial function; τ is defined as follows:

$$\tau(f(x^n, y^n, ...)) = f(x, y, ...),$$

where $x, y, ... \in \{0, 1\}$. For example, $\tau(x^2 + y + 1) = x + y + 1$.

Definition 6.2 Let L be the set of all functions satisfying $\tau(f) = f$. Then $L = \{f : \tau(f) = f\}$. L is the set of pseudo-linear real polynomial functions. In the case of two variables, $L = \{axy + bx + cy + d | a, b, c, d \in \mathbf{R}\}$.

Definition 6.3 L_1 is inductively defined as follows:

1. Variables are in L_1.
2. If f and g are in L_1, then $\tau(f \cdot g)$, $\tau(f + g - f \cdot g)$ and $\tau(1 - f)$ are in L_1. (We call these three calculations τ *calculation.*)
3. L_1 consists of all functions finitely generated by the (repeated) use of 1 and 2.

If f is in L_1, then f satisfies $\tau(f^2) = f$. That is, $f \in L_1 \Rightarrow \tau(f^2) = f$. Obviously, if $f \in L_1$, then $f \in L$.

2.2 A new model for classical logic

Theorem 6.4 Let the correspondence between Boolean algebra and τ calculation be as follows:

1. $F \wedge G \quad \Leftrightarrow \quad \tau(fg)$.
2. $F \vee G \quad \Leftrightarrow \quad \tau(f+g-fg)$.
3. $\overline{F} \quad \Leftrightarrow \quad \tau(1-f)$.

Then $(L_1, \tau$ calculation) is a model for classical propositional logic; that is, L_1 and τ calculation satisfy the axioms for Boolean algebra. Proofs can be found in (Tsukimoto 1991).

We show a simple example. $(X \vee Y) \wedge (X \vee Y) = X \vee Y$ is calculated as follows:

$$\begin{aligned} \tau((x+y-xy)(x+y-xy)) &= \tau(x^2+y^2+x^2y^2+2xy-2x^2y-2xy^2) \\ &= x+y+xy+2xy-2xy-2xy \\ &= x+y-xy. \end{aligned}$$

2.3 Extension of model

2.3.1 *Extension of truth value*

The truth value is extended from $\{0,1\}$ to [0,1]. By this extension, functions become continuous functions ($f : [0,1] \to \mathbf{R}$).

2.3.2 *Elimination of $\tau(f^2) = f$*

Let f be a function of one variable $(ax+b)$. Then $\tau(f^2) = f$ is $\tau((ax+b)^2) = ax+b$. The solutions are 0, 1, x, and $1-x$, which correspond to contradiction, truth, affirmation (of a proposition), and negation, respectively. These functions are classical logical functions of one variable. This property also holds for cases of many variables. Therefore, f in a subset of L, which satisfies $\tau(f^2) = f$, is a member of L_1 and $(f \in L) \wedge (\tau(f^2) = f) \Leftrightarrow f \in L_1$ follows from $(f \in L) \wedge (\tau(f^2) = f) \Rightarrow f \in L_1$ and $f \in L_1 \Rightarrow \tau(f^2) = f$ in Section 2.1. Thus, the subset of L which satisfies $\tau(f^2) = f$ is equal to L_1 (classical logical functions).

Here, $\tau(f^2) = f$ is eliminated in order to expand this space; then the space of pseudo-linear polynomial functions(L) is obtained. Elimination of $\tau(f^2) = f$ means the elimination of the idempotent law for formulas except variables (for example x, y,...). Therefore, this space is a model for a weak logic where the contraction rule holds only for variables. Therefore this space must related to Grisin's logic (Grisin 1982), where the contraction rule does not hold. The study of the relation between this model and Grisin's logic will be done in the future.

Hereinafter, this space(L) will be made into a Euclidean space (a finite-dimensional inner product space).

2.4 Euclidean space

2.4.1 *Inner product*

Definition 6.5 An inner product is defined as follows:

$$< f, g >= 2^n \int_0^1 \tau(fg)dx,$$

where f and g are in L, and the integral is generally a multiple integral.

This inner product has the following properties:

1. $< f, f >\geq 0, < f, f >= 0 \Leftrightarrow f = 0$.
2. $< af, g >= a < f, g >$, where a is a real number.
3. $< f + g, h >=< f, h > + < g, h >$.

Property 1 is proved in the case of one variable as follows. Let $f = f_1 x + f_0(1 - x)$:

$$\begin{aligned} < f, f > &= 2^1 \textstyle\int_0^1 \tau(ff)dx \\ &= 2^1 \textstyle\int_0^1 (f_0^2(1 - x) + f_1^2 x)dx \\ &= f_0^2 + f_1^2 \geq 0 \end{aligned}$$

and

$$< f, f >= 0 \leftrightarrow f_0^2 + f_1^2 = 0 \leftrightarrow f = 0.$$

$f_0^2 + f_1^2 = 0 \leftrightarrow f = 0$, because $f \in L$. This property is also proved in the same manner in the case of many variables.

2.4.2 *Norm*

Definition 6.6 A norm is defined as follows:

$$\| f \| = \sqrt{< f, f >}.$$

This norm has the following properties:

1. $\| f \| \geq 0, \| f \| = 0 \Leftrightarrow f = 0.$
2. $\| af \| = |a| \; \| f \|.$
3. $\| f + g \| \leq \| f \| + \| g \|.$

This norm is denoted by $N_r(f)$. N_r stands for relative norm, which depends on the dimension of space (the number of variables).

Theorem 6.7 L becomes an inner product space with the above norm. The dimension of this space is finite, because L consists of the pseudo-linear polynomial functions of n variables, where n is finite. Therefore, L becomes a finite-dimensional inner product space, namely a Euclidean space.

2.4.3 *Orthonormal system*

The orthonormal system is as follows:

$$\phi_i = \prod_{j=1}^{n} e(x_j) \; (i = 1, ..., 2^n, j = 1, ..., n),$$

where $e(x_j) = 1 - x_j$ or x_j. It is easily understood that these orthonormal systems are the expansion of atoms in Boolean algebra. In addition, it can easily be verified that the orthonormal system satisfies the following properties:

$$\begin{aligned} < \phi_i, \phi_j > &= 0 (i \neq j), \\ &= 1 (i = j), \end{aligned}$$

$$f = \sum_{i=1}^{2^n} < f, \phi_i > \phi_i.$$

For example, the vector representation of $x + y - xy$ of two variables (dimension 4) is as follows:

$$
\begin{aligned}
< f, xy > &= 2^2\int_0^1\int_0^1 \tau(x+y-xy)xy\,dx\,dy = 1,\\
< f, x(1-y) > &= 2^2\int_0^1\int_0^1 \tau(x+y-xy)x(1-y)dx\,dy = 1,\\
< f, (1-x)y > &= 2^2\int_0^1\int_0^1 \tau(x+y-xy)(1-x)y\,dx\,dy = 1,\\
< f, (1-x)(1-y) > &= 2^2\int_0^1\int_0^1 \tau(x+y-xy)(1-x)(1-y)dx\,dy = 0.
\end{aligned}
$$

Therefore, $f = 1\cdot xy + 1\cdot x(1-y) + 1\cdot(1-x)y + 0\cdot(1-x)(1-y)$ and the vector representation is $(1,1,1,0)$, where the bases are $xy = (1,0,0,0), x(1-y) = (0,1,0,0), (1-x)y = (0,0,1,0)$ and $(1-x)(1-y) = (0,0,0,1)$.

The space of the functions of n variables is 2^n-dimensional. The vector representation of classical logical functions is the same as the representation expanded by atoms in Boolean algebra. This vector representation is called a logical vector. This Euclidean space is an expansion of the Hasse diagram of classical propositional logic. Nonclassical logical vector (ex. $(0.8, 0.8)$) will be discussed in Section 3.2.

3 CORRESPONDENCE BETWEEN THE LOGICAL VECTOR AND THE PROBABILITY VECTOR

The following terms and notations are used. f stands for a logical function, $\mathbf{f}((f_i))$ stands for a logical vector and $\mathbf{p}((p_i))$ stands for a probability vector. Note that f stands for a logical function, while f_i stands for an element of a logical vector $\mathbf{f}$.

3.1 The information of a logical function

3.1.1 *A new norm*

Definition 6.8 A new norm is defined as follows:

$$N(f) = \sqrt{2^{-n}}N_r(f).$$

The values of the square of this norm for 1, x, xy and 0 are as follows:

$$
\begin{aligned}
(N(1))^2 &= 1,\\
(N(x))^2 &= 0.5,\\
(N(xy))^2 &= 0.25,\\
(N(0))^2 &= 0.
\end{aligned}
$$

As these figures show, this norm can be interpreted as the degree of satisfiability of propositions.

3.1.2 *The information of a logical function*

The information of a logical function which is called logical entropy is introduced.

Definition 6.9 The logical entropy H_L is defined as follows:

$$H_L(f) = -log_2(N(f))^2 (= -log_2(\int_0^1 \tau(f^2)dx)).$$

Examples are shown below.

1. $H_L(1) = 0$, which means that the information contained in a tautology is zero bit.
2. $H_L(x) = 1$, which means that the information contained in x (namely the affirmation of a certain proposition) is one bit.
3. $H_L(xy) = 2$, which means that the information contained in xy (namely the conjunction of a certain proposition and another proposition) is two bits.
4. $H_L(0) = \infty$, which means that the information contained in a contradiction is infinite.
   That the information contained in a contradiction is infinite can be explained as follows. Generally speaking, the information in a proposition is proportional to the number of propositional variables in the proposition. However, the probability of contradiction in the proposition is also proportional to the number of propositional variables in the proposition. Therefore, the information contained in the conjunction of infinite propositional variables should be infinite and could be false.

As the above examples show, the logical entropy is reasonably defined.

3.1.3 *The logical entropy is equal to the information of probability distribution*

The information of probability distribution I is as follows:

$$I = n - H(H = -\sum_{1}^{2^n} p_i log_2 p_i),$$

where n is the number of variables and p_i is probability.

Theorem 6.10 The logical entropy is equal to the information of probability distribution, that is, $H_L = I$.

Proof. Let a logical function be of n variables and let m elements of the logical vector be 1. The information of the logical function is calculated as follows:

$$\begin{aligned} H_L(f) &= -log_2(N(f))^2 \\ &= -log_2(\textstyle\int_0^1 \tau(f^2)dx) \\ &= -log_2(\textstyle\int_0^1 f dx) \\ &= -log_2 f(1/2) \\ &= -log_2(m(1/2^n)) \\ &= n - log_2 m. \\ (\int_0^1 \tau(f)dx &= f(1/2) \text{ can be easily verified}). \end{aligned}$$

The information of a probability distribution (I) is calculated as follows:

$$\begin{aligned} I &= n - H \\ &= n - (-\textstyle\sum_1^{2^n} p_i log_2 p_i) \\ &= n + \textstyle\sum_1^{2^n} p_i log_2 p_i \\ &= n + m \times (1/m log_2(1/m)) \\ &= n - log_2 m. \\ (p_i &= 1/m \text{ is Baysian-like}). \end{aligned}$$

As both values are calculated to $n - log_2 m$, the proof is completed. Therefore $H_L = I$.

3.2 Interpretation of nonclassical logical vectors

We discuss the case of one variable. Let (a, b) be a logical vector, where (a, b) means $ax+b(1-x)$. The logical vectors whose logical entropy is 1 are the arc whose radius is 1, as shown in Figure 6.2. $x(= (1, 0))$ corresponds to a proposition, while $1 - x(= (0, 1))$ corresponds to the negative proposition.

The vectors on the arc are perfect in information. That is, such vectors correspond to a concrete fact. The vectors on the arc represent an intermediate proposition between the proposition represented by $(1, 0)$ and the proposition represented by $(0, 1)$. For example, let the proposition represented by $(1, 0)$ be "The tree is tall." Then the proposition represented by $(0, 1)$ is "The tree is low." and the proposition represented by the point on the arc is "The tree is not tall and is not low." and so on.

The proposition represented by $(1, b)(0 < b < 1)$ is less than the proposition represented by $(1, 0)$ in the information, therefore $(1, b)$ represents an imperfect proposition such as "The tree may be tall." Similarly, $(a, 1)$ represents an imperfect proposition such as "The tree may be low." $(0.8, 0.8)$ represents an intermediate proposition, because it lies between $(1, 0)$ and $(0, 1)$. It also represents an imperfect proposition, because the information is less than 1.

3.3 Correspondence principle between the logical vector and the probability vector

First, the case of logical function 1 $(= x + (1 - x))$ of one variable, which means a tautology $X \vee \overline{X}$ is considered. The logical vector of 1 $(= x + (1 - x))$ is (1, 1). Then, this logical function means a tautology. Therefore, this logical function has no information, namely the information of the logical function is zero bit. The probability vector whose information is zero bit is (1/2, 1/2). Therefore, logical vector (1, 1) corresponds to probability vector (1/2, 1/2). Similarly, the following correspondences are obtained:

$$2 \text{ variables } (1,1,1,1) \Leftrightarrow (1/4,1/4,1/4,1/4),$$
$$n \text{ variables } (1,...,1) \Leftrightarrow (1/(2^{n}),...,1/(2^{n})).$$

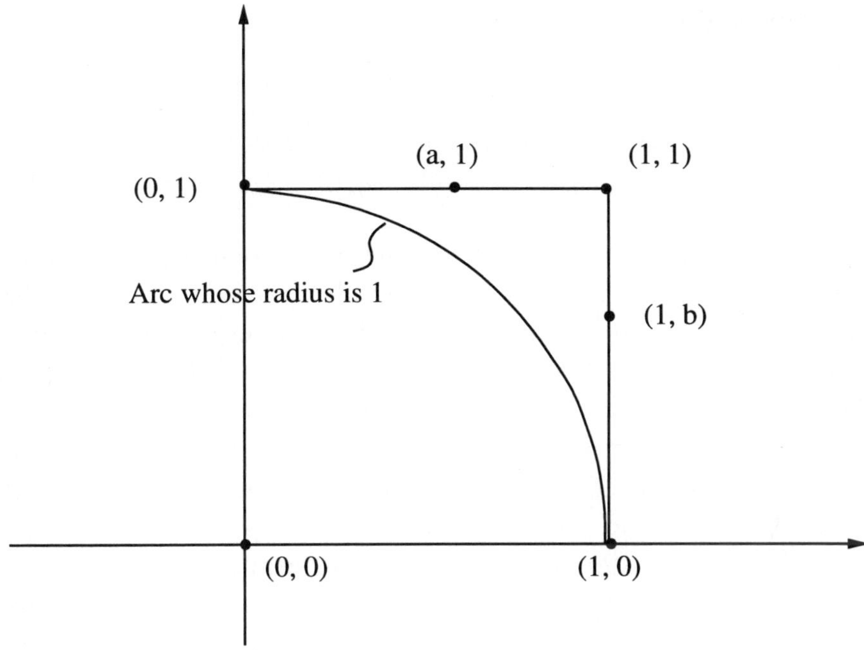

Figure 6.2. Interpretation of nonclassical logical vector

The following correspondences are obtained in the same manner:

$$x \quad \text{of 1 variable } (1,0) \Leftrightarrow (1,0),$$
$$x \quad \text{of 2 variables } (1,1,0,0) \Leftrightarrow (0.5,0.5,0,0).$$

The latter correspondence follows from the following two facts. The first fact is that x does not contain $1-x$, therefore the probability of events corresponding to $1-x$ must be 0. Therefore, the third element and the fourth element are 0. The other fact is that the logical vector (1, 1) corresponds to the probability vector (0.5, 0.5). Therefore, the first element and the second element are 0.5.

These correspondences can be generalized to the following principle:

Suppose m elements of (f_i) are 1.
If $f_i = 1$, then $p_i = 1/m$ and if $f_i = 0$, then $p_i = 0$.

Examples of two variables are shown below:

$$\begin{aligned}(1,1,1,0) &\Leftrightarrow (1/3,1/3,1/3,0),\\ (0,0,1,1) &\Leftrightarrow (0,0,1/2,1/2),\\ (1,0,0,0) &\Leftrightarrow (1,0,0,0).\end{aligned}$$

This correspondence principle means that a logical vector whose information is the smallest corresponds to a probability vector whose information is the smallest. The correspondence principle is Baysian-like, because when we have no information, the probability distribution is uniform. This correspondence principle means that the direction of the logical vector is the same as that of the probability vector and that they differ in the norm.

In classical logic, x of two variables corresponds to $(a, b, 0, 0)$, where a and b are probabilities. For example, $(a, b, 0, 0)$ is $(0.7, 0.3, 0, 0)$ or $(0.5, 0.5, 0, 0)$. Therefore x does not necessarily correspond to $(0.5, 0.5, 0, 0)$. However, we have constructed an extended model to represent nonclassical logics, so we can represent a logical function corresponding to (0.7, 0.3, 0, 0) as $cxy + dx(1-y) + e(1-x)y + f(1-x)(1-y)$, which is not equal to x. (As for the calculation of coefficients, refer to (Tsukimoto 1991).) Therefore, in this extended model, x corresponds to $(0.5, 0.5, 0, 0)$. In other words, the mapping from probability to L_1 is a homomorphism (n-to-one correspondence), while the mapping from probability to L is an isomorphism (one-to-one correspondence).

3.4 Transformation formula

Two important relations about the correspondence between the logical vector and the probability vector are obtained:

1. The information of the two vectors is the same.
2. The directions of the two vectors are the same.

Hereinafter, these relations are extended to nonclassical logics. In other words, we will use the above relations in nonclassical logics too. This extension is natural, so no problems are anticipated. However, further study of this extension will be done in future work.

The transformation from a probability vector to a logical vector is obtained from the above two relations. From 1 ($H_L = I$), $N_r = 2^{H/2}$ is obtained as follows:

$$\begin{aligned} H_L &= I \\ \rightarrow H_L &= n - H \\ \rightarrow -log_2(2^{-n/2}N_r)^2 &= n - H \\ \rightarrow 2^{-n/2}N_r &= 2^{H/2-n/2} \\ \rightarrow N_r &= 2^{H/2}. \end{aligned}$$

The directions are the same from 2. Therefore, the transformation formula is

$$\begin{aligned} \mathbf{f} &= (2^{H/2}/|\mathbf{p}|)\mathbf{p}, \\ \text{where } H &= -\Sigma_1^{2^n}(p_i log_2 p_i), \\ |\mathbf{p}| &= (\Sigma_1^{2^n} p_i^2)^{1/2}, \\ \mathbf{p} &= (p_1, ..., p_{2^n}). \end{aligned}$$

By this transformation formula, we can transform a probability vector to a logical vector. Since this formula also holds when attributes are non-binary, we also use this formula in such cases. (See Table 6.2.)

4 INDUCTIVE LEARNING FROM FREQUENCY DATA

4.1 The algorithm

The transformation formula from a probability vector to a logical vector has been obtained. Therefore, we can infer a proposition inductively from frequency data. The method is as follows.

1. The frequency data are transformed into a probability vector using the following method. Let the frequency data be (a_i), then the probability vector (p_i) is $(a_i/\sum a_i)$.
   Since the events whose frequency data is 0 can be ignored, only those whose frequency data is not 0 are computed. That is, the actual vector size is not exponential in the number of attributes. Therefore, the computational complexity does not depend on the number of events, but on the number of training examples. This case also holds for items 2 and 3.

2. The probability vector (**p**) is transformed into a logical vector (**f**) by the formula $\mathbf{f} = (2^{H/2}/|\mathbf{p}|)\mathbf{p}$.
3. The logical vector is approximated by the nearest classical logical vector. Let (f_i) be a logical vector which represents a nonclassical logical proposition, and let $(g_i)(g_i = 0, 1)$ be a classical logical vector whose elements are 0 or 1. Note that a classical logical vector does not necessarily represent a classical logical proposition, because when attributes are nonbinary, it does not represent a classical logical proposition. (See Section 3.3.) The nearest classical logical vector minimizes $\sum(f_i - g_i)^2$. Each term can be minimized independently and $g_i = 1, 0$. Therefore, the approximation method is as follows. If $f_i \geq 0.5$, then $g_i = 1$, otherwise $g_i = 0$.
4. This classical logical vector is transformed into a logical proposition. Propositions are used as the presentation method in this chapter, but decision trees can also be used for the representation.
5. The proposition is reduced to a minimum one. We can use several methods. We use a multivalued version of the Quine-McCluskey method, which has been developed by ourselves.

The algorithm is described as a batch processing. However, an incremental processing is also possible. (We will give a brief description below.)

The approximation by a classical logical vector corresponds to pruning the items whose value is less than 0.5. Some threshold in probability corresponds to 0.5 in logic, but it is very difficult to find the threshold in probability, because the threshold depends on the probability vector. This difficulty can be seen from the transformation formula $\mathbf{f} = (2^{H/2}/|\mathbf{p}|)\mathbf{p}$. It is clear that a fixed threshold in probability is not effective for the pruning (Fisher 1989).

As the above explanation shows, this algorithm uses no heuristics and has a rigid theoretical background. We briefly compare the algorithm with other methods in computational time. To make the comparison simple, an incremental processing for the algorithm is assumed.

The incremental processing is as follows:

1. Frequency data are modified by new data.
2. A classical logical vector is derived from the given frequency data.
3. The logical proposition (or decision tree) is modified; that is, the logical proposition or the decision tree is minimized only when the classical logical vector changes
4. These steps are repeatedly performed.

The pruning in this algorithm is independent of constructing trees or propositions and is not a search, while the pruning in other methods is not independent of constructing trees or propositions and is a search (Fisher 1989). Therefore, the pruning in this algorithm consumes less time than other methods, because the pruning in this algorithm is not a search, while the pruning in other methods is a search.

The number of objects for constructing trees or propositions in this method is smaller than that of other methods, because the objects for constructing trees or propositions in this method is the pruned data, which is smaller than the original data.

Therefore, if the number of necessary training examples is the same, the computational time of this method is shorter than other methods, because the pruning in this method consumes less time than other methods and the number of objects for constructing trees or propositions is smaller than other methods.

We give the results for the example in the Introduction:

The frequency data is (23, 2, 30, 45).
So, the probability vector is (0.23, 0.02, 0.3, 0.45).
The logical vector is (0.69, 0.06, 0.90, 1.35).
The nearest classical logical vector is (1, 0, 1, 1).
The proposition is $XY \vee \overline{X}Y \vee \overline{XY}$.
This proposition is reduced to $\overline{X} \vee Y (= X \rightarrow Y)$.

4.2 Experiments

We will give experimental analyses here. The first experiment consists of ten attributes with 5%, 10%, and 25% noise. We set a correct concept, which contains ten atoms. (Atoms correspond to events in Tables 1 and 2.) Figure 6.3 shows the results. After about 100 training examples, the correct prediction is greater than 90% under the effect of 5%, 10% and 25 % noise. This result means that this algorithm is noise proof.

The second experiment consists of the same condition as the first except that 16 attributes are used. As Figure 6.4 shows, the result is almost the same as the first.

The third experiment consists of the same conditions as the first except that the concept contains 100 atoms. As Figure 6.5 shows, the necessary training examples are more than for the ten-atom concept. This result is reasonable, because a 100-atom concept is more complicated than a ten-atom concept, so the learning of a 100-atom concept is more difficult than that of a ten-atom concept. The third experiment also shows that this algorithm is noise proof.

The fourth experiment consists of 16 attributes and the other conditions are the same as the third experiment. As Figure 6.6 shows, the results are almost the same.

These four experiments show that the number of necessary training examples does not depend on the number of attributes. In four experiments, the correct prediction decreases at a few points and increases again. This phenomenon happens when the threshold for pruning increases. To summarize these results, this algorithm is noise proof and the convergence of correct prediction is good. Experimental comparison with other methods will be undertaken in the future.

4.3 Description length of the proposition obtained by this algorithm

In this section, we calculate the description lengths of the propositions obtained by our algorithm to confirm that they basically satisfy MDL criteria. We use the example in (Nakakuki *et al.* 1990), because we can compare the results. Table 6.2 gives an example about the malfunctions of some devices. There are 16 events, each of which has two attributes. One is type { *a,b,c,d,e,f,g,h* }, the other is age$\{old, new\}$. Instance A contains the same data as in (Nakakuki *et al.* 1990). The obtained proposition is $(b \wedge old) \vee (b \wedge new) \vee (g \wedge new)$, which is reduced

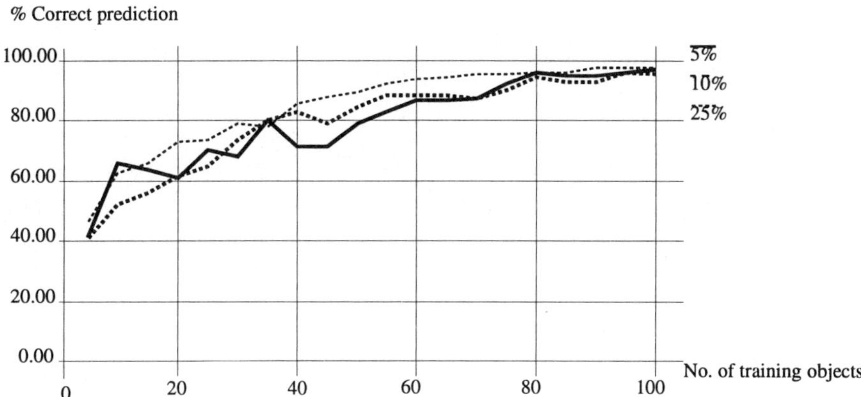

Figure 6.3. Learning curves

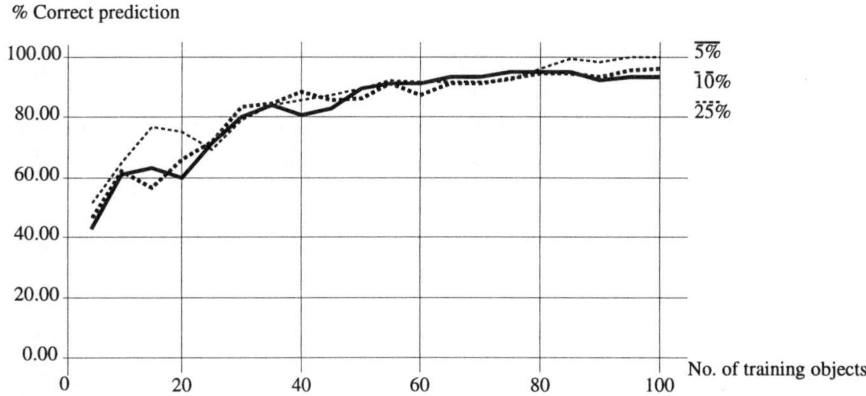

Figure 6.4. Learning curves

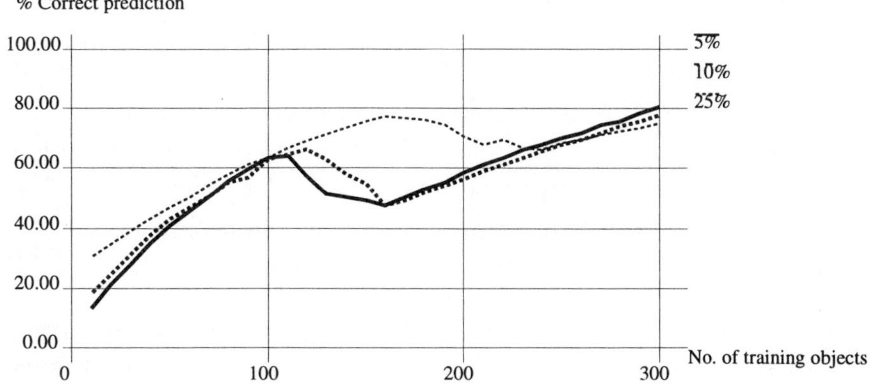

Figure 6.5. Learning curves

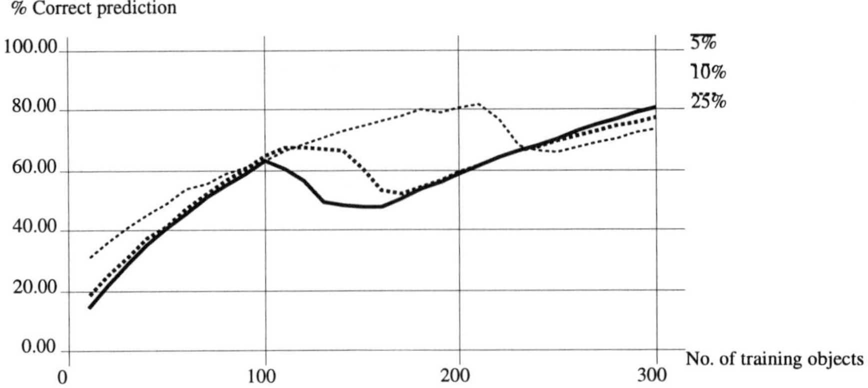

Figure 6.6. Learning curves

to the proposition $b \vee (g \wedge new)$. This result is the same as in (Nakakuki *et al.* 1990),which searches the optimal decision tree(proposition) in MDL criteria. Therefore, we can obtain the optimal proposition in MDL criteria using our algorithm. In instance B, the induced proposition is $(b \wedge new) \vee c \vee (g \wedge old)$. The description length is 147.1. In instance C, the induced proposition is $((b \vee c \vee h) \wedge old) \vee (c \wedge new)$. The description length is 124.6. These propositions also satisfy MDL criteria.

Other experiments also show that the propositions obtained by this algorithm satisfy MDL criteria. However, the propositions obtained by this algorithm will not always satisfy MDL criteria, because the result depends on the reduction method to the minimum proposition. In reality, in instance C, the proposition can be reduced to $(b \wedge old) \vee c \vee (h \wedge old)$ and this proposition's description length is 126.2, which is bigger than 124.6. It is plausible that the description length depends mainly on the threshold for pruning and only slightly on the method for reduction. Therefore, it is expected that, even when the description length of the proposition obtained by this algorithm is not the minimum, it would be close to the minimum.

In MDL, only methods such as brute-force searches are used. The computational complexity of this algorithm is much better than that of brute-force searches. Therefore, this algorithm is very efficient as a quasi-algorithm for MDL.

	Attributes		Number of observations		
Event	Type	Age	A	B	C
1	a	old	1	0	1
2	a	new	0	2	1
3	b	old	13	1	8
4	b	new	9	9	0
5	c	old	1	12	13
6	c	new	1	10	5
7	d	old	0	0	0
8	d	new	0	0	1
9	e	old	0	1	1
10	e	new	0	0	0
11	f	old	1	1	0
12	f	new	0	1	0
13	g	old	0	6	0
14	g	new	5	0	0
15	h	old	1	0	7
16	h	new	0	1	1
	Total		32	44	38

Table 6.2. Malfunctions of some devices

5 CONCLUSIONS

We have presented an algorithm to induce an appropriate proposition from frequency data. This algorithm is possible due to the combination of the vector representation of logical proposition (logical vector) and the correspondence between the probability vector and the logical vector. The most important feature is setting the optimal threshold for pruning. This pruning method does not use any heuristics and has a theoretical background. The pruning method is not a search and the number of objects for constructing trees or propositions is smaller than for other methods, therefore the computational time is shorter than for other methods when the number of necessary training examples is the same. Experimental results show that this algorithm performs well. Experimental results also show that the propositions obtained by the algorithm basically satisfy MDL. We are applying this algorithm to several problems to discover appropriate propositions or rules. The results using our algorithm are propositions, so if we want causalities, the propositions must be transformed to causalities. However, this transformation is another problem. Future work also includes a comparison study with other methods and the relationship between this algorithm and MDL.

REFERENCES

Fisher, D. H. (1987a). Conceptual clustering, learning from examples and inference. *Proceedings of the Fourth International Workshop on Machine Learning*, 38-49.

Fisher, D. H. (1987b). Knowledge acquisition via incremental conceptual clustering. *Machine Learning*, *2*, 139-172.

Fisher, D. H. (1989). Noise-tolerant conceptual clustering. *Proceedings of 11th IJCAI*, 825-830.

Grisin, V. N. (1982). Predicate and set theoretic calculi based on logic without contraction. *Math. USSR Izvestija*,*18*, 1982, pp41-59.

Mingers, J. (1989). An empirical comparison of pruning methods for decision tree induction. *Machine Learning*, *4*, 217-243.

Nakakuki, Y. , Koseki, Y. and Tanaka, M. (1990). Inductive learning in probabilistic domain. *Proceedings of 8th AAAI*, 809-814.

Quinlan, J. R. (1986). Induction of decision tree. *Machine Learning*, *1*, 81-106.

Quinlan, J. R. (1987). Simplifying decision trees. *International Journal of Man-Machine Studies*, *27*, 221-234.

Rissanen, (1978). J. Modeling by shortest description. *Automatica*, *14*, 465-471.

Tsukimoto, H. (1990). A topological model for propositional logics. *Transactions of Information Processing Society of Japan*,*31*, 783-791 (in Japanese).

Tsukimoto, H. (1991). A topological model for logics. working paper.

Watanabe, S. (1969). *Knowing and guessing.* John Wiley and Sons.

7

Learning Non-deterministic Finite Automata from Queries and Counterexamples

T. Yokomori

Department of Computer Science and Information Mathematics,
University of Electro-Communications

1 INTRODUCTION

In the recent theoretical research activity of inductive learning, in particular, of inductive inference, Angluin has introduced the model of learning called *minimally adequate teacher* (MAT), that is, the model of learning via membership queries and equivalence queries, and has shown that the class of regular languages is efficiently learnable using deterministic finite automata (DFAs) (Angluin 1987b). More specifically, she has presented an algorithm which, given any regular language, learns from MAT a minimum DFA accepting the target in time polynomial in the number of states of the minimum DFA and the maximum length of any counterexample provided by the teacher.

The MAT learning model is reasonably accepted for the following reasons. First, the limit of the learning capability from only given example data is well-recognized. Actually, Gold shows that the time complexity of learning consistent DFAs from given data is computationally intractable (Gold 1978). Hence, learning models from more than given data are required to study the feasible learnability. On the other hand, there is another motivation for introducing the MAT learning model which comes from a more practical viewpoint. Suppose one wants to construct an expert system (or knowledge system) and (s)he is trying to collect inference rules by interviewing human experts.

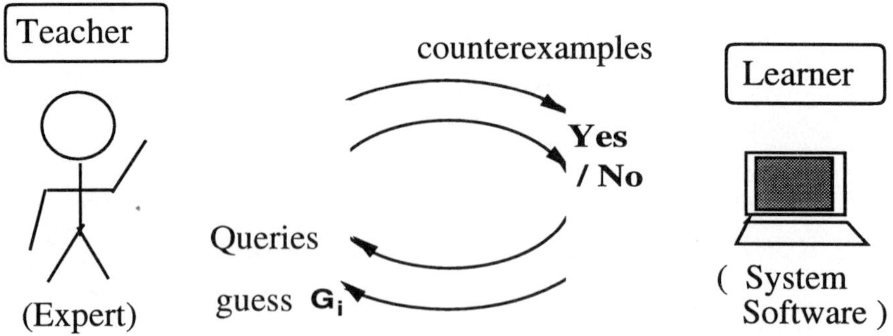

Figure 7.1. An Interactive Learning in MAT Model

The problem here is that human experts often do not retain their expert knowledge as a form of rules in a systematic fashion, and thus, no one can expect to obtain those expert rules directly from human experts. However, it is usually possible for experts to provide concrete example (knowledge) derived from those rules to be collected. Hence, it is of crucial importance to find a method for achieving rule acquisition from a large number of examples through queries and answering. Thus, this requirement nicely meets an interactive learning by the MAT model which is illustrated by Figure 7.1.

Following Angluin's work, several extended results about polynomial-time MAT learnability for subclasses of context-free grammars have been reported (Berman and Roos 1987; Ishizaka 1990; Shirakawa and Yokomori 1993). However, the polynomial-time learnability of the whole class of context-free grammars is still open and seems to be negative, which is strongly suggested by a recent result that under a certain cryptographic assumption the class of context-free grammars is not learnable in polynomial time from MAT (Angluin and Kharitonov 1991).

On the other hand, it is well recognized that *non-deterministic finite automata* (NFAs) are useful in many domains for both theoretical and practical reasons. We know, for example, that various theoretical properties of DFAs can be easily proved using the notion of NFAs. From a more practical viewpoint, we can pick up the *pattern matching problem* as a typical task for which the non-determinism works in a much more elegant man-

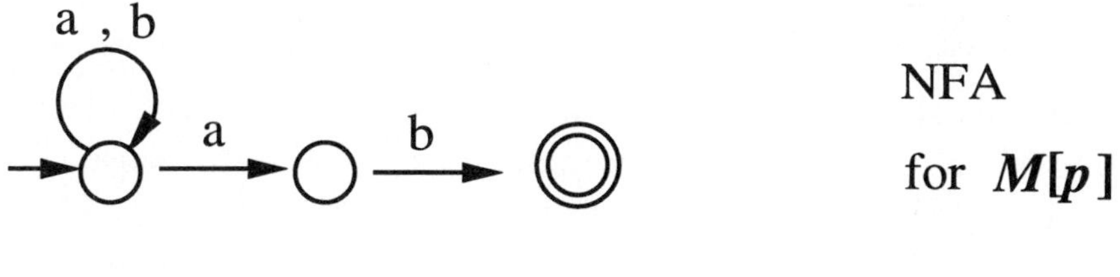

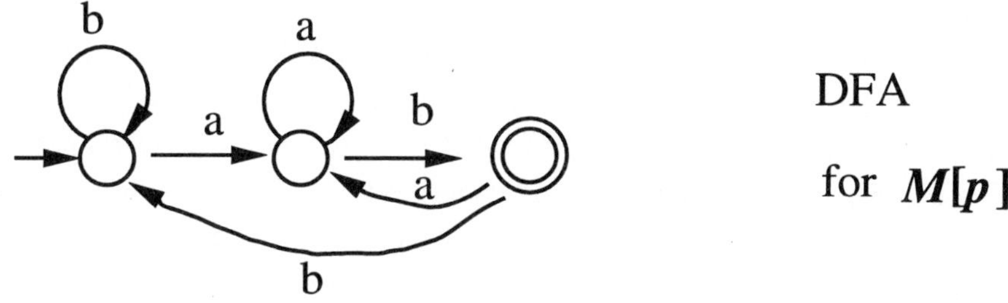

Figure 7.2. NFA and DFA for Pattern Matching

ner than the determinism.

Let t and p be strings over $\{a, b\}$, called *text* and *pattern*, respectively. Assuming that $p = ab$, the problem here is, given an input text t, to check if a text t ends with a pattern p or not. A finite automaton $M[p]$ is usually constructed for this purpose, and it is checked whether t is accepted by $M[p]$ which accepts $\{a, b\}^*p$. The non-deterministic version of $M[p]$ is given together with its deterministic version in Figure 7.2. It is seen that an NFA yields a simpler and more natural description for a concept in question than a DFA.

More generally, it may be said that human experts often retain their expert knowledge in the form of non-deterministic rules. (In the case above, an NFA description is much easier for human experts to recognize its correctness for their knowledge $M[p]$ than is a DFA description.)

Returning to the MAT-learnability issues, a recent study shows that under a certain cryptographic assumption the class of NFAs is not learnable in polynomial time from MAT. More exactly, under the assumption the class is not learnable from MAT in time polynomial in the number of states of a minimum

NFA equivalent to the target NFA and the maximum length of any counterexample (Angluin and Kharitonov 1991).

In this chapter we shall show that the class of NFAs is learnable in polynomial time from MAT in the following sense that given any regular language L an algorithm learns an NFA M accepting L in time polynomial in n and ℓ, where n is the number of states of a minimum DFA equivalent to M and ℓ is the maximum length of any counterexample provided during the learning process. This provides an alternative algorithm for learning regular languages in polynomial time from MAT.

The idea used in this chapter is roughly explained in two steps. First, all the necessary states and transition rules are introduced from positive counterexamples. Then, among these transition rules, wrong (incorrect) transitions are removed using negative counterexamples. A learning method based on this idea is employed by Angluin (1987a), Ishizaka (1989, 1990) and in fact, the learning algorithm proposed here is almost immediately obtained as a special case from one of our recent papers (Shirakawa and Yokomori 1993).

Using an example, we shall outline the basic idea used in the present chapter. Let $L = \{a^m b | m \geq 0\}$ be a target regular language. The goal is to find an NFA M accepting L in the MAT learning model. Suppose that, in response to the initial conjecture M_0 (accepting an empty set) from a learning algorithm, a string $w_1 = ab$ is given as the first (positive) counterexample. Then, we construct the set of candidate states $Q_1 = \{[\lambda], [a], [ab]\}$, and set $[\lambda]$ and $[ab]$ as the initial and final states, respectively. For simplicity, let us rename $q_0 = [\lambda], q_1 = [a], q_2 = [ab]$. By constructing a set of transition rules δ_1:

$$\begin{array}{llll}
(r_{00}:)\ q_0 \xrightarrow{a} q_0, & (r'_{00}:)\ q_0 \xrightarrow{b} q_0, & (r_{01}:)\ q_0 \xrightarrow{a} q_1, & (r'_{01}:)\ q_0 \xrightarrow{b} q_1, \\
(r_{02}:)\ q_0 \xrightarrow{a} q_2, & (r'_{02}:)\ q_0 \xrightarrow{b} q_2, & (r_{10}:)\ q_1 \xrightarrow{a} q_0, & (r'_{10}:)\ q_1 \xrightarrow{b} q_0, \\
(r_{11}:)\ q_1 \xrightarrow{a} q_1, & (r'_{11}:)\ q_1 \xrightarrow{b} q_1, & (r_{12}:)\ q_1 \xrightarrow{a} q_2, & (r'_{12}:)\ q_1 \xrightarrow{b} q_2, \\
(r_{20}:)\ q_2 \xrightarrow{a} q_0, & (r'_{20}:)\ q_2 \xrightarrow{b} q_0, & (r_{21}:)\ q_2 \xrightarrow{a} q_1, & (r'_{21}:)\ q_2 \xrightarrow{b} q_1, \\
(r_{22}:)\ q_2 \xrightarrow{a} q_2, & (r'_{22}:)\ q_2 \xrightarrow{b} q_2, & &
\end{array}$$

The first conjecture is $M_1 = (\{q_0, q_1, q_2\}, \{a, b\}, \delta_1, q_0, \{q_2\})$. Since $L(M_1)$ generates the set Σ^+, we expect a (negative) counterexample from MAT, say, $w_2 = ba$. Then, after parsing w_2 via M_1 we have a transition sequence: $q_0 \xrightarrow{b} q_1 \xrightarrow{a} q_2$. Then, check if $[a](= q_1) \xrightarrow{a} [ab](= q_2)$ is correct for L or not. This is done making a query that $aa \in L$? Since the answer is No, we decide that a transition rule $(r_{12} :)q_1 \xrightarrow{a} q_2$ is an incorrect rule for L and remove it from δ_1, making the second conjecture M_2 with $\delta_2(= \delta_1 - \{r_{12}\})$. This procedure is justified by the principle called *contradiction backtracing* (Angluin 1987a; Ishizaka 1989, 1990; Shapiro 1981), which will be discussed in detail later.

For the next (negative) counterexample $w_3 = bb$, using the current conjecture M_2 we have a transition sequence : $q_0 \rightarrow {}^{b}\ q_1 \xrightarrow{b} q_2$. To check if $[a](= q_1) \xrightarrow{b} [ab](= q_2)$ is correct for L, we make a query that $ab \in L$? Since the answer is Yes, this implies that $(r'_{01} :)q_0 \xrightarrow{b} q_1$ is incorrect for L and removed from δ_2. Thus, we have the third conjecture M_3 with $\delta_3(= \delta_2 - \{r'_{01}\})$.

In a similar way, we see that when negative counterexamples bb, a, abb, bab, bbb, ba are provided in this order, rule sets $\{r'_{00}, r'_{22}\}, \{r_{02}\}$, $\{r'_{10}, r'_{11}\}$, $\{r_{20}, r_{21}\}$, $\{r'_{20}, r'_{21}\}$, $\{r_{22}\}$ are, respectively, removed from the conjecture. As a result, we have a conjecture $M' = (\{q_0, q_1, q_2\}, \{a, b\}, \delta', q_0, \{q_2\})$ with

$$\delta' = \{q_0 \xrightarrow{a} q_0,\ q_0 \xrightarrow{a} q_1,\ q_0 \xrightarrow{b} q_2,\ q_1 \xrightarrow{a} q_0,\ q_1 \xrightarrow{a} q_1,\ q_1 \xrightarrow{b} q_2\}.$$

If we transform this NFA M' into an equivalent minimum DFA $M = (\{p_0, p_1\}, \{a, b\}, \delta, p_0, \{p_1\})$ with $\delta = \{p_0 \xrightarrow{a} p_0, p_0 \xrightarrow{b} p_1\}$, it is easy to see that M accepts the target L.

Thus, the above procedure works as a learning algorithm to identify an NFA accepting a target L from MAT.

This chapter is organized as follows. After providing basic definitions in Section 2, we formalize the above idea to show that the class of regular languages is polynomial-time learnable from MAT using NFAs in Section 3. A learning algorithm LA for NFAs is first described, and the worst-case analysis for the time complexity of the proposed algorithm as well as its correctness is then given. The main result provides an alternative efficient MAT learning algorithm for regular languages. Section

4 deals with related topics, in which a comparative analysis of two algorithms (i.e., LA and Angluin's one) for learning regular languages is discussed in Section 4.1. We also mention in Section 4.2 a corollary of the main result claiming that there exists a subclass of NFAs which is polynomial-time learnable from MAT, in contrast to the fact that it seems not to be the case for the whole class of NFAs. Further, a practical variant of LA is briefly discussed in Section 4.3.

2 PRELIMINARIES

2.1 Definitions and Notation

We assume the reader to be familiar with the rudiments of formal language theory (see, e.g., Harrison 1978 or Salomaa 1973).

For a given finite alphabet Σ, the set of all strings with finite length (including *zero*) is denoted by Σ^*. (An *empty* string is denoted by λ.) $lg(w)$ denotes the length of a string w. Σ^+ denotes $\Sigma^* - \{\lambda\}$. A *language* L over Σ is a subset of Σ^*. For any strings x, $y \in \Sigma^*$ and any languge L over Σ, let $x\backslash L = \{y \mid xy \in L\}$.

A *non-deterministic finite automaton* (NFA) is denoted by $M = (Q, \Sigma, \delta, p_0, F)$, where Q and Σ are finite sets of *states* and *terminals*, respectively, p_0 is the *initial state* in Q, $F(\subseteq Q)$ is a set of final states, and δ is a finite set of transition rules of the form : $p \overset{a}{\to} q$, where p, q are states and a is a terminal from Σ. M is *deterministic* iff for each $a \in \Sigma$ and $p \in Q$, there exists at most one $q \in Q$ such that $p \overset{a}{\to} q$ is in δ. For each $p, p', q \in Q$ and $a \in \Sigma, w \in \Sigma^*$, define $p \overset{\lambda}{\to} p$, and $p \overset{wa}{\longrightarrow} q$ iff $p \overset{w}{\longrightarrow} p'$ and $p' \overset{a}{\to} q$.

For each $p \in Q$, let $L(p) = \{w \in \Sigma^* | p \overset{w}{\longrightarrow} q,\ q \in F\}$. In particular, $L(p_0)$, equivalently denoted by $L(M)$, is called the language accepted by M. Two NFAs M and M' are *equivalent* iff $L(M) = L(M')$ holds.

By $|Q|$(the cardinality of Q) we define the size of an NFA M, denoted by $size(M)$. M is *minimum* iff $size(M)$ is minimum, that is, for any NFA M' that is equivalent to M, $size(M) \leq size(M')$ holds.

A language L is *regular* if there exists an NFA M such that $L = L(M)$.

Since we are concerned with the learning problem of regular languages, without loss of generality, we may restrict our consideration to only λ-free regular languages.

2.2 MAT Learning

Let L be a target language to be learned over a fixed alphabet Σ. We assume the following types of queries in the learning process.

A *membership query* proposes a string $x \in \Sigma^*$ and asks whether $x \in L$ or not. The answer is either *yes* or *no*. An *equivalence query* proposes an NFA M and asks whether $L = L(M)$ or not. The answer is *yes* or *no*, and in the latter case together with a counterexample w in the symmetric difference of L and $L(M)$. A counterexample w is *positive* if it is in $L - L(M)$, and *negative* otherwise.

The learning protocol consisting of membership queries and equivalence queries is called *minimally adequate teacher* (MAT). The purpose of the learning algorithm is to find an NFA $M = (Q, \Sigma, \delta, p_0, F)$ such that $L = L(M)$ with the help of the minimally adequate teacher.

3 MAIN RESULTS

3.1 Learning NFAs

Throughout this section, for a target regular language L, let $M_* = (Q_*, \Sigma, \delta_*, \tilde{p}_0\ F_*)$ be a minimum DFA such that $L = L(M_*)$.

3.1.1 *Introducing new states*

Given a *positive* counterexample w of L, let $Q(w)$ be a set of new states consisting of all the prefixes of w, that is,

$$Q(w) = \{\, [x] \mid w = xy \text{ for some } y \in \Sigma^*\}.$$

The following lemma obviously holds.

Lemma 7.1 Let w be in $L = L(M_*)$. Then, for any state $\tilde{p} \in Q_*$ that appears in the transition sequence $\tilde{p}_0 \xrightarrow{w} \tilde{q}$ (for some $\tilde{q} \in F_*$), there is an $[x]$ in $Q(w)$ such that $L(\tilde{p}) = x\backslash L$. $\square$

3.1.2 *Constructing new candidate rules*

Suppose that we have a conjectured NFA $M = (Q, \Sigma, \delta, p_0, F)$ at the current stage of learning process, where $p_0 = [\lambda]$. From the set of new states $Q(w)$ produced above and the current set of states Q, we newly construct the set of new candidate transition rules δ_{new} as follows:

$$\delta_{new} = \{p \xrightarrow{a} q \mid p, q \in Q \cup Q(w), a \in \Sigma, \text{ and at least one of } p \text{ and } q \text{ is in } Q(w)\}.$$

(As seen below, we set $M = (Q \cup Q(w), \Sigma, \delta \cup \delta_{new}, p_0, F \cup \{[w]\})$ to obtain new conjectured NFA for the next stage.)

3.1.3 *Diagnosing the set of transition rules* δ

Let $M = (Q, \Sigma, \delta, p_0, F)$ be a conjectured NFA, where $p_0 = [\lambda]$.

Let r be a transition rule $p \xrightarrow{a} q$ in δ, where $p = [x], q = [y]$. Then, r is *incorrect for L* iff there exists $w \in \Sigma^*$ such that $yw \in L$ and $xaw \notin L$. A rule is *correct* for L iff it is not incorrect for L.

Now, given a *negative* counterexample w', the algorithm has to prevent the conjectured NFA from accepting w' by removing wrong rules. In order to determine such wrong rules, the algorithm calls the diagnosing procedure whenever a negative counterexample is provided.

Let w' be in Σ^+. A transition sequence : $p' \xrightarrow{w'} q$ (for some $q \in F$) in M is denoted by $\text{Trans}(M, p', w')$.

The diagnosing algorithm is a modification of Shapiro's *contradiction backtracing algorithm* (Shapiro 1981). For a given transition sequence $\text{Trans}(M, p', w')$ as an input, this procedure outputs a transition rule r of M which is incorrect for L, where $w' \in L(M) - L$.

Let $\text{Trans}(M, p', w') : p' \xrightarrow{a} p \xrightarrow{w} q (\in F)$, where $w' = aw (a \in \Sigma,\ w \in \Sigma^*)$ and $p = [y]$.

procedure diag($\text{Trans}(M, p', w')$)
  If $w' = a \in \Sigma$, **then** output $r : p' \xrightarrow{a} q$ and halts
  else make a membership query $yw \in L$? ;
    if the answer is *no*

then call **diag**(Trans(M, p, w)) ;
else output $r: p' \overset{a}{\rightarrow} p$ and halts

We are now ready to prove the correctness of the diagnosing algorithm.

Lemma 7.2 Given a negative counterexample w' and a conjectured NFA M, the diagnosing algorithm always halts and outputs a rule incorrect for L.

Proof. It is clear that since the procedure **diag** is recursively called with a proper suffix of the initial input string w' in Tans(M, p_0, w'), the algorithm always halts and outputs some rule in δ. Let $p_0 \overset{u}{\longrightarrow} p' \overset{a}{\rightarrow}$
$p \overset{w}{\longrightarrow} q(\in F)$, where $w' = uaw$ and $p_0 = [\lambda]$, $p' = [x]$, $p = [y]$, $q = [z]$. By definition, we note that $z \in L$.

If w' is a terminal a in Σ, then a rule $r: p_0 \overset{a}{\rightarrow} q$ is returned, where $u = w = \lambda$ and $p_0 = p'$, $p = q$. Since $z \in L$ and $w' \notin L$, a rule r is incorrect for L.

Assume that a rule $r: p' \overset{a}{\rightarrow} p$ is returned by this algorithm. Then, since this is the first time the answer of a membership query is *yes*, at this moment we have that $yw \in L$. Further, from the property of procedure **diag**, it must hold that $xaw \notin L$. This implies that a rule r is incorrect for L. □

3.1.4 *Learning algorithm LA*

The following is a learning algorithm LA for NFAs:

Input : a regular language L over fixed Σ.
Output : an NFA M such that $L = L(M)$;
Procedure :
  initialize $M = (\{p_0\}, \Sigma, \emptyset, p_0, \emptyset)$, where $p_0 = [\lambda]$;
  repeat
    make an equivalence query to the current $M = (Q, \Sigma, \delta, p_0, F)$;
    If the answer is *yes* **then** output M and halts
    else if the answer is a *positive* counterexample w
    then introduce the set of new states $Q(w)$ from w;
        $Q := Q \cup Q(w)$;
        construct the set of new rules δ_{new};
        $\delta := \delta \cup \delta_{new}$;
        $F := F \cup \{[w]\}$;
    else (the answer is a *negative* counterexample w')

parse w' via the conjectured NFA M to obtain
the transition sequence $\mathrm{Trans}(M, p_0, w')$;
call **diag**($\mathrm{Trans}(M, p_0, w')$) to find an incorrect rule r;
$\delta := \delta - \{r\}$

A flowchart diagram for LA is given in Figure 7.3.

3.2 The Correctness and Time Analysis of LA

3.2.1 *Correctness*

For states $p = [x]$ in Q and $\tilde{p} \in Q_*$, we say that p is *well-corresponding to* $\tilde{p}$ iff $x \backslash L = L(\tilde{p})$ holds. A state p is *non-corresponding to* M_* iff there is no state $\tilde{p} \in Q_*$ to which p is well-corresponding. A transition rule $p \xrightarrow{a} q$ in δ is *well-corresponding to* $\tilde{p} \xrightarrow{a} \tilde{q}$ in δ_* iff p and q are well-corresponding to $\tilde{p}$ and $\tilde{q}$, respectively.

It is clear that every rule well-corresponding to a rule of M_* is correct for L.

We show that if a string w' which is not in L is accepted by the conjectured NFA M, then there exists at least one incorrect rule in δ, that is, if δ has no incorrect rule for L, then M accepts no string w' such that $w' \notin L$.

Lemma 7.3 Let $M = (Q, \Sigma, \delta, p_0, F)$ be a conjectured NFA and L be a target language. If all of the rules in δ are correct for L, then for all $p = [x] \in Q$, $L(p) \subseteq x \backslash L$ holds.

Proof. Let w be a string such that $w \in L(p)$. By the induction on the length i of the transition sequence for w, we show that w is also in $x \backslash L$. Suppose $i = 1$, that is, $p \xrightarrow{a} q$ for some $q \in F$. Let $q = [z]$, then $z \in L$. Further, suppose $xa \notin L$, then we have that $p \xrightarrow{a} q$ is incorrect for L, contradicting the assumption. Therefore, $a \in x \backslash L$.

Next, suppose that the claim holds for all w' such that $lg(w') < lg(w)$. Let $p \xrightarrow{a} p'(= [x']) \xrightarrow{w'} q$ and $q \in F$. Then, by the induction hypothesis, we have that $w' \in x' \backslash L$. Suppose $w = aw'$ is not a string in $x \backslash L$. Then, since $w' \in x' \backslash L$, we have that the rule $p \xrightarrow{a} p'$ is incorrect for L. This contradicts the assumption, and therefore, the lemma holds. □

Corollary 7.4 If all of the rules in δ are correct for L, then $L(M) \subseteq L$.

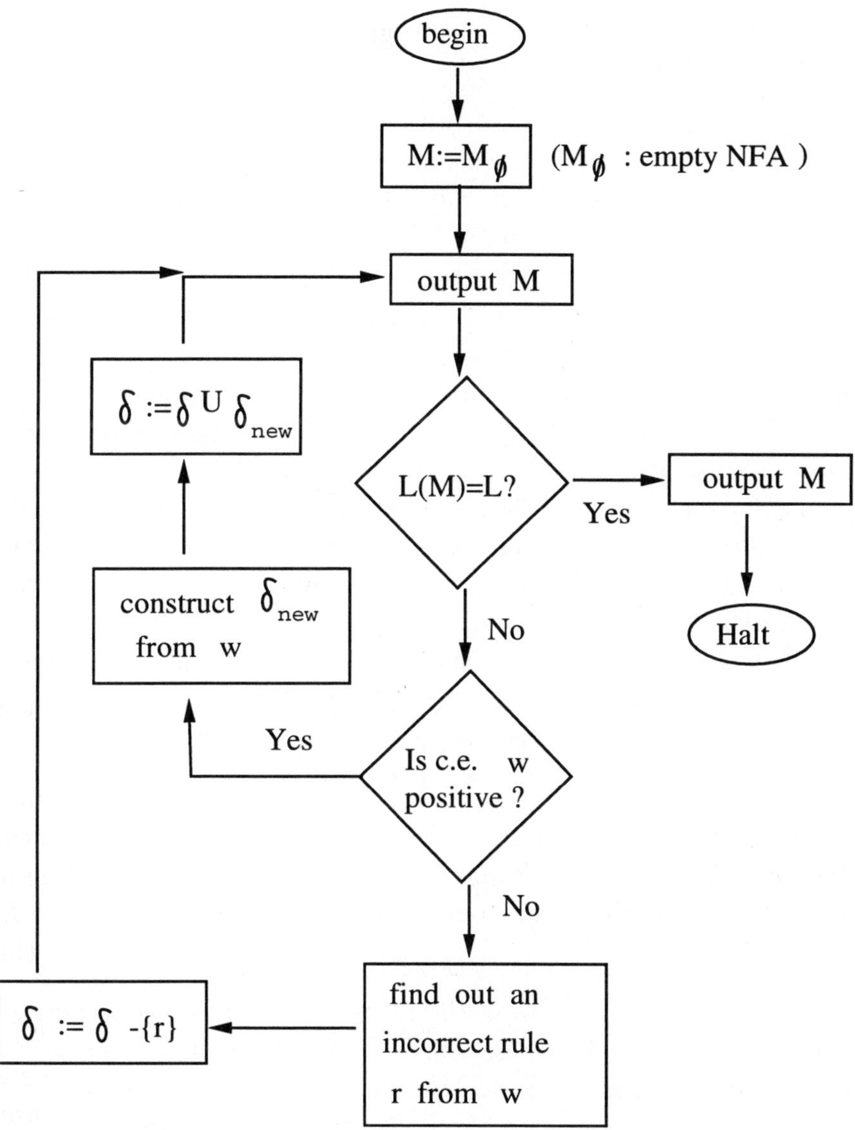

Figure 7.3. Flowchart Diagram for LA

Let $M_* = (Q_*, \Sigma, \delta_*, \tilde{p}_0, F_*)$ be a minimum DFA such that $L = L(M_*)$.

Lemma 7.5 The number of positive examples needed to identify a correct NFA by LA is at most $|Q_*|$ (the cardinality of the state set of M_*).

Proof. Let $M = (Q, \Sigma, \delta, p_0, F)$ be a conjectured NFA from LA, and suppose a positive counterexample w is given. Then, we claim that at least one new state p well-corresponding to some $\tilde{p}$ in Q_* is introduced.

For $w \in L$, let $Q_*(w)$ be the set consisting of all the states in Q_* used in the transition sequence of M_* for accepting w. By the nature of LA, if a positive counterexample w is given, then there exists at least one state p not contained in Q but needed to accept w. By Lemma 7.1, $Q(w)$ contains all states well-corresponding to the states in $Q_*(w)$ that appear in the transition sequence $\tilde{p}_0 \xrightarrow{w} \tilde{q}(\in F_*)$. Further, for each new state p in $Q(w)$, all the rules containing p are added to the rule set δ, and only incorrect rules in δ are removed by the diagnosing procedure.

To sum up, whenever a positive counterexample w is given, there exists at least one state p in $Q(w) - Q$ that is necessary for accepting w and is well-corresponding to $\tilde{p}$. Thus, this justifies the claim and, hence, completes the proof. □

Note that this lemma implies that after receiving at most $|Q_*|$ positive counterexamples, Q includes sufficient number of states to accept the target language L.

When $|Q_*|$ positive counterexamples are given, the set of transition rules δ includes all the rules that are well-corresponding to the ones in δ_*. In other words, any string in L is accepted by the conjectured NFA M. Thus, at that time it holds that $L \subseteq L(M)$, and hence no more positive counterexample is required and provided. By Lemma 7.2, each time a negative counterexample is given, one incorrect rule is determined and removed from δ of the current conjecture M. Further, we know that no correct rule is removed at any stage. Therefore, the number of required negative counterexamples is at most the maximum number of rules of conjectured NFA. By Corollary 7.4, if all incorrect rules are removed, the resulting conjectured NFA M accepts no string which is not in L, that is, $L(M) \subseteq L$.

From these facts described above, we conclude that LA always converges and outputs a correct NFA. Thus, we obtain the main theorem.

Theorem 7.6 For any regular language L, the learning algorithm LA eventually terminates and outputs an NFA M accepting L. □

3.2.2 *Time Analysis*

We note that for a positive counterexample w, the number of states newly introduced from w in LA is obviously bounded by $lg(w)$, i.e., it holds that $|Q(w)| \leq lg(w)$.

Let ℓ be the maximum length of any counterexample provided during the learning process. Futher, let $n(= |Q_*|)$ be the number of states of a minimum DFA accepting a target L.

Lemma 7.7 The total number $|\delta_{total}|$ of all rules introduced in the learning algorithm is bounded by $|\Sigma|n^2\ell^2$. □

Proof. Let $|Q_{max}|$ be the maximum number of states of any conjectured NFA. Let $\{w_1, ..., w_t\}$ be the set of positive counterexamples provided by LA in the entire process of learning. (Note that, by Lemma 7.5, t is at most $|Q_*|$.) Then, from the above observation, $|Q_{max}| \leq \sum_{i=1}^{t} lg(w_i) \leq |Q_*|\ell$ is obtained. Further, from the manner of constructing δ,

$$\begin{aligned} |\delta_{total}| &\leq |\Sigma||Q_{max}|^2 \\ &\leq |\Sigma||Q_*|^2\ell^2 \\ &= |\Sigma|n^2\ell^2 \end{aligned}$$

is obtained. □

Now, suppose a negative counterexample w' is given. Then, the learning algorithm LA constructs the transition sequence Trans(M, p_0, w') by parsing w' via the conjectured NFA $M = (Q, \Sigma, \delta, p_0, F)$ at that time. Since there exists an algorithm that constructs a transition sequence for w' in time proportional to $|\delta| \lg(w')^2$, each parsing procedure requires at most $|\delta_{total}|\ell^2$ times.

From these observations, the next lemma follows.

Lemma 7.8 The total time complexity of LA is bounded by $O(|\Sigma|^2 n^4 \ell^6)$, where n is the number of states of a minimum DFA for a target language, and ℓ is the maximum length of any counterexample provided.

Proof. It is obvious that the total time complexity of LA is dominated by those of parsing negative counterexamples and of obtaining the transition sequences for them.

Each time a negative counterexample is provided, the number of transition rules δ of the resulting conjecture from LA is reduced by exactly one. Further, by Lemma 7.7, the number of transition rules δ of any conjectured NFA M is at most k, where $k = |\Sigma| n^2 \ell^2$. Hence, the total time required for parsing is bounded by

$$k\ell^2 + (k-1)\ell^2 + \cdots + \ell^2 = \frac{1}{2}k(k+1)\ell^2.$$

□

Thus, we have the following theorem.

Theorem 7.9 The total running time of LA is bounded by a polynomial in ℓ the maximum length of any counterexample provided during the learning process and n the number of states of a minimum DFA for a target language. □

4 DISCUSSIONS

4.1 A Comparative Analysis

We will make a comparison of the main results in this chapter with that of Angluin (1987). The time complexity of her algorithm is dominated by the task of constructing the observation table whose size is at most $O(n^2\ell^2 + n^3\ell)$, and the algorithm makes at most n different guesses before terminating with a correct minimum DFA for the target, where n is the number of states of the minimum DFA and ℓ is the maximum length of any counterexample. As a result, the total time complexity is bounded by $O(n^3\ell^2 + n^4\ell)$. (See Table 7.1.) From these, in the worst-case analysis, Angluin's algorithm has a great advantage over LA presented in this article.

(1) Although Table 7.1 shows that LA does not seem better than Angluin's one in the worst case, we do not know whether or not the worst case can really occur in LA. Moreover, we should like to call one's attention to the following fact that there is a subclass of regular languages for which LA in the worst case runs faster than the other in the best case.

Let $L = \{w \in \{a,b\}^* | lg(w) \geq 2\} (= \{a,b\}^* - \{\lambda, a, b\})$ be a target language to be learned. As the first (positive) counterexample, suppose $w_1 = ab$ is given. Then, LA produces as the first guess M_1 pictured in Figure 7.4. At this moment, since there are only two negative counterexamples: a and b to M_1, suppose $w_2 = a$ is taken for the second counterexample. Then, after diagnosing the transition sequence for w_2, the second guess of LA will be M_2. Further, for the third counterexample $w_3 = b$, LA eventually outputs M_3 which is correct for L and terminates. (Note that this termination is independent of the choice of counterexamples, i.e., the other choice of $w_2 = b$ and $w_3 = a$ leads to the same correct NFA M_3.) It is important to note that no membership query is actually needed in this case, and that the choice for w_1 is completely arbitrary. Thus, we observe that only three counterexamples have been provided to successfully terminate even in the worst case.

On the other hand, Angluin's algorithm works as follows. As the initial guess, using three membership queries, it first produces M'_1 accepting the empty set in Figure 7.4. Then, given a positive counterexample w_1, the algorithm eventually produces a correct DFA M'_2 after 16 membership queries and terminates at its best. Thus, for a target language L, LA clearly terminates faster than Angluin's algorithm, provided that each operation involved equally takes one unit time.

This shows that the total time performance of the two algorithms strongly depends upon what class of regular languages is targeted.

(2) When we look at these two algorithms from the *human-*

Table 7.1. Performance Comparison in the Worst-case Analysis

	DFA learning	NFA learning
Number of guesses	$O(n)$	$O(n^2\ell^2)$
Total time	$O(n^3\ell^2 + n^4\ell)$	$O(n^4\ell^6)$
Number of membership queries per revison	$O(n\ell)$	ℓ

machine interface point of view, the most important factor is the response time between conjectures and, in particular, the time complexity the *teacher* (human) must get involved in. That is, the comparison of the number of membership queries required per one revison of conjecture shows that our algorithm LA is superior to Angluin's one in this respect. Although the total number of guesses of LA is greater than that of Angluin's one in the worst case, taking the empirical number of guesses into consideration, we believe that the former may provide better human-machine interface for (at least some type of) application systems than the latter. (See 4.3.)

Thus, it is an interesting open problem to investigate the empirical performance of LA over Angluin's one, and such a quantitative analysis will give clearer and more practical comparisons of the two algorithms.

4.2 A Subclass of NFAs

As strongly suggested by a recent result (Angluin and Kharitonov 1991), the *whole* class of NFAs seems not to be learnable in polynomial time from MAT. This leads us to consider a subclass of NFAs which allows us to have a positive result for learning the subclass.

We say that an NFA M is *polynomially deterministic* if there exists a DFA M' equivalent to M such that the number of states of M' is bounded by a polynomial in the number of states of M. It is trivially clear that any DFA is polynomially deterministic, and there is an NFA which is *not* polynomially deterministic. (For example, $L_n = \{a,b\}^*a\{a,b\}^n$ accepted by an NFA with $(n+2)$ states has no DFA with states less than 2^{n+1}.)

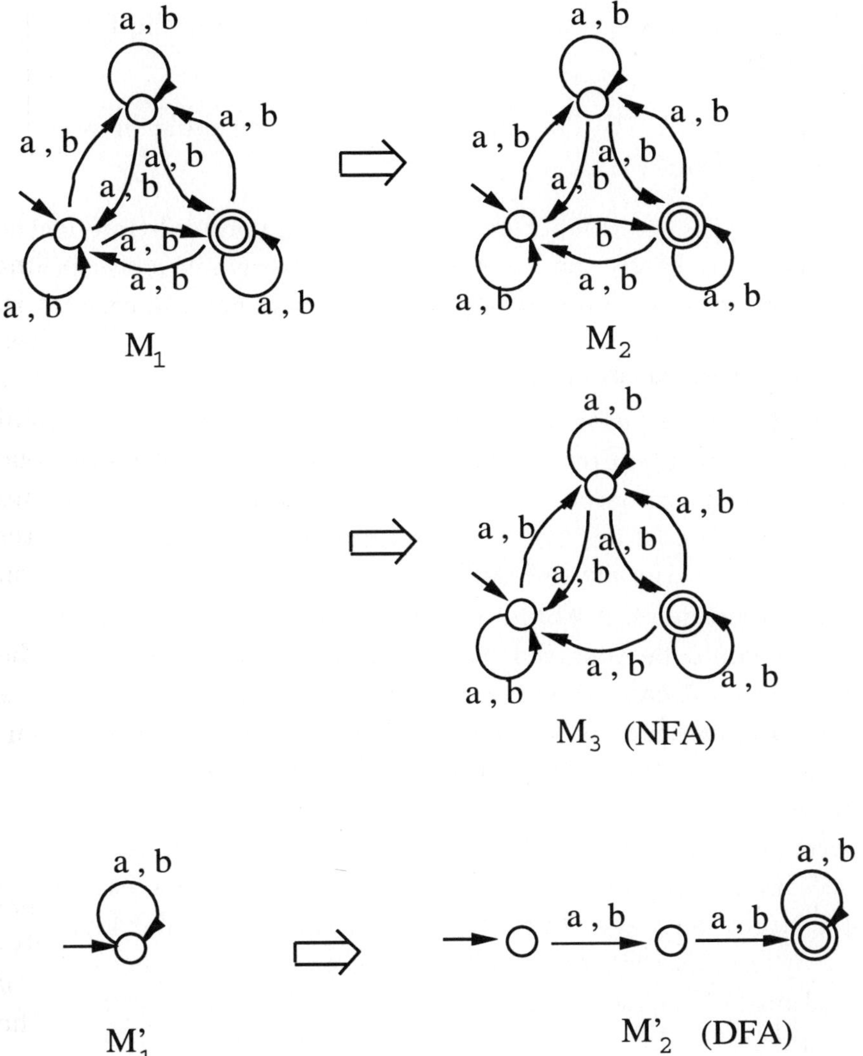

Figure 7.4. Example Runs

Let $\mathbf{NFA}_{poly}$ be the class of polynomially deterministic NFAs. Then, we have that $\mathbf{DFA} \subset \mathbf{NFA}_{poly} \subset \mathbf{NFA}$, where **NFA** (**DFA**) denotes the class of NFAs (DFAs).

From the main result, we immediately obtain the following proposition.

Proposition 7.10 The class $\mathbf{NFA}_{poly}$ is learnable in polynomial time from MAT.

It remains open whether or not the polynomial-time learnability from MAT can be extended to the whole class **NFA**, although it is strongly suggested that this is not the case.

4.3 A Practical Variant of LA

In a practical situation in performing the algorithm LA, there may be a problem on the feasibility of equivalence queries because, when compared with the membership queries, the equivalence queries cost too much to carry out, and they are sometimes even computationally infeasible.

In this respect, it would be very convenient if the equivalence queries could be replaced with an other device. Actually, the algorithm LA can induce in a straightforward manner its modified version which is viewed as a learning algorithm on the in the limit basis (Gold 1967) where no equivalence queries are required. That is, we can think of LA as an algorithm LA_m which learns a correct NFA *in the limit* from membership queries and the *complete presentation* (i.e., positive and negative examples) of a target language. Further, LA_m learns any regular language L *consistently, conservatively*, and *responsively* in the sense of Angluin (1980), and may be implemented to run in time polynomial in n (the number of states of a minimum DFA for L) and m (the total lengths of all examples provided so far). Note that Angluin's algorithm mentioned above also has its variation of this type. A flowchart diagram for LA_m is given in Figure 7.5.

Proposition 7.11 The class of regular languages is learnable in the limit using membership queries by an algorithm LA_m. Further, LA_m may be implemented to run in time polynomial

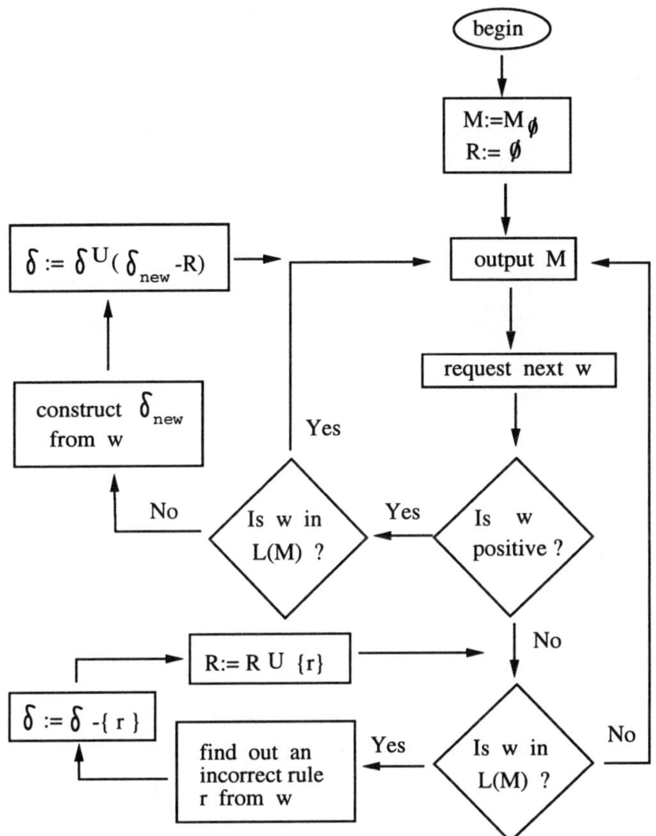

Figure 7.5. Flowchart Diagram for LA_m

in n (the number of states of a minimum DFA for a target language) and m (the total lengths of all examples provided so far).

Thus, the algorithm LA_m provides a reasonably practical, efficient algorithm for learning the class of regular languages. In a related work (Ishizaka 1989) Ishizaka proposes an algorithm for learning regular languages based on the model inference algorithm. His algorithm learns a correct DFA in a logic program formulation in the limit from membership queries and the complete presentation, and may be implemented to run in time polynomial in n, ℓ, and N (the total number of examples provided so far). Thus, this algorithm is close to LA_m but is different in that it learns DFAs in a logical formulation.

An interesting problem to be investigated is how far we can go along the line of research proposed here in the MAT learning paradigm. In fact, we have obtained a couple of extended results to superclasses of regular languages (Yokomori 1992; Shirakawa and Yokomori 1993).

Acknowledgments

The author would like to thank a referee for many constructive suggestions which greatly improved the quality as well as readability of this chapter.
This work is supported in part by Grants-in-Aid for Scientific Research No. 04229105 from the Ministry of Education, Science and Culture, Japan.

REFERENCES

Angluin, D. (1980). Inductive inference of formal language from positive data. *Information and Control*, *45*, 117–135.

Angluin, D. (1987a). Learning k-bounded context-free grammars. *Research Report*, *557*, Dept. of Computer Sci., Yale Univ.

Angluin, D. (1987b). Learning regular sets from queries and counterexamples. *Information and Computation*, *75*, 87–106.

Angluin, D. and Kharitonov, M. (1991). When won't membership queries help? *Proceedings of 23rd ACM Symposium on Theory of Computing*, 444–454.

Berman, P. and Roos, R. (1987). Learning one-counter languages in polynomial time. *Proceedings of 28th IEEE Symposium on Foundations of Computer Science*, 61–67.

Gold, E.M. (1967). Language identification in the limit. *Information and Control*, *10*, 447–474.

Gold, E.M. (1978). Complexity of automaton identification from given data. *Information and Control*, *37*, 302–320.

Harrison, M.A. (1978). *Introduction to Formal Language Theory.* Addison-Wesley, Reading, MA.

Ishizaka, H. (1989). Inductive inference of regular languages based on model inference. *International Journal of Computer Mathematics*, *27*, 67–83.

Ishizaka, H. (1990). Polynomial time learnability of simple deterministic languages. *Machine Learning*, *5*, 151–164.

Salomaa, A. (1973). *Formal Languages.* Academic Press, New York, NY.

Shapiro, E. (1981). Inductive inference of theories from facts. *Research Report*, *192*, Dept. of Computer Sci., Yale Univ.

Shirakawa, H. and Yokomori, T.(1993). Polynomial time MAT Learning of c-deterministic context-free grammars. *Transactions of Information Processing Society of Japan*, *34*, 380-390.

Yokomori, T. (1992). On learning systolic languages. *Proceedings of 3rd Workshop on Algorithmic Learning Theory*, 41–52.

SCIENTIFIC DOMAINS

8

Machine Learning and Biomolecular Modelling

M. J. E. Sternberg

R. A. Lewis

Biomolecular Modelling Laboratory,
Imperial Cancer Research Fund,
Lincoln's Inn Fields,
London WC2A 3PX, UK.

R. D. King

Department of Statistics, Strathclyde University,
Glasgow , G1 1XH, UK

S. Muggleton

Turing Institute, George House,
36 North Hanover Street,
Glasgow G1 2AD, UK

Abstract

Two problems in biomolecular modelling have been tackled by the machine learning program, Golem. The program, based on inductive- based logic programming, takes as input observations and combines them with background knowledge to yield rules. The first system tackled was modelling the quantitative structure-activity relationship (QSAR) of a series of trimethoprim analogues binding to E. coli dihydrofolate reductase. The Golem rules were a better model than standard regression approaches. More importantly, these rules described the chemical properties of the enzyme binding site that were in broad agreement with the crystallographic structure. The other system was the prediction of the secondary structure of proteins to identify regions that form α-helices. On a test set the prediction was 81% accuracy which is better than the results achieved by neural networks. The rules from Golem defined patterns of residues forming α-helices.

1 INTRODUCTION

Biomolecular modelling aims to understand the inter-relationships of chemical formula, three-dimensional structure and function of molecules of biological importance. In many area, rules are derived from a experimental observation of the properties of a series of compounds. This paper will describe two topics where the approach of inductive-based machine learning, implemented in a program Golem (Muggleton and Feng, 1990) , has been applied to map the experimental relationship. The first topic is to model quantitatively the structure activity relationship of a series of drugs. The second topic is the prediction of the local secondary structure of a protein from its sequence. In both systems rules are obtained that are better than conventional statistical approaches and also yield rules that provide insight into the stereochemistry of the system. All figures referred to in the text are to be found at the end of this chapter.

2 GOLEM

Golem (Muggleton and Feng, 1990) is a program for machine learning by inductive logic programming (ILP). In Golem the concepts are expressed in a language that is a subset of first-order predicate calculus. The input consists of: (1) observations coded as facts and negative counter-examples and (2) background knowledge that in both applications describes the relevant chemical principles. Golem then forms an inductive hypothesis by generalising the observations and background knowledge (Figure 8.1).

The basic algorithm implemented by Golem is:

1. take at random two examples from the observations;
2. computes the set of properties that are common to both examples;
3. generate a rule which will be true for both examples;
4. evaluate the accuracy of the rule on the remaining data;

5. repeat steps (1) to (4) several times (typically 10) and then keep the best rule;
6. choose a further example is at random and apply steps (2) and (3) to generate a rule;
7. test the rule from (6) on further observations;
8. if the rule continues to be accurate, store the rule and added it to the background knowledge;
9. start again from (1) and generate new rules until no further rules can be obtained.

In Golem, PROLOG is used to represent the observations, background knowledge and rules. The use of the predicate calculus in PROLOG is expressive enough for most mathematical concepts but is readily comprehensible and therefore makes an ideal representation for machine learning. The program Golem is written in the language C and runs on SUN Sparcstations and VAXes.

3 DRUG DESIGN

3.1 Background

In the design of a potent drug, the pharmaceutical chemist starts with a lead compound with some of the required activity. Modifications are then made introducing different chemical groups whose activities are assayed. To direct the choice of better compounds to synthesise, one requires a quantitative structure-activity relationship (QSAR) (see reviews (Goodford, 1985). One standard method is the Hansch equation (Hansch, 1969; Hansch, et al., 1962) in which an empirical equation based on linear regression links the properties of the substituents to their observed activity. More recently, neural networks (Andrea and Kalayeh, 1991) have also been applied to model QSAR. However neither approach yields insight into the chemistry of the drug / receptor interaction.

Here we report a comparison of modelling a QSAR by the Hansch method (Hansch, et al., 1982) and by Golem. The particular system is the activities of trimethoprim analogues on dihydrofolate reductase (DHFR). In most QSAR problems, the three-dimensional structure of the receptor in which the drug binds is not known. However for DHFR there is precise three-dimensional information at the atomic level about the stereochemistry of the interaction obtained by protein crystallography (Champness, et al., 1986; Matthews, et al., 1985) . Thus we have an ideal system to establish whether Golem can learn rules describing this stereochemistry. Full details of this study can be found in King et al. (1992).

3.2 Method

The study was performed with the same training set of 44 trimethoprim analogues (Figure 8.2) as used in the study on the Hansch linear regression (Hansch, et al., 1982). In addition a testing set of 11 drugs was obtained from the literature.

The input to Golem consisted of

1. The observed activities of the drugs expressed as paired examples of greater activity, e.g. *great(d20, d15)* which states that drug no. 20 has higher activity than drug no. 15.
2. The background knowledge in which the chemical structure of the drugs is represented in the form *struc(d35, NO2, NHCOCH3, H)* which states that drug no. 35 has: NO2 substituted at position 3 , NHCOCH3 substituted at position 4, and no substitution at position 5;
3. The background knowledge of the chemical properties of the substituents in which the properties were assigned heuristically. The particular properties are: polarity, size, flexibility, hydrogen-bond donor, hydrogen-bond acceptor, p donor, p acceptor, polarisability, branching and s effect. This was represented using different predicates for each property and value, e.g *polar(Br, polar3)* states that Br has polarity of value 3.
4. In-built arithmetic so information was explicitly given about the relative values of these properties for the substituent groups e.g. *great_polar(polar4, polar3)*. The input to Golem was 871 observations and 2976 items of background information. The run time was about 30 cpu minutes on a

SUN SparcStation 1 to generate one rule.

3.3 Results

Nine rules were obtained that predicted the relative activity of two drugs in terms of the chemical properties of the substituents (see Appendix A). An example of a rule as a PROLOG clause is:

```
great(A, B):-
        struc(A, D, E, F), struc(B, h, C, h),
        h_donor(D, h_don0), pi_donor(D, pi_don1),
        flex(D, G), less4_flex(G).
```

The interpretation in English is:

Drug A is better than drug B if
drug B has no substitutions at positions 3 and 5 and
drug A at position 3 has hydrogen donor = 0 and
drug A at position 3 has pi-donor = 1 and
drug A at position 3 has flexibility < 4.

The rank values predicted for the 44 drugs in the training set by Golem and by the application of the Hansch equation are tabulated in Figure 8.3 and shown graphically in Figures 8.5 and 8.6. The Spearman rank correlation with the observed order is 0.92 for Golem and 0.79 for the Hansch equation. The difference in these rank correlation was shown to be significance at the 5% level by FisherUs z transformation (Kendall and Stuart, 1977) .

A better test of a prediction method is its performance on data not used in developing the algorithm. Tabulation of the results for the 11 testing drugs are also shown in Figure 8.4. The rank correlation for the 11 drugs by Golem was 0.46 compared to 0.42 for the Hansch method. The Fisher z-value is 0.54 which is not significant reflecting the similar rank correlations obtained on a small test set. Further cross- validation trials selecting at random 44 training drugs and evaluating the success on the 11 remaining compounds gave similar results. In general, on the test set the machine learning approach is slightly more accurate than the regression method of Hansch.

We now consider whether the rules obtained by Golem describing the chemical properties of favourable drugs can provide information about the stereochemistry of drug / DHFR interactions. The second half of Figure 8.2 shows a cartoon of the crystallographically- observed binding of trimethoprim to E. coli DHFR in the complex (Champness, et al., 1986) with NADPH. Golem suggested that the general properties of a favourable substituent at the 3 position are: pi-donor of 1; neither a hydrogen bond donor nor an acceptor; flexibility < 4 and a polarity of zero. The crystal structure shows that the 3 position is constrained in size and is not exposed to solvent. Thus the substituents should not be polar or have hydrogen bond donor or acceptor character. In addition, Golem found that the properties that favour positions 4 are: polarity of 2; size of 3 or size of 2 with the potential to be a hydrogen bond acceptor. These properties are consistent with a site that is exposed to solvent and should be polar.

To summarise, the rules produced by Golem are at least as good as those obtained by the linear regression method of Hansch. More importantly, these rules can automatically provide a chemical description of the interaction of the drug with the receptor.

4 SECONDARY STRUCTURE PREDICTION

4.1 Background

Proteins are biological macromolecules whose activity, such as catalysis, results from its complex three-dimensional structure (called tertiary structure). However structure determination remains difficult so today there are around 500 different tertiary structures known. In contrast, the chemical structure of the protein, known as the amino-acid sequence, can readily be obtained from the gene sequence and today more than 50,000 sequences are known.

Experiments suggest that it should be possible to predict theoretically the three- dimensional structure of a protein from its sequence, for reviews see (Blundell, et al., 1987). This problem remains complex and a simpler, first step, has attracted much research. Most proteins have segments of the chain that adopt a regular local conformation, known as secondary structure. There are two major types, α-helices and β-sheets, and typically between 5 to 15 residues of the chain can be assigned to one local region of secondary structure. Thus a protein chain that ranges between 50 and 1000 amino-acid residues will have several regions of α and/or β structure linked by less regular (but still defined) conformation known as coil.

Many early methods of secondary structure prediction were based on simple statistical analyses and these achieved about 60% accuracy for a three state (α,β and coil) prediction (Kabsch and Sander, 1983) . Recently, approaches using larger data sets (Gibrat, et al., 1987) and/or sophisticated analyses such as neural networks (Kneller, et al., 1990) yielded only marginal improvements of up to 65% accuracy. One method of improving prediction is to restrict the algorithm to a subset of all proteins that have just α-helices and coil. Recently neural networks by Kneller et al (Kneller, et al., 1990) have been applied to this α/α class and yielded 76% accuracy. However a major problem with neural networks is that the weights do not provide insight into principles of folding. Here we report on the use of Golem to secondary structure prediction. This extends earlier work (King and Sternberg, 1990) and full details can be found in Muggleton et al (Muggleton, et al., 1992) .

4.2 Method

The structural data consisted of a training set of 12 protein with a testing set of 4 proteins that were not related (i.e. non-homologous). The absence of any homology between the training and testing set is essential for an evaluation of the power of any algorithm. We note that there are several studies where this requirement has been ignored.

The input to Golem was:

1. The observations of the location in the protein of the α-helices coded as: *alpha(protein_name,position)*. For example: *alpha(155C, 110)* states that residue 110 in protein 155C is in an α-helix.
2. The background information of the amino-acid sequence coded as: *position (155C,110,V)* that states residue at

position 110 in 155C is a valine.

3. The background knowledge about the chemical properties of each of the 20 different amino-acid residues. These properties were:
hydrophobic, very_hydrophobic, hydrophilic,
positive, negative, neutral,large, small, tiny,
polar, aliphatic, aromatic, hydrogen_bond_donor,
hydrogen_bond_acceptor, not_aromatic, small_or_polar,
not_polar,aromatic_or_very_hydrophobic,
either_aromatic_or_aliphatic,not_proline, not_lysine.
These were coded as: aliphatic(V).
4. Built-in arithmetic that explicitly coded information about the sequential relationship of residues.

4.3 Results

Golem was applied and yielded rules that led to a speckled prediction with individual residues not predicted as helical but within a run of helical residues. α-Helices are continual runs of at least 5 residues and this was then included into the algorithm. A bootstrapping procedure was followed where the predictions made by Golem were used as background knowledge and Golem learnt how to smooth the data. This was repeated twice and the resultant rules generalised by hand to include symmetry.

The accuracy of secondary structure prediction is conventionally expressed as the percentage of residues whose state is correctly predicted. The final prediction, after smoothing, was 78% and 81% on the training and testing sets. The better improvement on the testing set rather than the training data reflects the standard error in each of these values of about 2%. This result can be compared with the recent work on neural networks on α/α proteins that was 76% accurate (Kneller, et al., 1990) .

The rules described the properties of residues at different positions along the helix (e.g. Figure 8.7). Inspection showed that well-known chemical principles of α-helix formation are described in the rules, for example the preference for one face to be formed from oily (i.e. hydrophobic) residues. However new insights into protein folding which may be included in the patterns are not easy to discern.

5 CONCLUSIONS

There are many similarities in the methodology and the results from the two studies. First the system was described as simple observations- for drugs the relative activities of the two compounds and for proteins the observed secondary structure. The background knowledge had two parts. First the chemistry was defined, either the substituents of the drug or the residue type in the protein. Then chemical properties were assigned. This is the most critical step as the choice of representation is central to the success of the approach. Finally built-in arithmetic had to be supplied explicitly.

The results showed that machine learning was at least as good as the state-of-the- art statistical methods, neural networks or linear regression. The rules obtained did include some information about the stereochemistry of the system. However careful study of the rules is required to extract this information.

Two areas in biomolecular modelling have been tackled by machine learning. There are however many other problems that await input from the artificial intelligence community. For example, we have developed an algorithm that searches for the most probable association in three-dimensions of two proteins of known structure (Walls and Sternberg, 1992) . This docking problem was tackled by an exhaustive search to match the two surfaces. The algorithm was implemented on the single instruction multiple datastream architecture of an AMT DAP . The search took two days on the DAP but would require at least two orders of magnitude longer on a SparcStation 1.

In general, there is an explosion in the data in the biological sciences. Machine learning can provide a powerful tool to extract generalisations of predictive quality that yield new insights.

Acknowledgments

MJES is supported by the Imperial Cancer Research Fund; RAL by a fellowship from the Royal Commission of 1851; RDK by the Esprit "Statlog" project; SM by for an SERC research fellowship.

APPENDIX A

Rules for QSAR derived by machine learning

The rules are first given as Prolog clauses (in which ":-" is a definition and a "," is logical "and") and then in English. The rules have been classified into those primarily relating to substituent 3 (rule 3.1 to 3.4); to substituent 4 (4.1 and 4.2). The rules given are the more general and three other more specific are not given.

Rule 3.1 (coverage 119/0 train : 105/0 test)

```
great(A, B):-
        struc(A, D, E, F), struc(B, h, C, h),
        h_donor(D, h_don0), pi_donor(D, pi_don1),
        flex(D, G), less4_flex(G).
```

Drug A is better than drug B if
drug B has no substitutions at positions 3 and 5 and
drug A at position 3 has hydrogen donor = 0 and
drug A at position 3 has pi-donor = 1 and
drug A at position 3 has flexibility < 4.

Rule 3.2 (coverage 244/71 train : 248/4 test)

```
great(A, B):-
        struc(A, C, D, E), struc(B, F, h, G), not except3.2(A, B).
except3.2(A, B):-
        struc(A, C, D, h), struc(B, E, h, F), h_donor(E, h_don0).
```

Drug A is better than drug B if
drug B has no substitution at position 4 unless
drug A has no substitution at position 5 and
drug B at position 3 has hydrogen donor = 0.

Rule 3.3 (coverage 102/13 train : 33/0 test)

```
great(A, B):-
        struc(A, G, H, I), struc(B, C, h, D),
        pi_donor(C, pi_don0), polar(C, E),
        great0_polar(E), h_acceptor(C, F), great0_h_acc(F).
```

Drug A is better than drug B if
drug B has no substitutions at position 4 and

drug B at position 3 has p-donor = 0 and
drug B at position 3 has polarity > 0 and
drug B at position 3 has hydrogen acceptor > 0.

Rule 3.4 (coverage 129/2 train: 126/0 test)

```
great(A, B):-
        struc(A, C, D, E), struc(B, G, h, h), h_donor(C, h_don0),
        pi_donor(C, pi_don1), flex(C, F), less4_flex(F),
        polarisable(G, H), less3_polari(H).
```

Drug A is better than drug B if
drug B has no substitutions at position 4 and 5 and
drug B at position 3 has polarisability < 3 and
drug A at position 3 has hydrogen donor = 0 and
drug A at position 3 has p-donor = 1 and
drug A at position 3 has flexibility < 4.

Rule 4.1 (coverage 289/72 train: 99/0 test)

```
great(A, B):-
        struc(A, D, E, F), struc(B, h, C, h), not except4.1(A,B).
except4.1(A,B):-
        struc(B, h, C, h), size(C, size3).
except4.1(A,B):-
        struc(B, h, C, h), size(C, size2), h_acceptor(C,h_acc1).
```

Drug A is better than drug B if
drug B has no substitutions at position 3 and 5 unless
        drug B at position 4 has size = 3 or
        drug B at position 4 has size = 2 and hydrogen acceptor = 1.

Rule 4.2 (coverage 187/2 train: 193/2 test)

```
great(A, B):-
        struc(A, E, F, G), struc(B, C, D, h),
        not_h(E), polar(F, polar2).
```

Drug A is better than drug B if
drug B has no substitution at position 5 and
drug A has a substitution at position 3 and
        drug A at position 4 has polarity = 2.

REFERENCES

Andrea, T.A. and Kalayeh, H. (1991). Applications of neural networks in quantitative structure-activity relationships of dihydrofolate reductase inhibitors. *J. Med. Chem.* 34, 2824-2836.

Blundell, T.L., Sibanda, B.L., Sternberg, M.J.E. and Thornton, J.M. (1987). Knowledge-based prediction of protein structures and the design of novel molecules. *Nature* (London) 326, 347-352.

Champness, J.N., Stammers, D.K. and Beddell, C.R. (1986). Crystallographic investigation of the cooperative interaction between trimethoprim, reduced cofactor and dihydrofolate reductase. *FEBS Letters* 199, 61-67.

Gibrat, J.F., Garnier, J. and Robson, B. (1987). Further developments of protein secondary structure prediction using information theory. New parameters and consideration of residue pairs. *J Mol Biol* 198, 425- 443.

Goodford, P.J. (1985). A computational procedure for determining energetically favorable binding sites on biologically important macromolecules. *J. Med. Chem.* 28, 849-857.

Hansch, C. (1969). A quantitative approach to biochemical structure-activity relationships. *Acc. Chem. Res.* 2, 232-239.

Hansch, C., Li, R.-l., Blaney, J.M. and Langridge, R. (1982). Comparison of the inhibition of Escherichia coli and Lactobacillus casei dihydrofolate reductase by 2,4-diamino-5-(substituted-benzyl) pyrimidines: quantitative structure-activity relationships, X-ray crystallography, and computer graphics in structure-activity analysis. *J. Med. Chem.* 25, 777-784.

Hansch, C., Maloney, P.P., Fujita, T. and Muir, R.M. (1962). Correlation of biological activity of phenoxyacetic acids with Hammett substituent constants and partition coefficients. *Nature* 194, 178-180.

Kabsch, W. and Sander, C. (1983). How good are predictions of protein secondary structure. *FEBS Lett* 155, 179-182.

Kendall, M. and Stuart, A. (1977). *The Advcnced Theory of Statistics*, Griffen and Company, London.

King, R.D., Muggleton, S., Lewis, R.A. and Sternberg, M.J.E. (1992). Drug design by machine learning : the use of inductive logic programming to model the structure-activity relationship of trimethoprim analogues binding to dihydrofolate reductase. *Proc. Nat. Acad. Sci.*, USA. 89, 11322-11326.

King, R.D. and Sternberg, M.J.E. (1990). A machine learning approach for the prediction of protein secondary structure. *J. Mol. Biol.* 216, 441- 457.

Kneller, D.G., Cohen, F.E. and Langridge, R. (1990). Improvements in protein secondary structure prediction by an enhanced neural network. *J. Mol. Biol.* 214, 171-182.

Matthews, D.A., Bolin, J.T., Burridge, J.M., Filman, D.J., Volz, K.W., Kaufman, B.T., Beddell, C.R., Champness, J.N., Stammers, D.K. and Kraut, J. (1985). Refined crystal structures of Escherichia coli and chicken liver dihydrofolate reductase containing bound trimethoprim. *J. Biol. Chem.* 260, 381-391.

Muggleton, S. and Feng, C. (1990). Efficient induction of logic programs. *Proceedings of the first conference on algorithmic learning theory*, Arikawa, S., Goto, S., Ohsuga, S. and Yokomori, T., eds. (Japanese Society for Artificial Intelligence, Tokyo) pp. 368-381.

Muggleton, S., King, R.D. and Sternberg, M.J.E. (1992). Protein secondary structure prediction using logic. *Prot. Eng.* 5 647-657.

Walls, P.H. and Sternberg, M.J.E. (1992). New algorithm to model protein-protein recognition based on surface complementarity - Applications to antibody-antigen docking. *J. Mol. Biol.* 228, 277-297 .

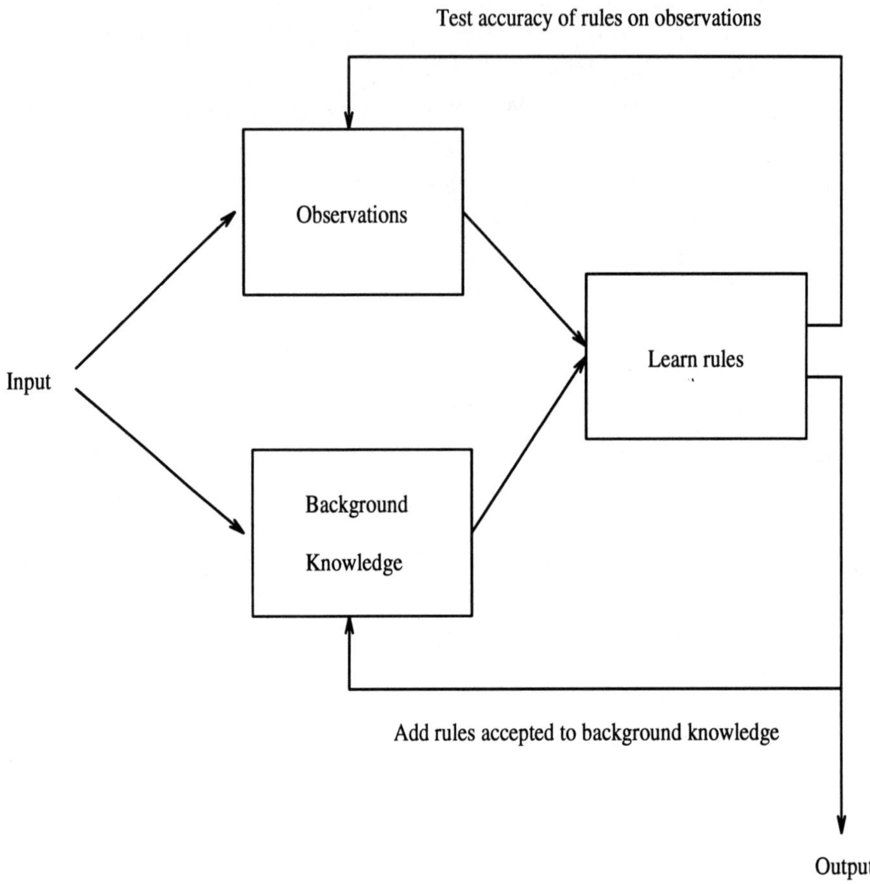

Figure 8.1. Machine learning by Golem.

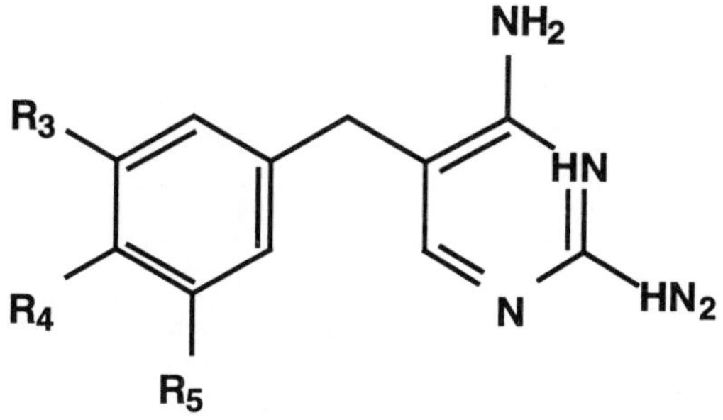

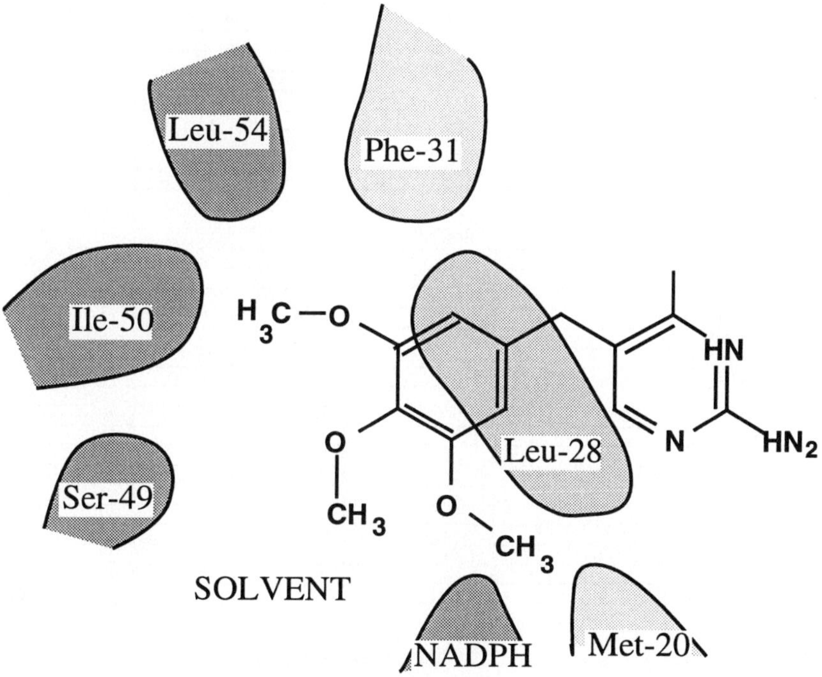

Figure 8.2. (a) the structure of trimethoprim analogues; (b) a cartoon of the interaction of trimethoprim with DHFR from X-ray structures (Champness, et al., 1986; Matthews, et al., 1985). Faint stippling indicates that the residue lies below the plane of the phenyl ring, darker stippling that the atoms are above.

X	$log(1/K_{iapp})$	Observed rank	rank by Golem	rank by Hansch
3,5-(OH)2	3.04	1	17	2
4-O(CH2)6CH3	5.60	2	4.5	4
4-O(CH2)5CH3	6.07	3	4.5	10
H	6.18	4	1	6.5
4-NO2	6.20	5	7.5	20.5
3-F	6.23	6	6	6.5
3-O(CH2)7CH3	6.25	7	15	6.5
3-CH2OH	6.28	8	2	16
4-NH2	6.30	9	7.5	3
3,5-(CH2OH)2	6.31	10	3	23
4-F	6.35	11	9.5	6.5
3-O(CH2)6CH3	6.39	12	16	11
4-OCH2CH2OCH3	6.40	13	20	20.5
4-Cl	6.45	14	12	17.5
3,4-(OH)2	6.46	15	18	1
3-OH	6.47	16	13	9
4-CH3	6.48	17	9.5	17.5
3-OCH2CH2OCH3	6.53	18	21	34
3-CH2O(CH2)3CH3	6.55	19	24	35.5
3-OCH2CONH2	6.57	20.5	14	13
4-OCF3	6.57	20.5	19	22
3-CH2OCH3	6.59	22	28	29.5
3-Cl	6.65	23	30.5	29.5
3-CH3	6.70	24	30.5	27.5
4-N(CH3)2	6.78	25	22.5	27.5
4-Br	6.82	26	11	24
4-OCH3	6.82	27	22.5	26
3-O(CH2)3CH3	6.82	28	29	32
3-O(CH2)5CH3	6.86	29	26	14
4-O(CH2)3CH3	6.89	30.5	27	15
4-NHCOCH3	6.89	30.5	25	12
3-OSO2CH3	6.92	32	33	25
3-OCH3	6.93	33	36	38
3-Br	6.96	34	37	35.5
3-NO2, 4-NHCOCH3	6.97	35	34	37
3-OCH2C6H5	6.99	36	35	31
3-CF3	7.02	37	32	19
3,4-(OCH2CH2OCH3)2	7.22	38	39	40
3-I	7.23	39	38	33
3-CF3, 4-OCH3	7.69	40	41.5	39
3,4-(OCH3)2	7.72	41	41.5	41
3,5-(OCH3)2, 4-O(CH2)2OCH3	8.35	42	43	43
3,5-(OCH3)2	8.38	43	40	42
3,4,5-(OCH3)3	8.87	44	44	44

Figure 8.3. Predicted and observed activity of trimethoprim analogues on training data. X gives the substituents. The observed value of the affinity is expressed as $log(1/K_{iapp})$. The first 44 drugs were used in the training set and the observed rank ranges from 1 to 44.

X	*log* $(1/K_{iapp})$	Observed rank	rank by Golem	rank by Hansch
3,5-(CH3)2, 4-OCH3	7.56	40 (1)	52.5 (9)	45.5 (4.5)
3-Cl, 4-NH2, 5-CH3	7.74	43 (2)	40 (2)	44 (3)
3,5-(CH3)2, 4-OH	7.87	44.5 (3.5)	40 (2)	41 (1)
3,5-Cl2, 4-NH2	7.87	44.5 (3.5)	40 (2)	45.5 (4.5)
3,5-Br2, 4-NH2	8.42	48 (5)	44 (4)	53 (10)
3,5-(OCH3)2, 4-OCH2C6H5	8.57	49 (6)	52.5 (9)	51 (9)
3,5-(OCH3)2, 4-CH3	8.82	50.5 (7.5)	47.5 (9)	54 (2)
3,5-(OCH3)2, 4-O(CH2)7CH3	8.82	50.5 (7.5)	52.5 (5.5)	42 (11)
3,5-(OCH3)2, 4-O(CH2)5CH3	8.85	52 (9)	52.5 (9)	48.5 (7)
3,5-I2, 4-OCH3	8.87	54 (10.5)	52.5 (9)	50 (8)
3,5-I2, 4-OH	8.87	54 (10.5)	47.5 (5.5)	47 (6)

Figure 8.4. Predicted and observed activity of trimethoprim analogues on test data. The final 11 drugs are the testing set and the first number is the rank in the 55 drugs and the second the rank for the 11.

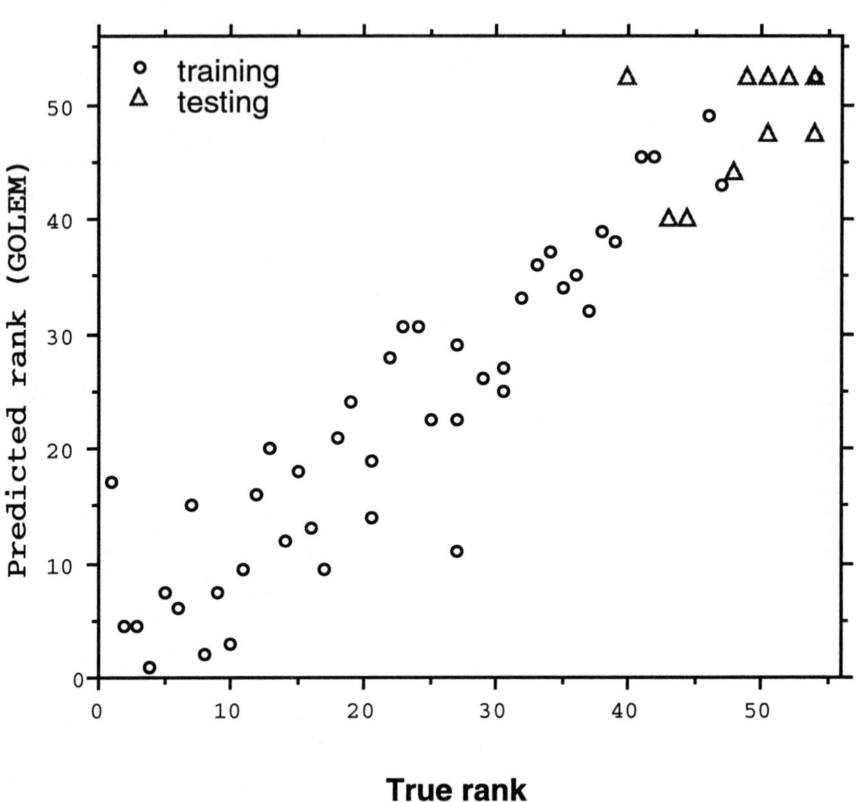

Figure 8.5. Scattergram of the observed rank versus that predicted by Golem.

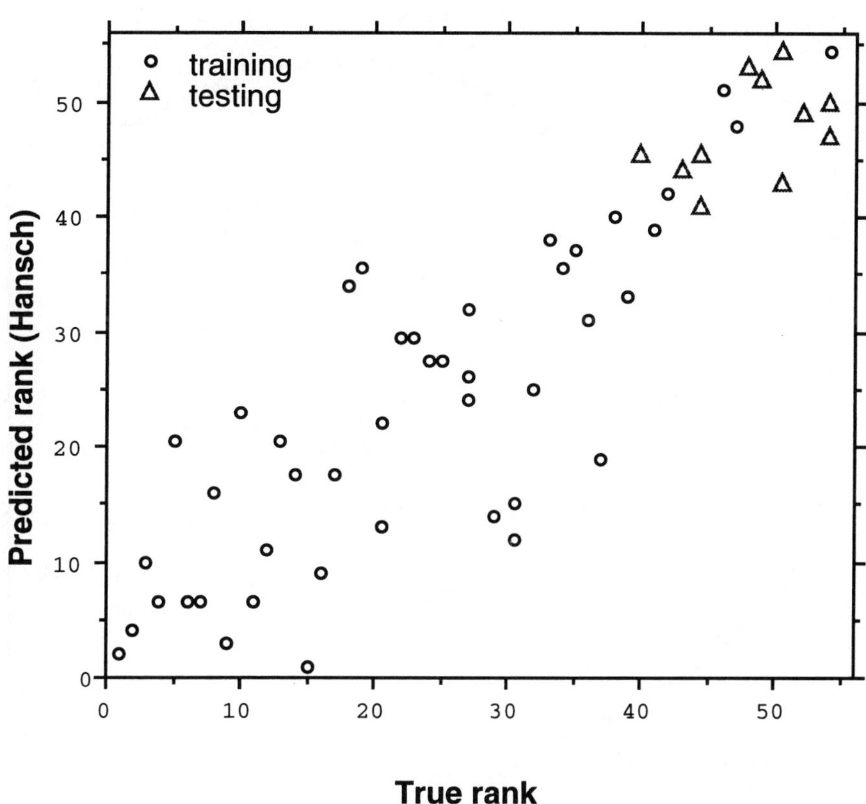

Figure 8.6. Scattergram of the observed rank versus that predicted by Hansch.

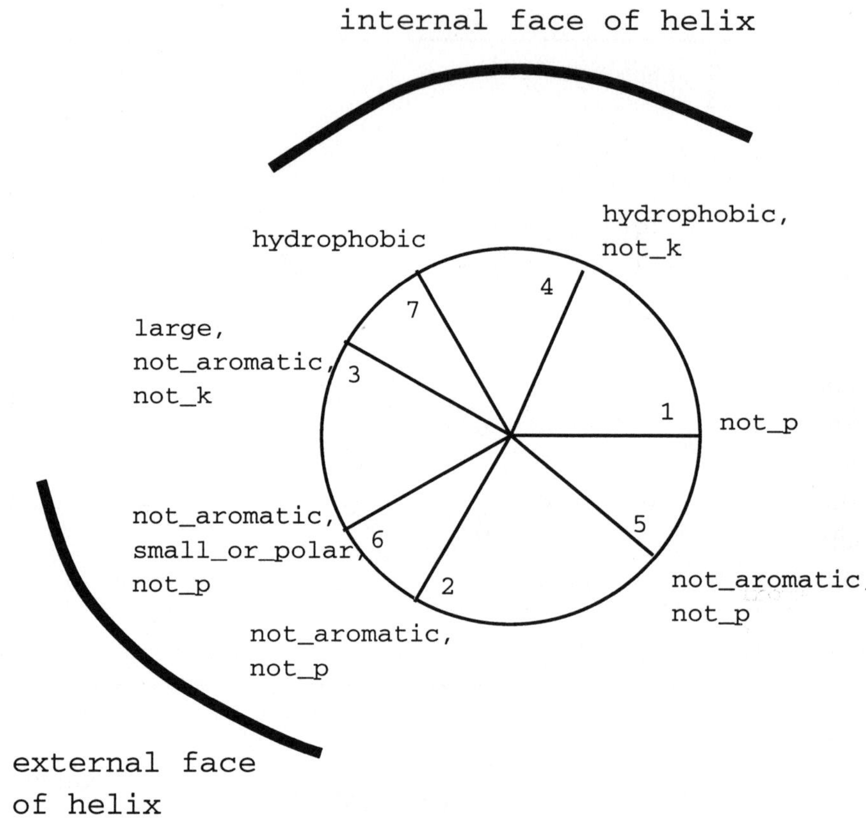

Figure 8.7. Schematic representaion a rule for secondary structure prediction.

9

More Than Meets the Eye: Animal Learning and Knowledge Induction

E. J. Kehoe

School of Psychology,
University of New South Wales,
Kensington, NSW 2033 Australia

Abstract

Animals share many of the same learning problems as faced by untutored machine systems. In the ordinary environment, animals must continuously discover associations between potential signals and subsequent events that are crucial to their wellbeing. Moreover, the animals must build this world model from a highly limited training set, without having prior knowledge or verbal instructions. Experiments using conditioned reflex methods have examined issues fundamental to all learning systems. These include:

- Credit assignment, e.g., animals capitalize on any sequence of reliable signals for subsequent events.
- Representation, e.g., the XOR problem can be solved by animals.
- Productivity, e.g., the rate of learning improves across successive problems that are structurally similar but differ in their superficial features.
- Anticipation, e.g., conditioned responses are finely timed so that they are neither too early nor too late during a warning signal.

These behavioural phenomena can be modelled by small connectionist networks and/or single-unit real time models. Thus, the study of learning in animals provides a biological counterpart to nonsymbolic learning systems.

1 INTRODUCTION

Animals share many of the same learning problems as faced by untutored machine systems. Among other things, animals must continuously discover associations between potential signals, actions, and subsequent events that are crucial to their wellbeing. Animals must build this world model from a highly limited training set, without having prior knowledge or verbal instructions. In addition, the relationships among events are changeable; both opportunities and threats can change in place and time. Over the last century, a large body of research concerning the fundamental principles of learning has emerged from laboratory experiments with animals. More recently, it has become increasingly clear that these principles may be useful in designing adaptive machines. However, the traffic is not all one way; developments in machine intelligence and adaptive systems have proved increasingly valuable in explaining animal learning. This chapter will describe how the principles of animal learning correspond to key issues in designing algorithms for learning. The chapter will be divided into five sections. First, the major procedures for the study of animal learning will be described and classified with respect to more general learning paradigms. The remaining four sections will describe the study of animal learning as it pertains to the issues of credit assignment, representation, productivity, and anticipation. Each of these latter sections will include a description of the issue, examples of well-substantiated findings, and examples of algorithms relevant to those findings. All figures referred to in the text are to be found at the end of this chapter.

2 ISSUES AND DEFINITIONS

2.1 Basic Procedures

Animal learning has been studied in a variety of ways in a variety of situations using a variety of species, ranging from flatworms to elephants. However, the variety of experiments can be grouped into two basic categories, namely, *classical conditioning* and *operant conditioning.*

2.1.1 *Classical conditioning*

The prototype for classical conditioning is Pavlov's (1906, 1927) method of conditioned reflexes. Pavlov's method entails the presentation of two stimuli to a dog. First, the *conditioned stimulus* (CS) is usually a relatively innocuous event, for example the well-known ringing of a bell. Second, the *unconditioned stimulus* (US) is an event of biological significance to the dog, for example the consumption of food. When presented in close succession, the CS acts as a signal for the US. Pavlov's method takes advantage of the innate ability of the US to elicit strong reactions known as the *unconditioned responses* (URs). Pavlov measured one of those responses, namely, salivation. Across repeated CSPUS pairings, the CS itself comes to elicit a response that broadly resembles the UR. This learned response to the CS is known as the *conditioned response* (CR).

2.1.2 *Operant conditioning*

Operant conditioning is also known as instrumental learning and sometimes as trial-and-error learning. It is commonly identified with Skinner (1938) but was started by Thorndike (1898). In the prototypic experiment, a cat is placed in a puzzle box, really a modest-sized cage. Just outside the cage, a bowl of food is visible. The cat is allowed to explore the box without constraint, but only a single action Q pressing a certain latch Q opens the cage door and permits access to the food reward. Across repetitions of the problem, the cat progressively refines its actions so that it smoothly and quickly opens the necessary latch. The distinguishing feature of this and all operant procedures is the *contingency* between the designated action and its reward, which is known technically as a *positive reinforcer.*

2.2 Perspectives

At various times, classical and operant conditioning have been thought to reflect different learning processes (see Gormezano and Kehoe, 1975). Classical and operant conditioning are usually distinguished in terms of the relationship between the learned response and the final significant event. In classical conditioning, the final event Q the unconditioned stimulus Q is presented on a preprogrammed basis irrespective of the conditioned response. In contrast, in operant conditioning, the delivery of the reward depends on whether or not the designated action occurs. However, from the perspective of adaptive systems, a different distinction is possible. Specifically, classical and operant conditioning can be viewed as two species of supervized learning. According to this distinction, it is the instructive or noninstructive nature of the final event itself that is important, not its programmed or contingent relationship to the prior action.

From an adaptive systems' perspective, classical conditioning is an example of supervized learning with *instructive feedback.* Instructive feedback contains explicit instructions as to the desired response. For example, in a classification task, such feedback would include the name of the appropriate category. If the learner classifies an item correctly, such feedback confirms the response. If the learner makes an incorrect classification, then the same feedback provides an explicit corrective. Likewise, in classical conditioning, the elicitation of the UR by a US effectively acts to either confirm an existing CR or to instruct the subject as to the nature of the CR to be acquired.

In contrast to classical conditioning, operant conditioning usually involves *critical* feedback. That is to say, the feedback only indicates whether or not the preceding action matched the desired output. Critical feedback gives no explicit guidance as to what a more appropriate response might be. In the case of a classification task, the learner would only be told whether a particular classification was 'right' or 'wrong'. Where there are many possible categories, being told that one's choice is 'wrong' is minimally informative. In operant conditioning, the delivery or absence of a reward serves the same critical function as 'right' and 'wrong'.

2.3 The Rabbit Eyeblink Preparation

The remainder of this chapter will focus on classical conditioning in one particular version, namely, classical conditioning of the protective eyelid closure in the rabbit. This preparation was chosen because it has generated a large and well-established body of findings relevant to key issues in adaptive systems. Moreover, the rabbit preparation is now the most widely used preparation for the neurophysiological study of learning in mammals (Gormezano, 1966; Gormezano, Thompson, and Prokasy, 1987; see also Gabriel, 1988). In this preparation, closure of the rabbit's inner, third eyelid Q the nictitating membrane (NM) Q is the measured response. The CSs have been mild auditory or visual stimuli that themselves do not innately elicit eyelid closure. Conversely, the US is a tactile stimulus Q either a puff of air to the eye or a low, electrical current delivered to the skin around the eye. These USs innately elicit closure of the eyelids as URs (see Figure 9.1). As is true of Pavlov's preparation, repeated presentation of a CS and US in close succession yields the acquisition of a CR, namely closure of the eyelids during the warning period provided by the CS in advance of the US (see Figure 9.2).

3 CREDIT ASSIGNMENT

Reliable signals for biologically significant events are usually embedded in a continuous and multifaceted stream of sensory events. An adaptive organism must possess mechanisms for discovering associations among reliable sequences and rejecting spurious associations among accidental sequences. From the perspective of adaptive systems, identification of reliable sequences is an example of credit assignment (Sutton, 1984). In game-playing, the credit-assignment problem is one of determining which single move or combination of moves is responsible for ultimately winning or losing a given game (Minsky, 1961). In classical conditioning, credit is assigned to stimuli rather than actions (Kehoe, 1990). CSs that reliably signal the US receive credit in the form of associative strength that drives the CR. Unreliable CSs receive no associative strength. CSs that signal long periods without the US receive credit in the form of negative associative strength that inhibits expression of the CR. In its simplest form, credit assignment has been examined by manipulating the interval between a CS and US. More complex forms of credit assignment have been examined by using a sequence of two or more CSs prior to the US.

3.1 CSPUS Interval

Aristotle's Law of Contiguity has served as the historic principle for credit assignment in associative learning. That is to say, events that occur close together in time become strongly associated. Conversely, events more widely separated gain proportionally less associative strength. Since the inception of conditioning research, there have been numerous studies examining CR acquisition across manipulations of the interval between the CS and US (e.g., Gormezano and Moore 1969; Hall 1976). Figure 9.3 summarizes the results from several studies using the rabbit NM preparation in which the CSPUS interval was manipulated. Each point on the diagram represents the mean CR likelihood achieved for a given CS-US interval used in training. As can be seen, the function is not monotonic but is concave. No CR acquisition occurs when the CSPUS interval is zero, that is to say, when CS onset and US onset occur simultaneously. As the CSPUS interval approaches values between 200 and 400 ms, CR acquisition reaches progressively higher levels. CSPUS intervals longer than 400 ms produce progressively less CR acquisition. The longest CS-US interval that produces any discernible CR acquisition appears to be less than 6000 ms. Concave functions of this sort are common in other conditioning procedures. These functions reveal that there is an optimal warning period for anticipating the US; the animal does not waste its resources on signals that give too short or too long a warning period. Beyond their adaptive significance, these concave functions have been used to map the time course of the *trace* of the CS, that is, it's internal representation. According to real-time models of learning (e.g., Kehoe 1990; Sutton and Barto 1981, 1990), it is the intensity of the trace at the time of US delivery that determines the rate of conditioning. Thus, the concave function reveals the moment-by-moment intensity of the CS trace as it recruits and then decays.

3.2 Serial CSs

Although the nature of credit assignment with a single, reliable CS is relatively straightforward, a sequence of discrete CSs engages additional mechanisms. The rabbits do not just use the CS that has the most optimal CSPUS interval but capitalize on the relationships among CSs. Studies with the rabbit preparation have commonly used a sequence of just two CSs such as a tone and a light to signal the US (CSAPCSXPUS) (Kehoe, 1982a). Using this minimal sequence, there have been two paradoxical findings.

3.2.1 *Facilitation of Remote Associations*

Facilitation of remote associations occurs when CSA is presented several seconds in advance of CSX and the US. The acquisition of the CR to CSA proceeds more quickly and reaches a higher level than would otherwise be the case. In fact, control experiments have revealed that, without CSX in the sequence, virtually no CR acquisition to CSA occurs (Kehoe, Gibbs, Garcia, and Gormezano, 1979; Kehoe, Marshall-Goodell, and Gormezano, 1987; Kehoe and Morrow, 1984) (see Figure 9.4).

3.2.2 *Impairment of Proximal Associations*

Impairment of proximal associations occurs when CSA precedes CSX by a short interval, say, less than a second. CSA has a deleterious effect on CR acquisition to CSX. That is to say, the level of responding to CSX is low and considerably less than that of a control group given only CSXPUS training (e.g., Kehoe, 1979; Kehoe *et al.*, 1979; Kehoe, 1983) (see Figure 9.5).

3.3 Basic Principles

Two processes appear to underlie the findings obtained with serial compounds.

3.3.1 *Associative Transfer*

The facilitation of responding to CSA in a serial compound appears to rely heavily on transfer of associative strength from CSX (Gibbs, Cool, Land, Kehoe, and Gormezano, 1991a; Gibbs,

Kehoe, and Gormezano, 1991b). Such associative transfer has been demonstrated clearly in 'second-order conditioning' (Pavlov, 1927; Rizley and Rescorla, 1972). In that procedure, the subjects are given separate 'second-order' CSAPCSX trials and 'first-order' CSXPUS trials. As CRs are acquired on the CSX-PUS trials, CRs are also acquired to CSA on the CSAPCSX trials (Kehoe, Feyer, and Moses, 1981). Associative transfer occurs in a serial compound (CSAPCSXPUS), because it contains both the CSAPCSX and CSXPUS relationships of second-order conditioning (Gormezano and Kehoe, 1989) (see Figure reffig:kehoe6).

3.3.2 *Competitive Learning*

Concurrent CSs appear to compete for a limited resource such as associative strength or attention (e.g., Rescorla and Wagner, 1972; Sutherland and Mackintosh, 1971). That is to say, one CS loses access to the key resource, because the other CS captures the resource very rapidly or has already captured all of it. Such factors as the intensity of the two CSs, their prior associative strengths, and their temporal order are commonly thought to influence their relative competitive advantage (Kehoe, 1987). In a serial compound, CSA may capture the subject's attention before the onset of CSX, thus precluding CSX from gaining associative strength.

3.4 The Sutton-Barto Model

It might seem paradoxical that CSA both gains strength through associative transfer from CSX but also competes with CSX for a key resource. However, associative transfer and stimulus competition have been woven together by Sutton and Barto (1981, 1990). Their model assumes that the animal is continuously updating associative strengths by comparing recent values with current values. Thus, changes in the associative strength of a CS occur whenever there is a discrepancy between the net associative strength at one moment versus the next. Such discrepancies can arise through presentation of another CS as well as the US. In a CSAPCSXPUS compound, CSA's associative strength (VA) would gain an increment at the onset of CSX from any associative strength previously gained by CSX (VX). Thus, as quickly as CSX gains associative strength, it would be transmitted to CSA. Later within a trial, at the point of US onset, both CSA and CSX gain associative strength on a competitive basis according to their relative salience. Thus, VA increases at the onsets of both CSX and the US, while VX increases only at the onset of the US. As VA approaches the asymptotic value, the summated associative strengths (VA + VX) could exceed that sustainable by the US. Consequently, at US onset, there would be decrements in both VA and VX. On the next trial, VA would regain its lost value at CSX onset, while VX would continue to suffer decrements at US onset until the sum VA + VX stabilized at a level sustainable by the US.

4 REPRESENTATION

Among its many aspects, representation concerns the learning of arbitrary mappings from stimulus inputs to response outputs. Representation problems of fundamental importance can be created with as few as two stimulus inputs. Specifically, a solution to the exclusive-OR (XOR) problem requires a response to each of two inputs presented separately but not to their joint occurrence (Barto, 1985, p. 35; Rumelhart, Hinton, and Williams, 1986, p. 319). Thus, it is impossible to generate the appropriate reaction, namely no response, to the joint stimulus inputs by summing the responses attached to the two separate inputs. The XOR problem has been studied in classical conditioning by using combinations of two CSs, most commonly tone and light. These studies have revealed two mapping processes that may operate in parallel. One is a linear process in which the associative strengths of the two CSs add together in a simple way. The other is a nonlinear process that can counteract the linear process and permit the animals to solve the XOR problem.

4.1 Summation in Conditioning

The left-hand side of Figure 9.7 shows evidence for the summation of associative strengths in determining the performance of a CR. The training procedure is known as stimulus compounding. The animals received a mixture of two types of trials: (a) a tone CS by itself consistently signalled the US, and (b) a light CS also consistently signalled the US. To determine how the animals would combine the associative strengths of the tone and light CSs, they were occasionally given a third type of trial on which the two stimuli were presented together to form a compound, which was never followed by the US. Inspection of the left-hand side of Figure 9.7 reveals that the animals showed CR acquisition to the separate tone and light CSs. The likelihood of the CR to each CS rose steadily to an asymptote around 80% over ten days of training. Moreover, on the compound test trials, the likelihood of a CR was even greater. This relationship had a very precise quantitative character. Specifically, the percentage CRs to the compound (Pc) equalled the sum of the percentage CRs to the separate CSs (Pt, Pl) as combined by the formula for statistically-independent events, namely Pt + Pl - (Pt x Pl) (Kehoe, 1982b; Kehoe, 1986; Kehoe and Graham, 1988).

4.2 Nonlinear, Configural Learning

Notwithstanding the strong evidence for linear summation in conditioning, it is also possible to engage a nonlinear process when required. The middle and right-hand portions of Figure 9.7 show the acquisition of XOR behaviour in a procedure known as *negative patterning.* In negative patterning, the animals continued to receive trials in which the tone signalled the US and the light signalled the US. However, the number of compound trials was increased dramatically so that they outnumbered the separate trials with the tone or the light. The compound was never followed by the US. Eventually, the rabbits were able to overcome summation and solve the XOR problem. That is to say, the rabbits still showed CRs on tone trials and light trials but gradually suppressed responding on compound trials. By the end of training on Day 28, the levels of responding on tone trials and light trials exceeded 80% CRs, while responding on compound trials had dropped to 40% CRs.

4.3 A Layered Network Model

In conventional psychological theory, processes of perceptual fusion have been postulated to account for nonlinear phenomena such as negative patterning (see Kehoe and Gormezano, 1980; Pearce, 1987). However, thanks to interdisciplinary developments, it is now well known that layered, neural networks can, in principle, solve any linear or nonlinear problem. What remains to be determined is whether a given network with a given set of parameter values can realistically duplicate both the linear and nonlinear outcomes like those seen in classical conditioning. For a small set of inputs, the answer appears to be 'yes'. Kehoe (1988, 1989) has presented a small network that can simulate both stimulus compounding and negative patterning. Figure 9.8 shows the architecture of that network. The network has three 'sensory' inputs, one each for the tone (T), the light (L), and the US. The inputs for the tone and the light have adaptive connections with the two hidden units (X, Y). In turn, the X and Y units have adaptive connections with a response generation unit (R). At the start of training, all these connections have zero weights. These adaptive connections can be altered during training according to a learning rule known variously as the RescorlaPWagner rule, the WidrowP Hoff rule, and the least-mean square rule. The US input acts as a teacher input and has fixed, highly weighted connections with X, Y, and R. Depending on the pattern of inputs and the operation of the learning rule, the adaptive connections can be tuned so that the units respond selectively to the tone, the light, or their compound as required.

Using a fixed set of parameters, the model can duplicate the results shown in Figure 9.7. Figure 9.9 shows the results of the simulations for stimulus compounding and negative patterning. The simulated learning curves are not perfect, but they are adequate to a first approximation. Thus, for a simple environment, a correspondingly simple network has considerable explanatory power.

5 PRODUCTIVITY

Productivity denotes the application of knowledge in flexible ways to novel situations (Agostino, 1984, p. 86; Reynolds and Flagg, 1977, p. 241). The most widely recognized example of productivity in adult humans is their ability to utter novel sentences as the occasion arises (Chomsky, 1964, pp. 7P8). Despite the widespread occurrence of productivity, its emergence in naive creatures is not well understood. Research in cognitive psychology has focused largely on the already highly-productive performance by highly-experienced human subjects. Thus, the emergence of productivity is taken as an accomplished fact, whatever its origins may be. In the study of both animal and human learning, the application of previous learning to new situations falls under the heading of 'transfer of training'. Transfer to novel situations has been explained partially through *stimulus generalization.* In stimulus generalization, responses acquired to one stimulus are applied to new stimuli depending on their physical similarity to the original stimulus. However, stimulus generalization does not explain readily the more abstract relationships that can arise between old and new situations. Such abstract relationships for applying knowledge seem to require the application of rules or schemas rather than the direct transfer of specific reactions. One means for tracing the emergence of productivity in its more abstract form is *learning to learn.* Learning to learn denotes a progressive increase in the rate of learning across a series of tasks that are similar in their abstract structure but differ dramatically in their superficial stimuli (Ellis, 1965, p. 32). Another researcher concluded, '... learning to learn transforms the organism from a creature that adapts to a changing environment by trial and error to one that adapts by seeming hypothesis and insight' (Harlow, 1949). There is growing evidence that learning to learn can be obtained in a clear but simplified form in classical conditioning by conducting transfer between CSs from different sensory modalities, such as tone and light. Figure 9.10 shows the details of learning to learn as it has been observed in conditioning of the rabbit NM response (Holt and Kehoe, 1985; Kehoe and Holt, 1984; Kehoe, Morrow, and Holt, 1984; 1991c). In Stage 1, one group of rabbits (Group E) received presentations of a CS (e.g., tone) which was the signal for a US presented 400 ms later. A second group (Group R) served as a 'no learning' control. The control rabbits were restrained in the cubicles but did not receive presentations of either the CS or US. As can be seen in Figure 9.10,

Group E showed rapid CR acquisition to the CS that reached a terminal level of 98% CRs, while Group R showed only a few spontaneous responses that never exceeded 2% during the measurement periods (Kehoe and Holt, 1984, Experiment 2). In Stage 2, both groups received CS-US training with a new CS (e.g., light). Learning to learn was evident almost as soon as training began with the new CS. As can be seen in the acquisition curves for Stage 2, the experimental group (E) showed extremely rapid CR acquisition to the new CS. For example, Group E achieved a mean CR likelihood of 46% CRs within the first block of CSPUS trials. In comparison, the control group (R) achieved a mean CR likelihood less than 10% CRs within the first block of trials. In this and other studies, the positive transfer between tone and light was symmetric.

Learning to learn can be explained by the same layered network used to explain summation and negative patterning. In fact, only one hidden unit (X) is needed. Nothing else about the model is changed in any way. Figure 9.11 shows simulations of learning to learn. CR acquisition with the initial CS, tone, requires the strengthening of both the TPX and XPR connections. In subsequent training with a new CS (light), only the LPX connection needs to be strengthened. As the LPX connection starts to strengthen, the earliest firings of X by L are translated immediately into CRs via the previously-established XPR connection, thus facilitating CR acquisition to L. In Figure 9.11, this facilitation can be seen in the lower right-hand panel, specifically in the simulated acquisition curve labelled as PRE, which denotes pretraining. For purposes of comparison, a learning curve for a naive rest control condition is also displayed, labelled as REST.

6 ANTICIPATION

One of the keys to behavioural adaptation is anticipation. It entails the ability to use signals to predict future events and to make an appropriate response. However, in taking anticipatory action, an organism faces a fundamental conundrum. On the one hand, early anticipation maximizes the amount of time available for planning and preparation. On the other hand, premature action can be wasteful and even deleterious to the organism. Take eyelid closure as an example of a protective response. Closure in anticipation of a threat maximizes protection of the eye but, at the same time, blinds the organism. Thus, the timing of such a response must be a compromise between protection of the eye and the current need for vision. In classical conditioning, examination of the time course of the CR is one means for discovering how a biological system meets the constraints on its anticipatory responses. Contrary to textbook depictions of 'reflexes' as stereotyped actions, CRs are graded in a manner that is finely attuned to the CSPUS interval. Specifically, the CRUs time of initiation moves away from the US, but the maximal extent of the CR Q the CR peak Q remains near the time of US onset. Figure reffig:kehoe12 shows two examples of CR timing. In the top example, rabbits were given training with two separate CSs, namely, a light and a tone. The light onset signalled that the US would occur 250 ms later, and the tone onset signalled that the US would occur 500 ms later. The diagram shows the time course of the CR on the separate tone and light 'test' trials, which were presented without the US. The upward excursion in each line represents eyelid closure. As can be seen in the diagram, the CRs began around 125 ms after CS onset. However, the CRs reached maximum closure at different times, corresponding to the expected time of US delivery.

The lower portion of Figure 9.12 shows the time course of the CR when there was some uncertainty about the arrival of the US. Specifically, the rabbits were given training with one CS. In half the trials, the US occurred 400 ms after CS onset. In the other half of the trials, the US occurred 900 ms after CS onset. Inspection of the diagram reveals that the rabbits covered both possibilities by blinking twice. Each peak in the CR coincided with one or the other of the two points of US delivery.

Certain 'real-time' models have been aimed at explaining the time course of such CRs (Desmond, 1990; Desmond and Moore, 1988; Desmond and Moore, 1991; Grossberg and Schmajuk, 1989; Sutton and Barto, 1990). These models assume that each CS generates a cascade of stimulus elements, each with its own time course. To illustrate how these real-time models build a CR from a cascade of stimulus elements, Figure 9.13 shows a schematic version of such models. Panel A shows the time course of ten stimulus elements (Xit) generated by the CS. The associative strength (Vi) of each element is proportional to its intensity at its point of contiguity with US delivery. Panel B shows the moment-by-moment product of each element's associative strength and its momentary intensity (ViXit). Panel C shows the simulated time course of the CR that would be generated by summing across the products for each element (SUM[ViXit]) at each time point *t*. The simulated CR is crude but nevertheless displays the key features of real CRs. First, the simulated CR is anticipatory. Because the time course of each element stretches over a substantial portion of the CS, the associative strength of an element is not confined to the point of US delivery but is smeared around that point. Second, the peak of the CR tends to occur near US delivery where the elements with the greatest associative strength coincide with each other.

Although the real-time models have successfully explained the time course of CRs, they have a curious character. They assume that anticipation is largely an accident related to the spread of associative strength along each stimulus element. Recently, Sutton and Barto (1990) have attempted to recast real-time models such that the CR reflects a predictive process. That is to say, the moment-by-moment amplitude of the CR reflects the animal's prediction concerning the forthcoming US weighted by its imminence. Thus, in at least a primitive way, the animal may be said to be planning for its future. In this way, Sutton and Barto are promoting a convergence between the processes of animal learning and those of machine systems that attempt to act in a proactive way.

7 CONCLUSION

The reader should be aware that the experiments and theories described in this chapter are only a small sample drawn from what is known about animal learning. The research using classical conditioning is much greater than described here, and this paper barely mentions the vast literature concerning operant conditioning. Moreover, there are conditioning studies using human subjects; these studies are often concerned with the interaction between language-based cognition and nonsymbolic learning systems. At the risk of being overly bold, there is probably at least one study in conditioning on any variable that one could imagine. And researchers in animal learning are still pushing the boundaries of what is known. From the perspective of adaptive systems and machine learning, the results of animal learning should be heartening. At a minimum, the ability of animals to show credit assignment, representation, productivity, and anticipation reveals that these issues are as fundamental in the biological realm as in machine systems. Hence, we can have confidence that we are on the right track in both fields. Moreover, the richness of the phenomena seen in animal learning demonstrates what can be done in a relatively simple system, one with little prior experience and no language ability. The brain of a rabbit is perhaps only simple when compared to that of a human. However, there is growing evidence that animals with very simple neural networks, namely varieties of sea slugs, can show some basic features of classical conditioning. There is undoubtedly a limit to how much learning a simple system can show. However, over the past 20 years, that limit has been pushed downward and outward. So, there may be more surprises.

Acknowledgments

Preparation of this manuscript was supported by Australian Research Council Grant AC89322441. The author is grateful to Amanda Horne, Peter Horne, and Michaela Macrae for their assistance in preparing this manuscript.

REFERENCES

Agostino (1984). Chomsky on creativity. *Synthese*, *58*, 85-117.

Barto, A. G. (1985). *Learning by statistical cooperation of self-interested neuron-like computing elements. (COINS Tech. Rep. 85- 11)*. Amherst, MA: University of Massachusetts.

Chomsky, N. (1964). *Current issues in linguistic theory.* The Hague: Mouton.

Desmond, J. E. (1990). Temporal adaptive responses in neural models: the stimulus trace. In M. Gabriel & J. W. Moore (Eds.), *Learning and computational neuroscience* (pp. 421- 456). Cambridge, MA: MIT Press.

Desmond, J. E., & Moore, J. W. (1988). Adaptive timing in neural networks: the conditioned response. *Biological Cybernetics*, *58*, 405-415.

Desmond, J. E., & Moore, J. W. (1991). Altering the synchrony of stimulus trace processes: tests of a neural-network model. *Biological Cybernetics*, *65*, 161-169.

Ellis, H. (1965). *The transfer of learning.* New York: Macmillan.

Gabriel, M. (1988). An extended laboratory for behavioral neuroscience: A review of classical conditioning (3rd Edition). *Psychobiology*, *6*, 79-81.

Gibbs, C. M., Cool, V., Land., T., Kehoe, E. J., & Gormezano, I. (1991a). Second-order conditioning of the rabbit's nictitating membrane response: Interstimulus interval and frequency of CS-CS pairings. *Integrative Physiological and Behavioral Science*, *26*, 282-295.

Gibbs, C. M., Kehoe, E. J., & Gormezano, I. (1991b). Conditioning of the rabbit's nictitating membrane response to CSA-CSB-US serial compound: Manipulations of CSB's associative character. *Journal of Experimental Psychology: Animal Behavior Processes*, *17*, 423-432.

Gormezano, I. (1966). Classical conditioning. In J. B. Sidowski (Ed.), *Experimental methods and instrumentation in psychology* (pp. 385-420). New York: McGraw-Hill.

Gormezano, I., & Kehoe, E. J. (1975). Classical conditioning: some methodological-conceptual issues. In W. K. Estes (Ed.), *Handbook of learning and cognitive processes.* NY: Lawrence Erlbaum Associates.

Gormezano, I., & Kehoe, E. J. (1989). Classical conditioning with serial compound stimuli. In J. B. Sidowski (Ed.), *Conditioning, cognition, and methodology: Contemporary issues in experimental psychology* (pp. 31-61). Lanham, MD: University Press of America.

Gormezano, I., & Moore, J. W. (1969). Classical conditioning. In M. H. Marx (Eds.), *Learning Processes*. New York, NY: Macmillan.

Gormezano, I., Thompson, R. F., & Prokasy, W. F. (Ed.). (1987). *Classical conditioning III.* Englewood Cliffs, NJ: Lawrence Erlbaum.

Grossberg, S., & Schmajuk, N. A. (1989). Neural dynamics of adaptive timing and temporal discrimination during associative learning. *Neural Networks*, *2*, 79-102.

Hall, J. F. (1976). *Classical conditioning and instrumental learning: a contemporary approach*. Philadelphia, PA: J. B. Lippincott Co.

Harlow, H. F. (1949). The formation of learning sets. *Psychological Review*, *56*, 51-65.

Holt, P. E., & Kehoe, E. J. (1985). Cross-modal transfer as a function of similarities between training tasks in classical conditioning of the rabbit. *Animal Learning & Behavior*, *13*, 51-59.

Kehoe, E. J. (1979). The role of CS-US contiguity in classical conditioning of the rabbit's nictitating membrane response to serial stimuli. *Learning and Motivation*, *10*, 23-38.

Kehoe, E. J. (1982a). Conditioning with serial compound stimuli: Theoretical and empirical issues. *Experimental Animal Behaviour*, *1*, 30-65.

Kehoe, E. J. (1982b). Overshadowing and summation in compound stimulus conditioning of the rabbit's nictitating membrane response. *Journal of Experimental Psychology: Animal Behavior Processes*, *8*, 313-328.

Kehoe, E. J. (1983). CS-US contiguity and CS intensity in conditioning of the rabbitUs nictitating membrane response to serial and simultaneous compound stimuli. *Journal of Experimental Psychology: Animal Behavior Processes*, *9*, 307-319.

Kehoe, E. J. (1986). Summation and configuration in conditioning of the rabbitUs nictitating membrane response to compound stimuli. *Journal of Experimental Psychology: Animal Behavior Processes*, *12*, 186-195.

Kehoe, E. J. (1987). RSelective associationS in compound stimulus conditioning with the rabbit. In I. Gormezano,W. F. Prokasy, & R. F. Thompson (Eds.), *Classical conditioning* (pp. 161-196). Hillsdale, NJ: Erlbaum.

Kehoe, E. J. (1988). A layered network model of associative learning: Learning-to-learn and configuration. *Psychological Review*, *95*, 411-433.

Kehoe, E. J. (1989). Connectionist models of conditioning: A tutorial. *Journal of the experimental analysis of behavior*, *52*, 427-440.

Kehoe, E. J. (1990). Classical conditioning: Fundamental issues for adaptive network models. In M. Gabriel & J. W. Moore (Eds.), *Learning and computational neuroscience*(pp. 389-408). Cambridge, MA: MIT Press.

Kehoe, E. J., Feyer, A., & Moses, J. L. (1981). Second-order conditioning of the rabbit's nictitating membrane response as a function of the CS2- CS1 and CS1-US intervals. *Animal Learning & Behavior*, *9*, 304-315.

Kehoe, E. J., Gibbs, C. M., Garcia, E., & Gormezano, I. (1979). Associative transfer and stimulus selection in classical conditioning of the rabbit's nictitating membrane response to serial compound CSs. *Journal of Experimental Psychology: Animal Behavior Processes*, *5*, 1-18.

Kehoe, E. J., & Gormezano, I. (1980). Configuration and combination laws in conditioning with compound stimuli. *Psychological Bulletin*, *87*, 351-378.

Kehoe, E. J., & Graham, P. (1988). Summation and configuration in negative patterning of the rabbit's conditioned nictitating membrane response. *Journal of Experimental Psychology: Animal Behavior Processes*, *14*, 320-333.

Kehoe, E. J., & Holt, P. E. (1984). Transfer across CS-US intervals and sensory modalities in classical conditioning in the rabbit. *Animal Learning & Behavior*, *12*, 122-128.

Kehoe, E. J., Marshall-Goodell, B., & Gormezano, I. (1987). Differential conditioning of the rabbit's nictitating membrane response to serial compound stimuli. *Journal of Experimental Psychology: Animal Behavior Processes*, *13*, 17-30.

Kehoe, E. J., & Morrow, L. D. (1984). Temporal dynamics of the rabbit's nictitating membrane response in serial compound conditioned stimuli. *Journal of Experimental Psychology: Animal Behavior Processes*, *10*, 205-220.

Kehoe, E. J., Morrow, L. D., & Holt, P. E. (1984). General transfer across sensory modalities survives reductions in the original conditioned reflex in the rabbit. *Animal Learning & Behavior*, *12*, 129-136.

Kehoe, E. J., & Napier, R. M. (1991c). Temporal specificity in cross-modal transfer of the rabbit nictitating membrane response. *Journal of Experimental Psychology: Animal Behavior Processes*, *17*, 26-35.

Minsky, M. L. (1961). Steps toward artificial intelligence. *Proceedings of the Institute of Radio Engineers*, *49*, 8-30.

Pavlov, I. P. (1906). The scientific investigation of the psychical faculties or processes in the higher animals. *Lancet*, *4336*, 911-915.

Pavlov, I. P. (1927). *Conditioned reflexes* (G. V. Anrep, Trans.). London: Oxford University Press.

Pearce, J. M. (1987). A model for stimulus generalization in Pavlovian conditioning. *Psychological Review*, *94*, 61-73.

Rescorla, R. A., & Wagner, A. R. (1972). A theory of Pavlovian conditioning: Variations in the effectiveness of reinforcement and nonreinforcement. In A. H. Black & W. F. Prokasy (Eds.), *Classical conditioning II* (pp. 64-99). New York: Appleton-Century-Crofts.

Reynolds, A. G., & Flagg, P. W. (1977). *Cognitive psychology*. Cambridge, MA: Winthrop.

Rizley, R. C., & Rescorla, R. A. (1972). Associations in second-order conditioning and sensory preconditioning. *Journal of Comparative and Physiological Psychology*, *81*, 1-11.

Rumelhart, D. E., Hinton, G. E., & Williams, R. J. (1986). Learning internal representations by error propagation. In D. E. Rumelhart & J. L. McClelland (Eds.), *Parallel distributed processing: Explorations in the microstructures of cognition* (pp. 318-362). Cambridge, MA: MIT Press.

Skinner, B. F. (1938). *The behavior of organisms*. New York, NY: Appleton-Century-Crofts.

Sutherland, N. S., & Mackintosh, N. J. (1971). *Mechanisms of animal discrimination learning*. New York: Academic Press.

Sutton, R. S. (1984) *Temporal credit assignment in reinforcement learning*. Unpublished doctoral dissertation, University of Massachusetts, Amherst.

Sutton, R. S., & Barto, A. G. (1981). Toward a modern theory of adaptive networks: Expectation and prediction. *Psychological Review*, *88*, 135-171.

Sutton, R. S., & Barto, A. G. (1990). Time-derivative models of Pavlovian reinforcement. In M. Gabriel & J. W. Moore (Eds.), *Learning and computational neuroscience* (pp. 497-537). Cambridge, MA: MIT Press.

Thorndike, E. L. (1898). Animal intelligence: An experimental study of the associative processes in animals. *Psychological Monographs*, *2*,(pp. 1-100).

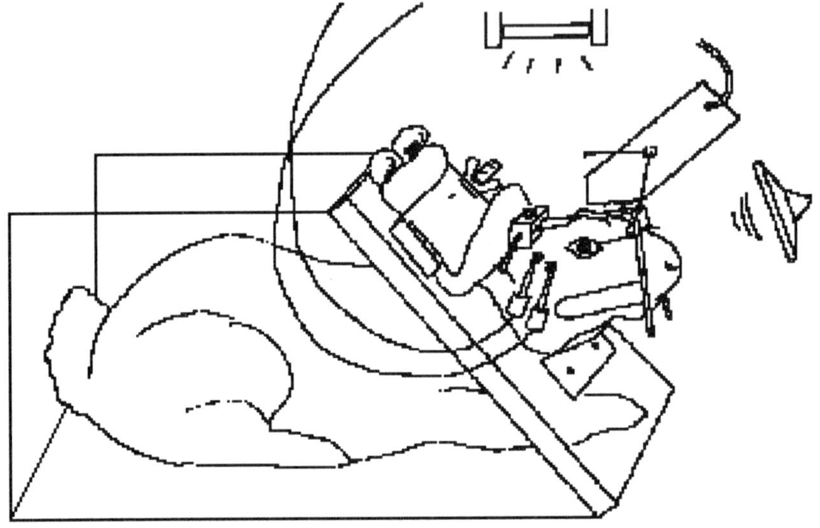

Figure 9.1. Drawing of a rabbit prepared for classical conditioning. The conditioned stimuli (CSs) can be either flashing of a neon light and/or presentation of pure tones through a speaker. The unconditioned stimulus (US) is a superficial tactile stimulus delivered by a brief, low-level electrical pulse delivered through wire leads to the skin near the eye. The learned, conditioned response (CR) and innate, unconditioned response (UR) entail movements of the rabbit's eyelids as measured by a mechanico- photo-electric transducer mounted over the rabbit's head.

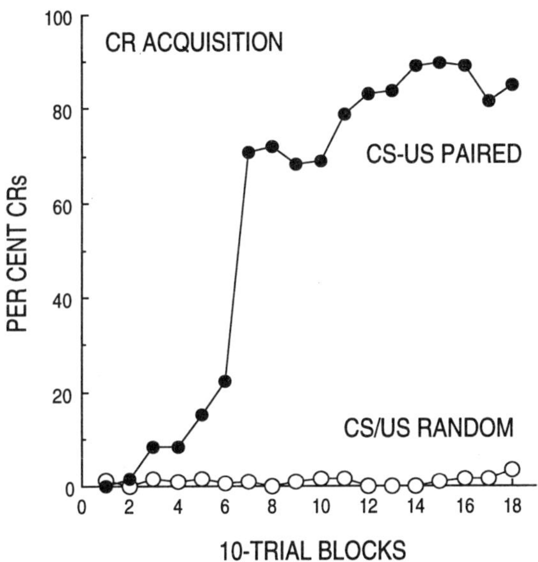

Figure 9.2. Example of a learning curve in classical conditioning of the rabbit's eyeblink. The abscissa represents blocks of 10 trials and the ordinate represents the proportion of trials on which the rabbits made a CR, that is, a blink during the warning period provided by the CS. The figure shows two curves. One curve shows successful CR acquisition in rabbits that received training in which the CS was always paired with the US at an interval of 250 ms. The other 'curve', which never rises from the line of the abscissa represents a control group of rabbits in which the CS and US were presented in a random order and never closer together than 30 sec.

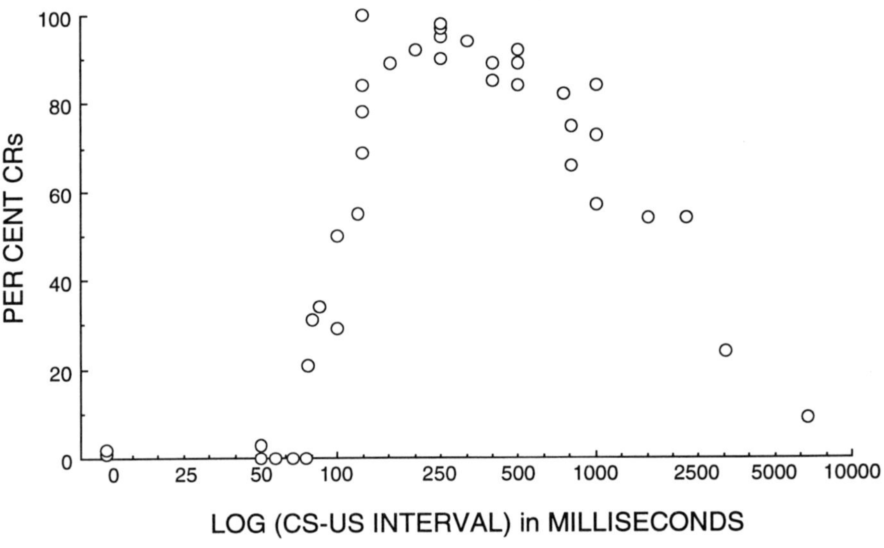

Figure 9.3. CR likelihood as a function of CSPUS interval. Each point represents the mean for a group of 8 to 12 rabbits trained at a given CSPUS interval.

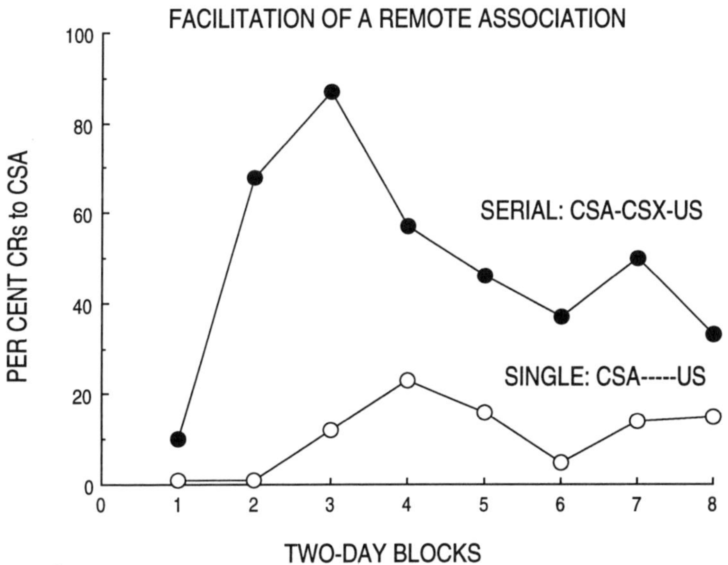

Figure 9.4. Facilitation of a remote association. Learning curves for CR acquisition to CSA in two groups of rabbits. One group (Serial) received training in which a sequence of two CSs signalled the US (CSAPCSXPUS). The other group (Single) had only CSA as a signal for the US (CSAPUS). The CSA preceded the US by 2800 ms in both groups. In Group Serial, the CSXPUS interval was 400 ms.

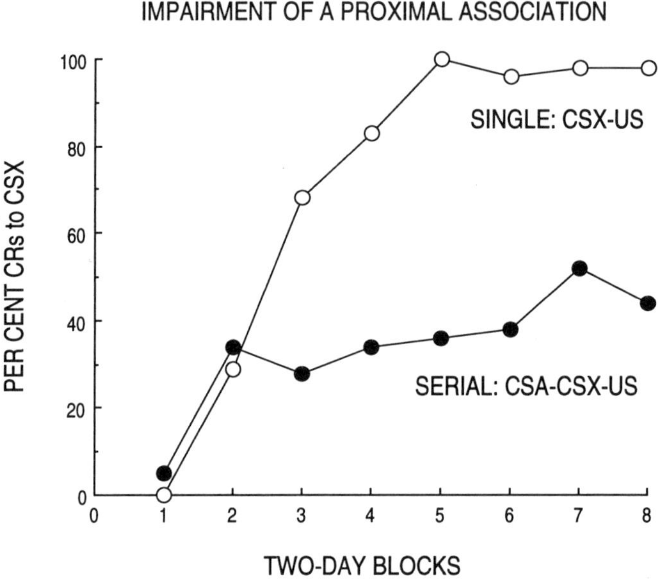

Figure 9.5. Impairment of a proximal association. Learning curves for CR acquisition to CSX in two groups of rabbits. One group (Serial) received training in which a sequence of two CSs signalled the US (CSAPCSXPUS). The other group (Single) had only CSX as a signal for the US (CSXPUS). CSX preceded the US by 400 ms in both groups, and CSA preceded the US by 800 ms in Group Serial.

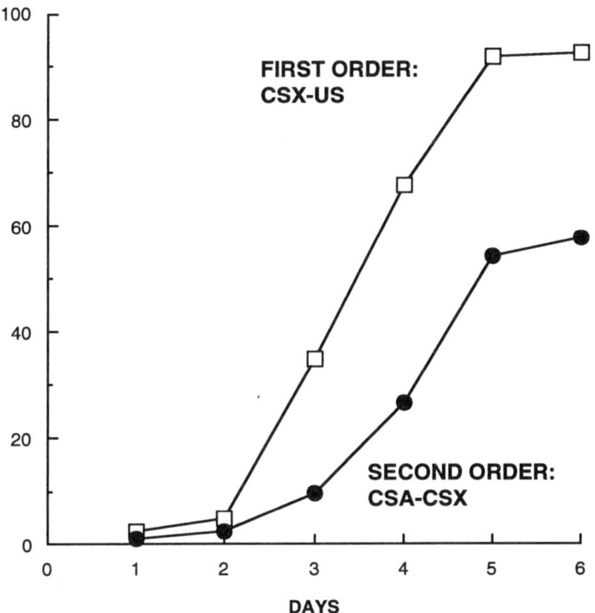

Figure 9.6. First- and second-order conditioning. One group of rabbits received training with two types of trials, namely, first-order trials (CSXPUS) and second-order trials (CSAPCSX). The curve for the first-order trials shows CR acquisition to CSX, and the curve for the second-order trials shows CR acquisition to CSA. The ordinate of the figure denotes percentage CRs.

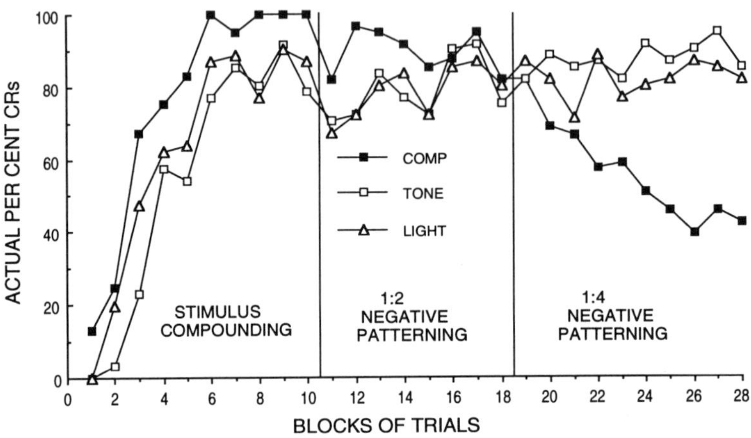

Figure 9.7. Learning curves that show linear and nonlinear outcomes. In the first phase ('stimulus compounding'), one group of rabbits received training with three types of trial: (1) pairings of a tone CS with the US, (2) pairings of a light CS with the US, and (3) occasional test presentations of a compound stimulus in which the tone and light were presented together but without the US. In this phase, response to the compound was a linear sum of the separate responding to the tone and light. In phases two and three ('negative patterning') the frequency of compound stimulus trials was increased so that there were two compound trials and later four compound trials, still without the US, for every separate pairing of the tone or light with the US. During these phases, the rabbits showed acquisition of XOR behaviour: they sustained responding to the separate tone and light but learned to suppress responding to the compound.

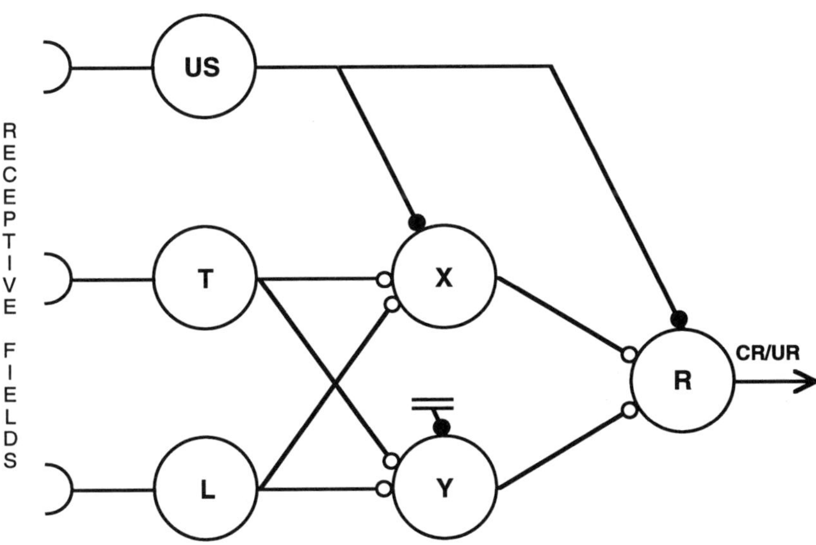

Figure 9.8. Architecture of a layered network that can produce linear and nonlinear behaviour in classical conditioning. The T and L units represent inputs for tone and light CSs, respectively. They have adaptive connections with hidden X and Y units, which in turn have adaptive connections with a response generator unit (R). The US unit represents the input for the unconditioned stimulus, which has fixed connections to the X, Y and R units.

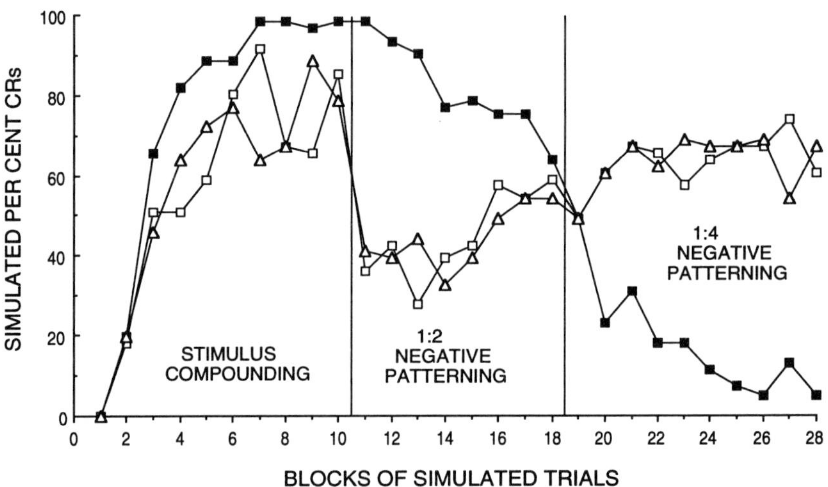

Figure 9.9. Simulations of stimulus compounding and negative patterning based on a layered network.

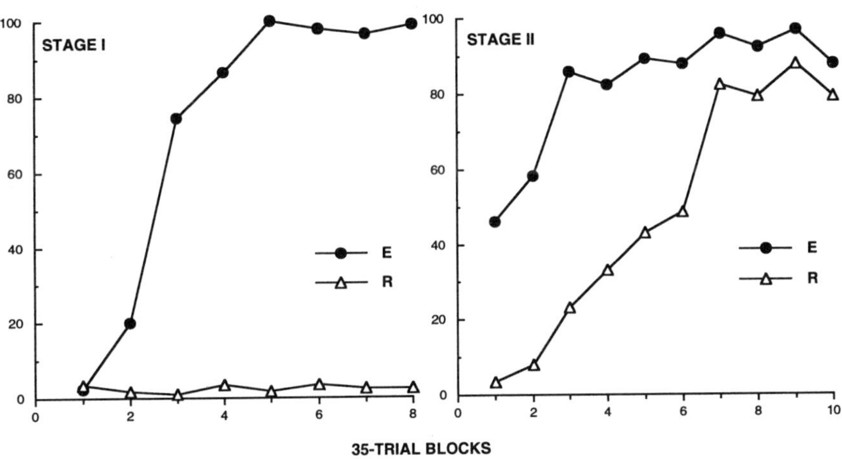

Figure 9.10. Demonstration of learning to learn. In Stage I, Group E received CSPUS pairings, while Group R received no exposure to either the CS or US. In Stage II, both groups received CSPUS pairings using a different CS from that used in Stage I. For example, if the CS in Stage I was tone, the CS in Stage II was light. The ordinate of the figure denotes percentage CRs.

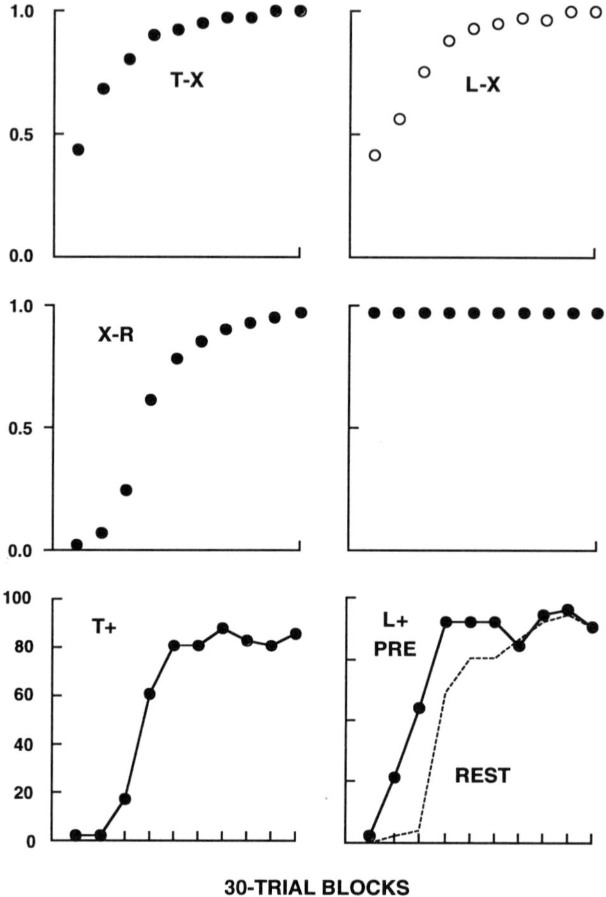

Figure 9.11. Simulations of learning to learn based on a layered network. Top: growth of the TPX and LPX connections in Stages I and II. Middle: growth and maintenance of the interior XPR connection. Lower: simulated acquisition curves to the tone and light in the 'PRE condition' (Group E). Simulated acquisition curve for the light is also shown for the 'REST condition' (Group R).

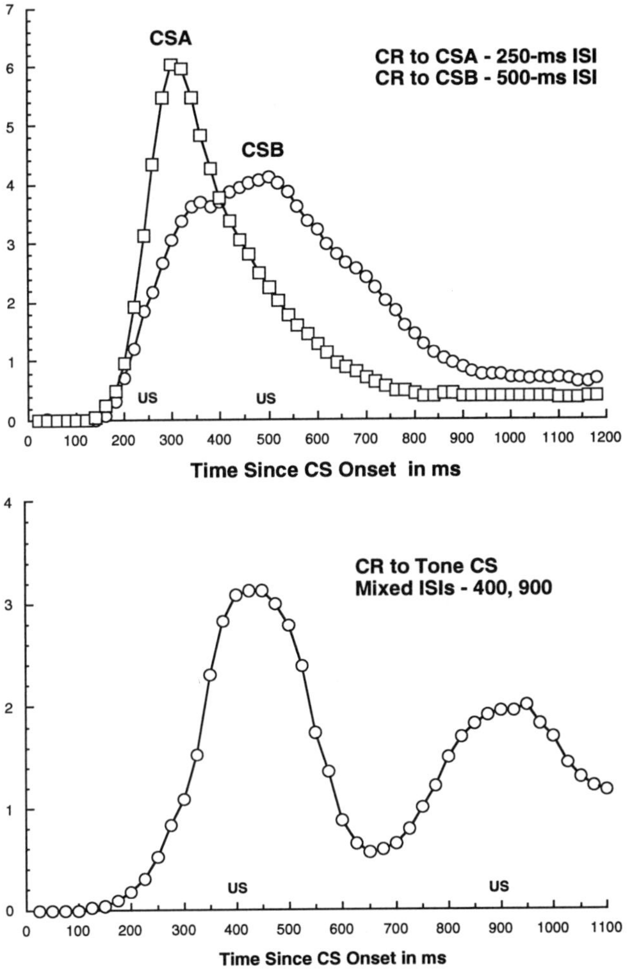

Figure 9.12. Moment-by-moment time course of conditioned responses. Upward excursions represent closure of the rabbit's eyelid. Top: two different CRs that evolved when stimulus CSA signalled the US at 250 ms and stimulus CSB signalled the US at 500 ms. Bottom: CR that evolved when one CS was followed unpredictably by the US at either 400 or 900 ms. Ordinate denotes CR magnitude in mm.

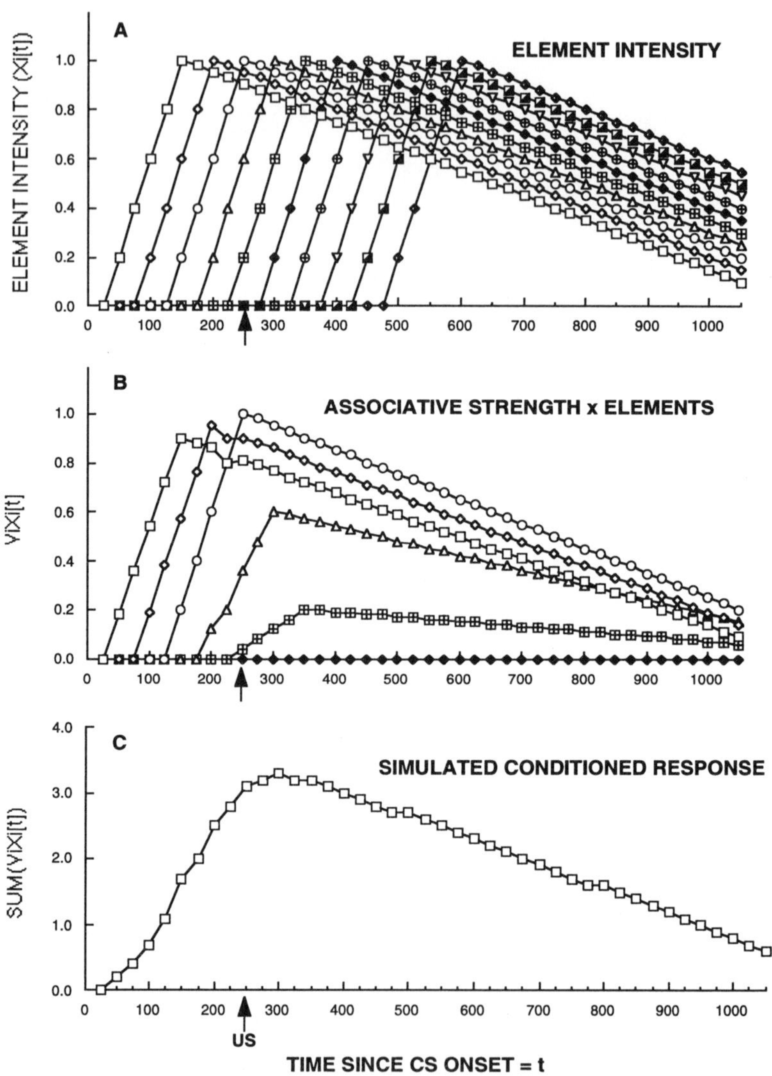

Figure 9.13. Generic real-time model. Top: momentary intensities of traces for ten stimulus elements following CS onset. Middle: time course of the traces weighted by their respective associative strengths. Bottom: simulated conditioned response obtained from summing the weighted traces. Arrow on the abscissa shows the locus of US delivery at 250 ms after CS onset.

10

Regulation of Human Cognition and its Growth

C. Trevarthen

The Edinburgh Centre for Research in Child Psychology,
Department of Psychology,
The University of Edinburgh

I will present a psychobiologist's view, citing three directions of research that have been fruitful in the past 25 years.

1 GENESIS OF CONSCIOUSNESS

Unlike technical structures which are constructed from without, living organisms grow from within. Cells dividing from the fertilized egg receive their genes and interact under the regulation of gene-produced molecules that determine the cells' cohesivness, mobility, multiplication, and capacity to survive in competition for essential substances. Messenger molecules pass between cells, changing gene transcription and directing populations of cells towards different life functions in specialized organs. Out of the undifferentiated multicellular mass an integrated individual forms—in the case of an animal, a perceptuomotor individual that will displace, orient, search, and choose objects, making selective use of the environment to maintain and develop its adaptive activities and structure. These activities are formed in a nervous system that carries nerve cell action potentials throughout the body. Molecular messengers (neurohormones), related to those that pattern gene expression, are transmitted at synapses to target cells, to control the intensity and selectivity of behaviour, patterning the animal's motivation.

At any given stage of development, because of the way it develops, the organism and its component parts show evidence of preparation for future selective use of its organization and the environment. The phenomenon of prefunctional morphogenesis is ubiquitous in the growth of organisms. Birds have wings, feathers, and beaks when in the egg. The eye, a camera-like optical instrument, is formed in the dark. As indicated, prefunctional organization also prevails at the physiological and bio-chemical levels.

Psychological activities of the brain, including consciousness, are no exception to this rule. Brain systems that will sense environmental situations and objects, and others that generate movements to change an animal's relations with the world in adaptive ways, were formed at a stage when no such experience or activity was possible. The brain forming in the embryo of all vertebrates becomes the central coordinator of behaviour. It integrates internal visceral and autonomic regulations with the requirements of behaviour; it incorporates basic reward and punishment systems that guide adaptive learning, controls body support and coordination in the field of gravitation, directs sensory orientations to the world, and integrates movement of the whole body. New components that increase the power and adaptive learning capacity of these functions, principally the cerebral hamispheres and cerebellum, are added in higher vertebrates in a foetal period.

In humans there are clear preparations for cultural intelligence in the way the cerebral hemispheres of a foetus are formed in utero. The cerebral cortex, which develops most of its significant functions after birth, some very slowly over years, already has organization related to the learning of speech half way through gestation, at 20 weeks post-conception. Its development is regulated by activity of other parts of the brain, especially neurones on the reticular core of the brainstem that started its development in the embryo, before there were any connections between the brain and sense organs or muscles.

It is always difficult to distinguish in the mature animal creature what is learned from what is prepared for in the morphogenesis of the brain before any learning could have occurred. We need a comprehensive theory of the strategy by which the motivation for learning is derived from the epigenetic elaboration of neural systems. Rival theories of how mature coordination and regulation of intelligence are achieved are of two main kinds. Embryologists, anatomists, and ethologists emphasize the autopoesis or self-creation of the brain before it has functional commerce with the environment outside the body. After all, the brain's component nuclei, tracts, and cortical mantles are clearly differentiated in their correct locations at birth, and animals show elaborate instinctive behaviours that link their life activities with each other and with the environment in ways that could not have been learned. On the other hand, psychological, physiological, and molecular-biological theories tend to a constructivist position. They hold that integrated psychological control of behaviour is a consequence of selective reinforcement, or selective retention, of neuronal assemblies that have the capacity to retain traces of the effects of environmental stimuli. These theories presume that the newborn brain must have little psychological organization, that its primary elements are detached from one another and require integration by plastic adaptation of redundant populations: of messenger molecules or cell adhesion molecules, of synaptic contacts or nerve circuits or reflexes, etc., that is, by selective retention determined from outside. They emphasize the role of learning.

In fact, the integrated and specialized behaviours and preferences of newborn infants indicate that prenatally formed brain mechanisms direct, gate, and choose the environmental effects that will be learned. Innate motive systems have a voice in deciding what kind of experience will be sought and what will implant in the brain new adaptive categories for awareness and new adaptive forms of action vis a vis the environment. Once this point is grasped, it becomes clear that psychological development throughout a person's life, including the construction in the child of a cultural consciousness mediated and transmitted by symbols, is guided by interaction and mutual influence of components of mental activity that were differentiated in the brain before birth. The brain system of our cultural intelligence is the key component of our evolutionary inheritance.

There are several different research strategies for exposing the original organization of perceptuomotor systems before the experience-dependent cortical structures are formed. I will here sketch three that have made significant strides in recent decades.

1. Developmental anatomy and experimental embryology of the nervous system.
2. Neuropsychological analysis of the effects of disconnections of key elements of the CNS.
3. Psychological analysis of newborn intelligence to determine its motives and learning capacities and their changes through early childhood.

2 BRAIN DEVELOPMENT FROM EMBRYO TO YOUNG CHILD

The neural structures of intelligence that develop to coordinate perceptions and voluntary action of one subject's integrated mind are formed before birth. They are the products of multiplication, migration, differential death, selective grouping, and interlinking of embryo brain cells into cooperative assemblies. The brain of an embryo is an integrated tissue system, one coactive fabric of closely coupled cells in which clusters and sheets emerge that are endowed with a dynamic form. They bunch, migrate, and fold under the influence of adhesive forces created by molecular interactions at the surfaces between them, or with non-neural tissues. Their activity, the way they adhere and move in relation to each other and surrounding substrates, within the brain itself and in interaction with the body, determines the ground plan of the future mind's awareness and its intentional response to the outside world, even before recognizable neurones are formed. The many brain systems identified in a newborn baby's brain are the product of intricate cell-to-cell communications that regulate the expression of genes in millions of multiplying neurones. For nine months this process has had no instructive input from the environment outside the body, except for diffuse physiological influences from the mother's body into the foetus attached to her as a benign parasite.

Neurones appear in the reticular core of the embryo brain that will generate clocks setting a common time base for brain activity. By axons distributed throughout the brain, they subsequently transmit neuromodulator substances (biogenic amines and others) that direct the activity, receptivity, and plasticity of the cortical neurone net. They also direct motor gating of sensory input (movements of selective attention and exploration by head, eyes, mouth, respiratory system, and hands). They, with diencephalic and limbic systems, constitute the motivations of the adult brain. From them are generated periodic expressive movements that, when perceived by a partner, can co-regulate motives and transfer learning. Such coupling, activated through prenatally-aimed person-seeking cells (in frontal and temporal cortex?) that detect expressive parts of another person's body, has unique developmental potential in humans. Subcortical regulators in the human brain induce an asymmetry in emotions and cortical development which affects growth of complementary cognitive functions and memories in the hemispheres.

A most important feature of brain morphogenesis is the construction of a mechanism for generating a Behavioural Field in which experience and action will be coordinated. Brain tissue has a specified place in the dorsal midline of the embryo body, and body and brain share the same morphogenetic axes of symmetry and polarity. They are bi-laterally mirror-symmetric with two orthogonal asymmetric or polarized anterio-posterior and dorso-ventral axes. Long before nerve conduction lines are formed or are active in transmitting action potentials, regions of the Central Nervous System (CNS) become committed to formation of neuronal arrays or networks that will serve different aspects of the body's self-regulation. Thus the active individual is endowed, from before the first movement in relation to an out-of-body event is perceived, with ground plans for distinct special purpose neural regulators of internal state. These specify a set of ideal sensory-motor relations to the outside world, ready to act in relation to its temporo-spatial arrays.

There are many body-shaped (somatotopic) maps in the newborn brain. These are further elaborated and more closely interconnected in postnatal development into a coherent control mechanism for behaviour that has many levels of complexity and refinement. They appear to be generated as subdivisions

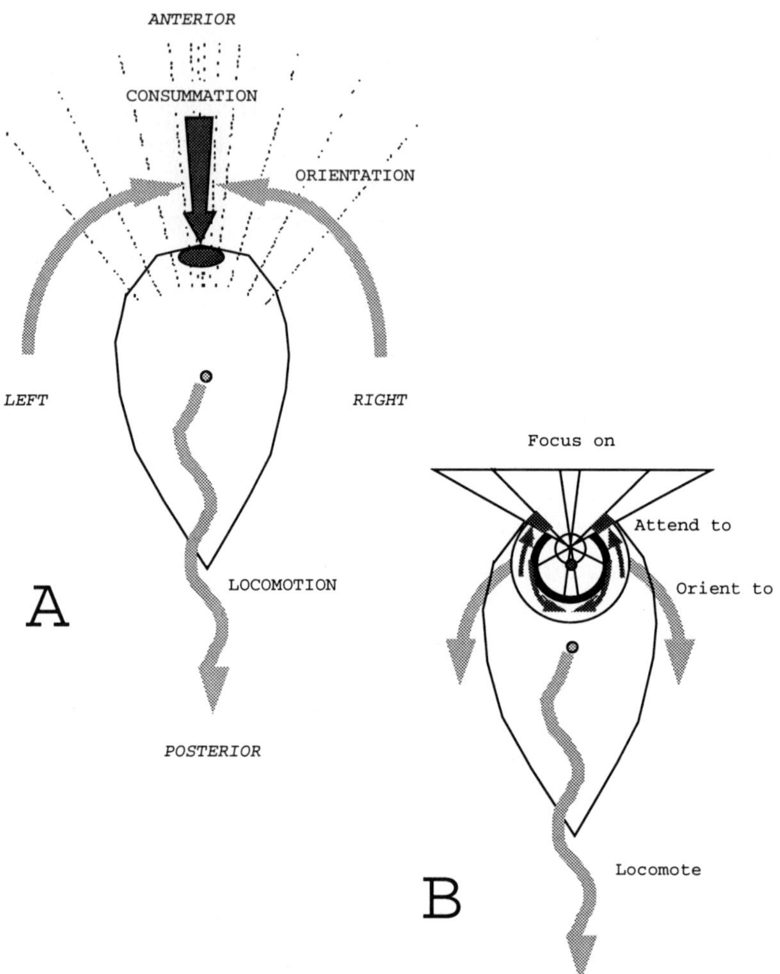

Figure 10.1. Simple Schema for a Visuo-motor Animal. Kinds of visuo-motor action regulated in the brain: **A:** A simple swimming animal with an anterior convex visual receptor and a mouth opening. Polarized Behaviour Field with the bi-symmetry of the body. **B:** A cyclopean with a lensed eye that can focus an image on a concave retina and turn to aim its optic axis independently of the head.

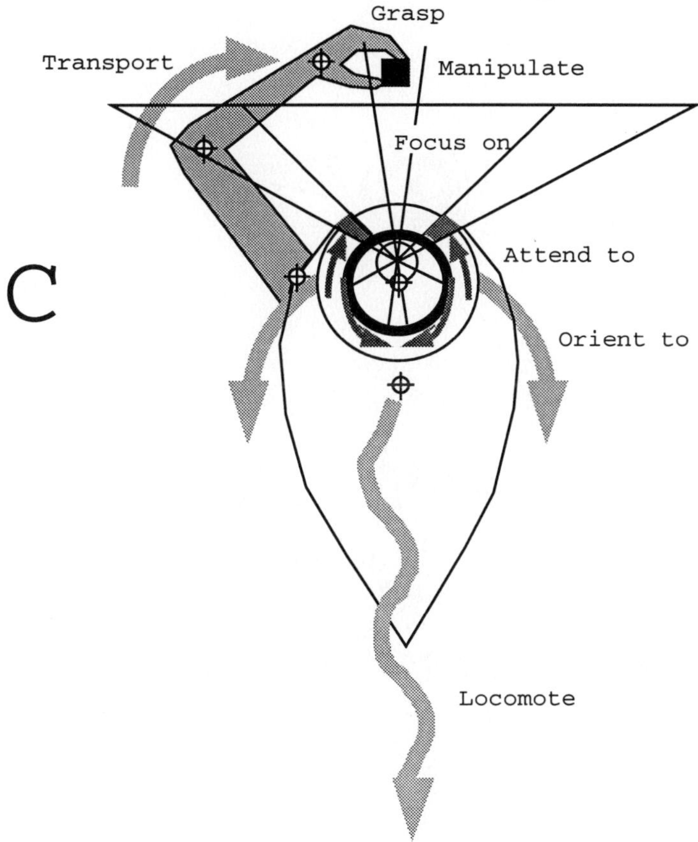

Figure 10.1. **C:** A right-armed cyclopean that can deploy a jointed member to grasp and manipulate under the visual guidance of a moveable eye.

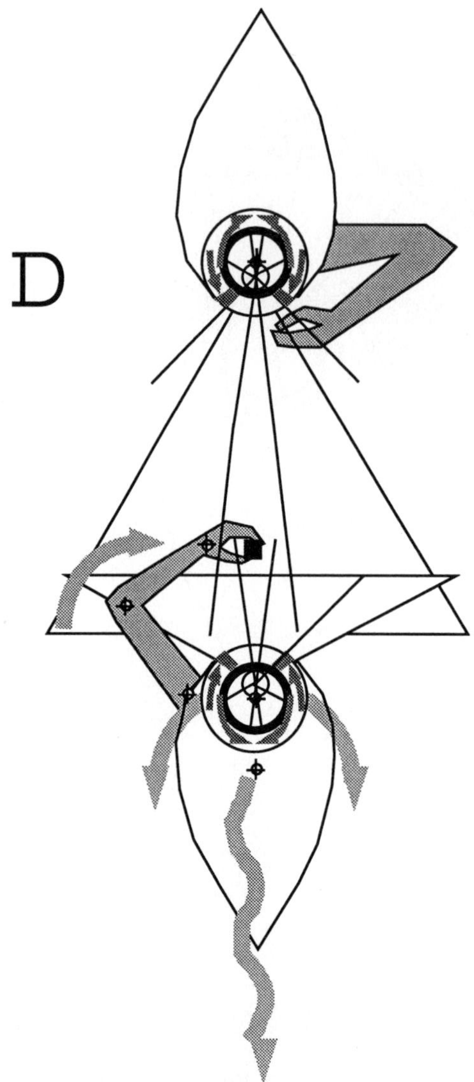

Figure 10.1. **D:** One right-armed cyclopean watches another, assimilating the other's orientations and fixations of interest, potentially imitating or cooperating.

of a body-shaped whole brain map that is formed first in register with the body of the early embryo. The future capacity of the brain to move one body coherently through information fields of many separate sensory modalities, and to have integrated mechanical effects by movements of the mouth, separate limbs, and digits, depends on the systematic interconnection or registration of the brain maps according to the common body-related plan. Cells of the primordial maps in the embryo are coded by chemical differentiation of their powers of adhesion so that links are established between corresponding points in the different body representations when nerve cell axons grow out and penetrate far in the CNS and the body. Distribution of one centre of attention in several modalities at once to different loci in one space–time frame, and the capacity to selectively attend with one specialized modality at a time (e.g. to see, touch or hear) both depend on this coherent somatotopy.

We cannot understand the development of articulate learned intelligence without a psychobiological analysis of the preparation for awareness and voluntary action that takes place in prenatal brain growth. This growth generates core regulator mechanisms that specify the parameters of a time–space behaviour field in terms of what the body can do and what it can be sensitive to. Motor activity and perception then create a partnership with the environment. Cognitive processes generate "perceptuo-motor cycles", and are regulated within them.

3 LEVELS AND DIRECTIONS OF CONSCIOUSNESS IN SPLIT BRAINS

The contribution of brain stem systems to consciousness and cognition has been revealed in a new way by the study of individuals in which the cortical mechanisms have been divided by neocortical commissurotomy. This operation, performed in humans to control devastating epilepsy, cuts the very large bundles of fibres in the corpus callosum and anterior commissure that provide the only direct links between the two halves of the cerebral cortex (Figure 10.2). According to latest estimates, the forebrain commissures of a human adult comprise nearly 1000 million (10^9) fibres. Because of the way somatotopic maps and neural projection systems are wired up prenatally, each cerebral hemisphere of a monkey or human receives its main sensory inflow from one half of the body-centered extracorporal space. The left hemisphere has full visual input pertaining to the right half of the subject's visual world. The right hemisphere likewise sees the left half of the world. Perceptual testing of split brain subjects with careful control of orientations shows that all vivid object-identifying consciousness is divided down the middle of body-centred space, for touch, hearing, seeing, and for smelling (note that, because of a peculiar overlap of the anatomy of the olfactory projection and the commissures of the brain, the sense of smell is represented for the left nostril in the left hemisphere, and for the right nostril in the right hemisphere in a commissurotomy patient).

In a monkey with a split brain (i.e. after commissurotomy between the cerebral cortices of the two hemispheres and division of the chiasma or cross-over of optic fibres underneath the brain so each eye is connected only to the visual centres on the same side of the brain) visual object-awareness and learning is divided in two. The animal learns separately what it sees with left and right eyes. Consciousness in each side is found to be contingent upon preparation of the action of response with one hand or the other. That is, the motive for using one hand makes awareness and learning active in the hemisphere of the opposite side. Meanwhile, the other hemisphere does not register what it receives from the eye to which it is connected. One can switch learning or recall from one eye to the other by making the split brain monkey change hands to respond.

These results fit the known anatomy of the most conspicuous visual projection, through the lateral geniculate nucleus of the thalamus to the striate cortex. However, some kinds of visual judgement, less dependent on high resolution of hue, brightness, and spatial detail, show "leakage" accross the surgical divide. Finding this lead me to identify a contribution to visual uptake of information that unsplit parts of the monkey's brain use for orienting and preparing whole-body or postural adjustments, such as simply reaching to a location outside the body, without taking account of the identity and reward-significance of

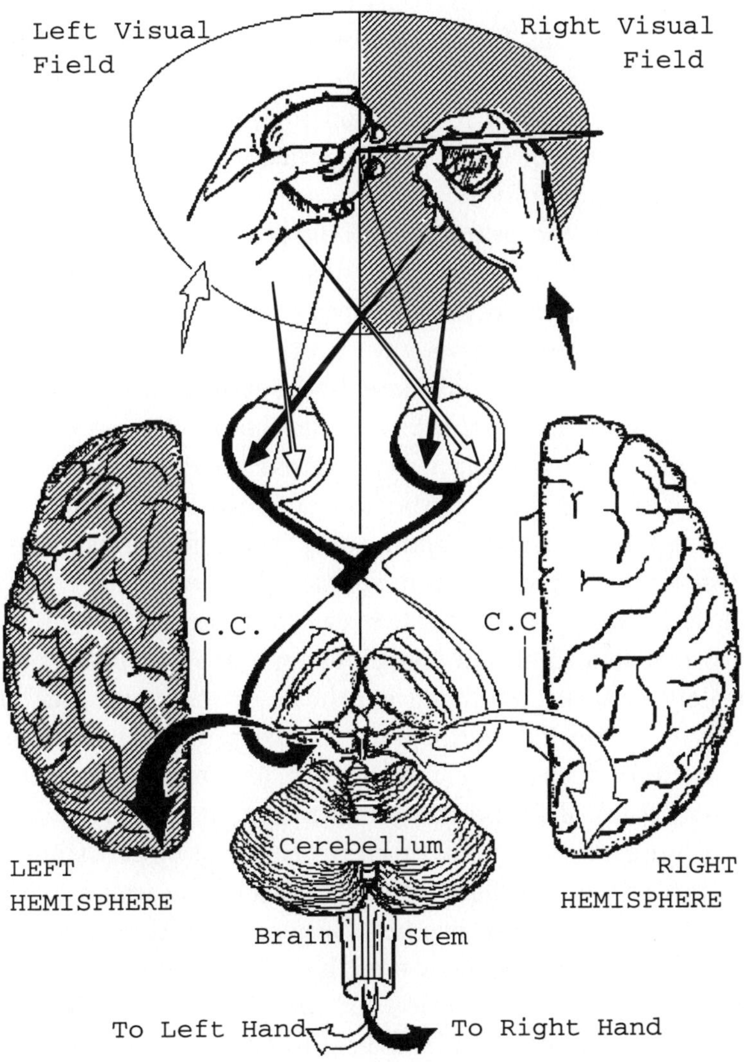

Figure 10.2. Anatomy of a Split Brain (C. C. = Corpus callosum).

the goal. This at least partly undivided Ambient Vision, which has relatively coarse spatial and light-intensity discrimination but high sensitivity to motion cues, is probably processed in the midbrain and cerebellum. These brain stem centres have their own visual inputs, partly direct and partly relayed by descending projections from the cerebral cortex. The completely divided Focal Division of detailed features that is needed for precise identification of forms, textures, hues, etc., that distinguish objects, and for refined manipulation and precise bimanual coordination, is wholly dependent on neocortical systems, and hence is split by forebrain commissurotomy.

The most striking feature of consciousness in a human commissurotomy patient, revealed when orientations of their receptors and cross-cuing across the midline of the body are well-controlled, is that they only speak about what they experience in the left hemisphere, the brain that holds the representation of the right half of their body-centred space. With rare exceptions in which the distribution of brain functions has been changed by early brain injury, only the left hemisphere of a commissurotomy patient can speak. Proper testing by nonverbal techniques reveals, however, that the right hemisphere has a versatile and skilful consciousness. It is actually superior to the left in tasks requiring synthetic or "appositional" mental activity. The left hemisphere, in keeping with its superior linguistic and declarative intelligence, is superior in itemizing serial strategies of propositional thinking. It quickly recognizes familiar nameable features, often overlooking circumstantial organization or arrangement of ensembles of features or objects. In the intact brain these two hemispheric cognitive systems, one more analytical and declarative, the other more holistic, pragmatic, and aesthetic, function as complements. Indeed, their differences are augmented in development through the selective removal of duplicate connections between and within the two halves of the cerebral cortex. Cortical centres differentiate in relation to each other under the direction of brain stem attentional and emotional systems which have inherent asymmetry. All the above effects of disconnection of the two halves of the cerebral neocortex, with all their acquired memories of conscious and culturally codified meanings, confirm the pre-eminence of the cortex in consciousness. However, when one examines the fundamental sensory-motor coordinations of locomotion and attentional orienting, and the self-regulatory functions of emotion, the surgery appears not to have divided the human mind. Because the subcortical circuits of the brain stem, cerebellum, and spinal cord are still fully integrated, the commissurotomy patient, like the split-brain monkey, retains the ability to orient his senses and react emotionally (at least in more direct ways, without cognitive mediation) as one being. He can turn his eyes and head and distribute the movements of his limbs and guide his body in locomotion much as before the operation.

The undivided subcortical mechanisms of attention and emotional feeling can also exercise powerful editorial influences over consciousness, even turn consciousness off when the cortical level has full information about phenomena to be perceived. For example, consciousness of something being looked at may be erased completely within one hemisphere by an effort of attention that the other hemisphere has initiated when preparing "its" hand, the contralateral one that has to cross the midline of the visual array to respond. It is even possible for this kind of directing of mind work to lead to futile attempts by one hemisphere to do a task for which it is not skilled. The hemisphere lacking the cognitive mechanism or memories required offers an inadequate answer or guess to the question posed, at the same time as the other more appropriately gifted hemisphere, given all the information required for a ready solution to the task, fails to respond. Levy and I called this capacity to direct, or misdirect, intelligence "metacontrol". It is undoubtedly necessary to have a capacity to recruit memory, imagination, and conscious discrimination in the most effective direction in relation to what one, as one mind, is motivated to do. In the normal fully-interconnected brain, this cognitive allocation process does not lead to prolonged or repeated misuse of the different resources of the hemispheres.

The mind-dividing effects in commissurotomy subjects have encouraged theories of consciousness as the work of a modular system, with differing quasi-independent parts of knowledge and cognitive specialism held together by external information loops, by expression through one body with its receptors. It is supposed that reasoning and the solution of motor problems may often be a debate or even a blind competition for solutions between different channels for processing environmental information. There is certainly abundant physiological and anatomical evidence that the cortex has specialized territories. Their transitory differential activation when a person engages in different kinds of mental activity is now dramatically demonstrated with the techniques of PET scan, regional cerebral blood flow monitoring, etc. But looking at the shifting patterns of nerve energy in the surface mantle of the brain of an immobile, observing, thinking, talking subject, who is kept still seated or lying down and performing a boring and repetitive task, may be misleading. We should not forget that the brain is designed for a freer existence. The modules of consciousness are products of the enterprise of the whole brain designed to coordinate and plan the activities of a complex body with many special senses, and to take up experience of one phenomenal field of existence. Mental activity is normally unified in whole-body activity. When thinking and apparently doing nothing else we borrow experience from more active states of being, including memories of interacting with other persons in dialogue or cooperative activity.

All these innate psychological aspects of central coordination in awareness and intention come to the fore when we consider consciousness in an infant, and how it develops through education of the child.

4 EMOTIONS OF INFANTS AND MOTHERS, AND DEVELOPMENT OF THE BRAIN

Detailed analysis of the expressive and orienting behaviours of infants has revealed that they are born with a consciousness that seeks communication with the emotions, interests, and actions of other minds. Mental interaction between the mother and her infant goes far beyond what is required to give the highly dependent young organism food, protection, rest, and comfort

Human newborns are, of course, adapted to receive physiological support from a caretaker in a protected environment, but the human needs more. Mother love involves giving human emotion that seeks a partnership with the emotions of the other. It is communicated in intricate protoconversational and imitative exchanges.

Neonates imitate face expressions, vocalizations, and hand movements, sometimes within a few minutes of birth. A foetus can learn to recognize, and prefer, the mother's voice, so that when the baby is born it will choose a recording of her and not one of another woman. Ultrasound movies show that fetuses in utero perform many coordinated movements, including patterns for communication — smiles, eye movements, hand gestures, lip and tongue movements, and, if injected air fills the lungs, vocalizations.

At two months after birth infants eagerly join in face-to-face "protoconversations" with their mothers, responding to the rhythms and intonation of a babytalk called "intuitive motherese", the melody of which transcends cultural conventions and language (Figure 10.3. A two-month-old can communicate through a double-television system by audition and vision alone. When the recordings are analysed, it is found that intercoordinations between mother and infant may be timed to a few tens of milliseconds and "intersynchrony" is achieved. Mother and infant both use vision, hearing, body-position, and touch to pick-up emotional and motivational invariants in the partner, and both make expressions with eyes, face, mouth-and-tongue, vocal organs, hands, and head. The infant's movements, as well as cortical electrical activity evoked by maternal signals, tend to be asymmetric, which shows that the newborn human brain has one-sided mechanisms for perceiving and generating expressions of communication, precursors of the territories that organize speech and language in the adult brain.

Evidence for a complex field of emotions in the two-month-old that respond to and influence the mother's expressions in lawful ways, comes from perturbation experiments, in which the mother's reactions are changed, as, for example, when she is asked to keep her face still and not to speak for a minute or two.

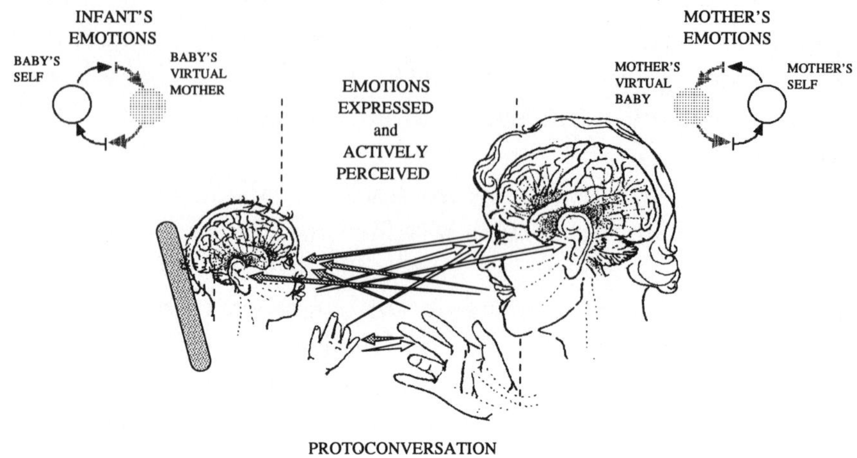

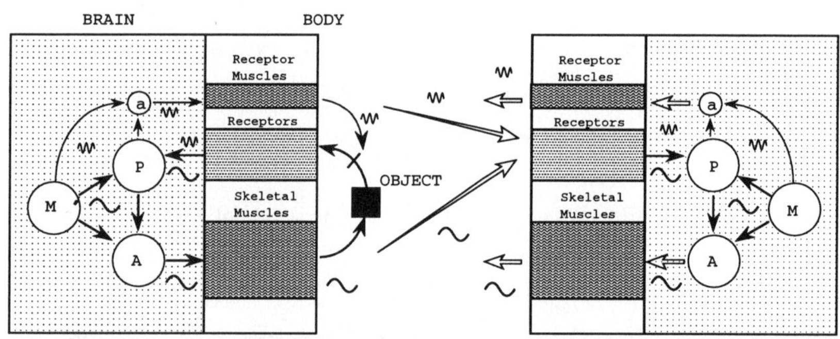

Figure 10.3. **Above**: Protoconversation mediated by eye-to-eye orientations, vocalizations, hand gestures, and movements of the arms and head, all acting in coordination to express interpersonal awareness and emotions. **Below**: Motor "gating" of perceptual information and the signals of mental activity that pass between two subjects, an actor and an observer. Motives M direct acts A and motor adjustments of the receptors a. At the same time, they set perceptual systems P for uptake of information from receptors. Large acts of the body A, including walking and reaching and grasping objects, have lower temporo-spatial periodicities than anticipatory adjustments of receptors a, such as saccadic eye movements or finger movements to feel objects. All movements can give information to an observer about the motives in an actor.

This causes immediate distress to the infant. Infants' reactions to strangers also show their emotional attachment to familiar persons. When a mother suffers from post-natal depression, the effects on the infant's behaviour and learning confirm the developmental importance of this communication with a familiar and affectionate person.

After three months, as an infant develops more effective support of the body against gravity and begins to actively seek objects for exploration and manipulation, the infant's communication becomes more obviously "self-conscious". It is increasingly monitored through others' reactions, by "emotional referencing" that checks on what others feel about the baby's "Me". Emotional negotiations of games, joking, and teasing lead to transfer of feelings to "objects of contemplation" or to "objects of use", and these things become animated and are given meaning in rhythmic and repetitive games with other persons. Mothers, fathers, and siblings, are uniquely playful and instructive when they are with an infant. In baby-songs and rhythmic body-games the mother solicits intense attention and cooperative movements and vocalizations from her infant. Six-month-olds are capable of participating in and learning a litany in poetic or musical form that lasts many seconds. Musical terms (beat, rhythms, pitch, intervals, harmony, timbre, melody) give exact description, of the innate forms and timing of the vocal and gestural communications in action games and songs for infants, and these are similar in all the world's languages. Evidently infants are equipped to learn in a particular style of dramatic or emotionally-transforming narrative or performance.

When they acquire referential content, baby-talk narratives in song and rhyme become stories, the primary vehicle of linguistic learning that is so readily adapted to comment on and explain pragmatic tasks of culture, such as building a construction, making a meal, delivering a gift, dressing up, performing a symbolic act or a role, and so on. Before speech, communication becomes integrated in one shared space for orientations; it becomes cooperative and conventionalized and the infant displays "protosigns". At the threshold of language, play is richly imaginitive, seeking expressions in imitated and socially approved roles, attitudes, and humorous displays that other persons in the familiar community readily recognize and appreciate.

The precocious development of conversational play in infancy has clear implications for brain growth. At least on the baby's side, the earliest communication is almost certainly largely generated and adjusted in subcortical systems of the brain, because for several weeks the neocortical circuits in the human are in a very rudimentary condition. The interconnections between cortical modules, including the links of the corpus callosum, have yet to differentiate. In spite of its relative immaturity, the brain that will learn language and traditional conventions for cooperative use of objects is born seeking an identified partner and teacher who "converses".

We must conclude that the infant, too immature to regulate exploratory or performatory contact with physical objects, and lacking focal perception, refined motor control, and "object memory", has dedicated cognitive and intentional systems seeking "operational closure" in communication with an affectionate partner. The Norwegian sociologist and cybernetician Stein Bråten has proposed, on the evidence of early communication, that the human brain is born with a dual personality; with a "virtual other" that can be fully active only when supported by another person's reactions.

Basic human communication establishing "dialogic closure" by "affect attunement" is used in teaching and in therapeutic communication. The "virtual other" is a key component of the innate intersubjectivity out of which cultural learning grows. Human social learning is in relationships identifying individual others as distinct sources of psychological or intersubjective teaching, and of cooperative initiative. Correspondingly, a parent is more than a caregiver, protector or scaffold for action. The infant seeks, in succession: an identified, emotionally available and responsive partner in communication of basic motives and emotions; an opponent in affectionate play; a companion and guide in emotionally evaluated experience with objects and events; a helper in task-performance; an audience, admirer, and critic, whose feelings convey the value of shared experiences and provide guidance towards greater competence and facility. These are the characteristics of the cognitive system that has employed symbols to transmit knowledge cumulatively over thousands of years.

REFERENCES

Bogen, J. E. (1969). The other side of the brain, II An appositional mind. *Bulletin of the Los Angeles Neurological Society* , 34: 135-162.

Bråten, S. (1988). Dialogic mind: The infant and adult in protoconversation. In: *Nature, Cognition and System* , ed. M.Cavallo. Dordrecht: Kluwer Academic Publications, pp. 187-205.

Changeux, J.-P. (1985) *Neuronal Man: The Biology of Mind* . New York: Pantheon.

Edelman, G.M. (1987). *Neuronal Darwinism: The Theory of Neuronal Group Selection.* New York: Basic Books.

Fodor, J. (1983). *The Modularity of Mind.* Montgomery VT: Bradford.

Gardner, H. (1984). *Frames of Mind.* London: Heinemann.

Gazzaniga, M. (1970). *The Bisected Brain.* New York: Appleton-Century-Crofts

Hunt, R. K. and Cowan, W. M. (1990). The chemoaffinity hypothesis: an appreciation of Roger W. Sperry's contributions to developmental biology. In: C. Trevarthen (ed.), *Brain Circuits and Functions of the Mind: Essays in honor of Roger W. Sperry* . New York: Cambridge University Press.

Innocenti, G. M. (1986). General organization of callosal connections in the cerebral cortex. In: E. G. Jones and A. Peters (eds.), *Cerebral Cortex* , Volume 5. New York: Plenum, 291-353.

Levy, J. (1990). Regulation and generation of perception in the asymmetric brain. In: C. Trevarthen (ed.), *Brain Circuits and Functions of the Mind: Essays in honor of Roger W. Sperry* . New York: Cambridge University Press.

Murray, L. (1992). The impact of postnatal depression on infant development. *Journal of Child Psychology and Psychiatry* , 33 (3): 543-561.

Murray, L. and Trevarthen C. (1985). Emotional regulation of interactions between two-month-olds and their mothers. In: T. Field and N. Fox (eds.), *Social Perception in Infants* . Norwood, N.J., Ablex, 177-197.

Singer, W. (1987). Activity dependent self-organization of synaptic connections as a substrate for learning. In: J.-P. Changeux and M. Konishi (eds.) *The Neural and Molecular Basis of Learning.* Wiley: New York.

Sperry, R. W. (1982). Some effects of disconnecting the cerebral commissures. *Science*, 217(4566): 1223-1226.

Sperry, R. W. (1984). Consciousness, personal identity and the divided brain. *Neuropsychologia*, 22(6): 661-667.

Stephan, M. (1990). *A Transformational Theory of Aesthetics.* London: Routledge.

Stern, D. N. (1985). *The Interpersonal World of the Infant: A View from Psychoanalysis and Development Psychology.* New York, Basic Books.

Trevarthen, C. (1984). Biodynamic structures, cognitive correlates of motive sets and development of motives in infants. In: W. Prinz and A.F. Saunders (eds.), *Cognition and Motor Processes.* Berlin-Heidelberg-New York: Springer Verlag, 327-350.

Trevarthen, C. (1985a). Facial expressions of emotion in mother–infant interaction. *Human Neurobiology* , 4, 21-32.

Trevarthen, C. (1985b). Neuroembryology and the development of perceptual mechanisms. In: *Human Growth* (Second Edition), eds. F. Falkner and J.M. Tanner. New York: Plenum, 301-383.

Trevarthen, C. (1986). Brain Science and the human spirit. *Zygon. Journal of Religion and Science* , 21(2) June 1986: 161-200.

Trevarthen, C. (1987a). Brain development. In: R.L. Gregory and O.L. Zangwill (eds.), *Oxford Companion to the Mind* . Oxford, New York: Oxford University Press, 101-110.

Trevarthen, C. (1987b). Split brain and the mind. In: R.L. Gregory and O.L. Zangwill (eds.), *Oxford Companion to the*

Mind . Oxford, New York: Oxford University Press, 740-747.

Trevarthen, C. (1987c). Subcortical influences on cortical processing in "split" brains. In: D. Ottoson (ed.) *Duality and Unity of the Brain* . (WennerGren International Symposium Series, Vol. 47). Basingstoke, Hampshire: Macmillan/New York: Stockton Press, 382-415.

Trevarthen, C. (1987d). Infancy, mind. In: R.L. Gregory and O.L. Zangwill (eds.) *Oxford Companion to the Mind* . Oxford, New York: Oxford University Press, 362-368.

Trevarthen, C. (1989). Development of early social interactions and the affective regulation of brain growth. In: *Neurobiology of Early Infant Behaviour* , eds. C. von Euler and H. Forssberg. (Wenner-Gren International Symposia Series) Basingstoke, Macmillan.

Trevarthen, C. (1990a). Growth and education of the hemispheres. In: C. Trevarthen (ed.), *Brain Circuits and Functions of the Mind: Essays in honor of Roger W. Sperry* . New York: Cambridge University Press.

Trevarthen, C. (1990b). Integrative functions of the cerebral commissures. In: F. Boller and J. Grafman (eds.), *Handbook of Neuropsychology* , Vol. 4. Amsterdam; Elsevier Science Publishers BV (Biomedical Division), 49-83.

Trevarthen, C. (1990c). Grasping from the inside. In: M. A. Goodale (ed.) *Vision and Action: The Control of Grasping* . Norwood, N.J.:Ablex, 181-203.

Trevarthen, C. (1993a). The function of emotions in early infant communication and development. In: J. Nadel and L. Camaioni (Eds.) *New Perspectives in Early Communicative Development* . London: Routledge. (in press)

Trevarthen, C. (1993b). The self born in intersubjectivity: An infant communicating. In: U. Neisser (Ed.) *Ecological and Interpersonal Knowledge of the Self* . New York: Cambridge University Press (in press).

Tucker, D. M. (1991). Developing emotions and cortical networks. In: M. Gunnar and C. Nelson (Eds.) *Minnesota*

Symposium on Child Psychology , Vol. 24: Developmental Behavioral Neuroscience. Hillsdale, NJ: Erlbaum.

Willshaw, D.J. and von der Malsburg, C., (1976). How patterned neural connections can be set up by selforganization. *Proceedings of the Royal Society of London, Series B.* , 194: 431-445.

Wilson, A. C., (1991). From molecular evolution to body and brain evolution. In: J. Campisi (ed.) *Perspectives on Cellular Regulation: From Bacteria to Cancer* , New York:Wiley-Liss, 331-340.

Zaidel, D. (1990). Long-term semantic memory in the two cerebral hemispheres. In: C. Trevarthen (ed.), *Brain Circuits and Functions of the Mind: Essays in Honour of Roger W. Sperry* . New York: Cambridge University Press.

Zaidel, E. (1990). The saga of right hemisphere reading. In: C. Trevarthen (ed.), *Brain Circuits and Functions of the Mind: Essays in honor of Roger W. Sperry* . New York: Cambridge University Press.

11

Large Heterogeneous Knowledge Bases

E. Tyugu

Royal Institute of Technology,
Dept. of Teleinformatics, Electrum/204, S-16440 Kista, Sweden

Abstract

This chapter discusses large knowledge bases as software development tools which support the creativity of programming in the large. User requirements, architecture and internal knowledge representation language of large knowledge bases are considered. Higher order constraint networks are proposed for representing knowledge about computability.

1 SOFTWARE REUSABILITY

An important characteristic of the software development process is the degree of reusability of software. Simply speaking, knowledge once encoded in the form of programs must be reusable every time it could be needed in programming new problems. A natural way to reuse programs is to apply large software libraries. It is expected that this increases the productivity of software development and reliability of the software produced. However, with the exception of a small number of specific applications, the software libraries of today tend to be very difficult to use. They lack comprehensive user interface, and require from the users too much effort in studying of documentation. One can use the following analogy. From a usability standpoint, a software library is like an ordinary library of literature containing a large number of books, except that it has no comprehensive catalogue, the books don't have title pages, and they are stored in a random order and are accessible only by numbers which

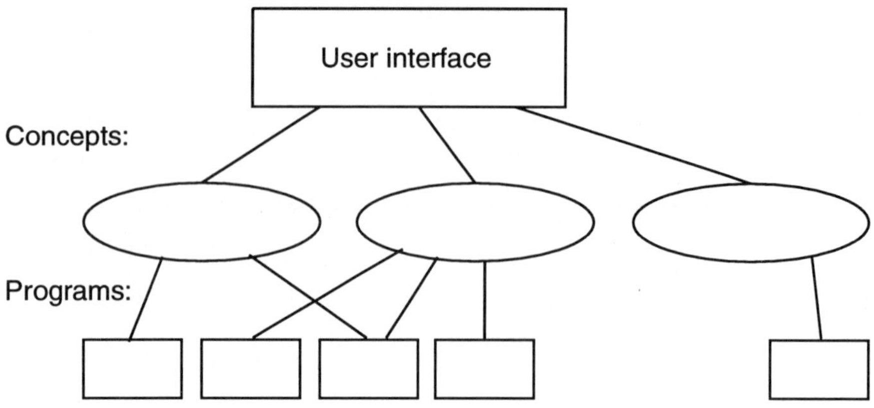

Figure 11.1. Layers of programming knowledge

are their formal addresses. Attempts are being made to build knowledge bases which could provide a guidance in selecting suitable software from software libraries (Devanbu et al. 1991).

The goal of the present work is to propose a design for a knowledge base which would support automatic construction of large programs from their declarative specifications. Roughly speaking, we shall build a software library which contains two layers of knowledge, Figure 11.1.

The lower layer is a repository of programs. These programs are not directly visible in the software development process. They are covered by the layer of knowledge about their applicability for solving different problems. This knowledge is visible to users (software developers) and it is represented in terms of concepts of problem domains, not in terms of programs. Besides that, we distinguish between the internal knowledge representation language, which is essentially a language for representing knowledge about computability, and a user language which is a high level knowledge representation language for specifying concepts and for reasoning about them.

This kind of kowledge representation has already been successfully used in several knowledge-based programming environments (Tyugu 1991). The novelty of the present work is the approach to a knowledge base as a general purpose software tool, not a narrowly problem oriented tool. This approach changes the requirements to the design of a knowledge base, as will be shown in section 3. The knowledge tools which worked satisfactorily in restricted problem domains such as engineering applications of computers for circuit analysis, design of machine parts, or selecting fits and tolerances, are not applicable for developing a large general purpose knowledge base.

We start with the explanation of the representation of knowledge about computability which is the key issue of software knowledge bases. Thereafter we present requirements to large knowledge bases and present some architectural solutions of large knowledge bases. We shall not discuss here the user language built on top of the internal knowledge representation language, as it has been represented in full details in several papers (Tyugu 1991).

2 REPRESENTATION OF KNOWLEDGE ABOUT COMPUTABILITY

In this section we describe briefly the internal knowledge representation language. This is a language used for representing semantics of concepts of the 'real world' in terms of programs available in the knowledge base. This language itself is divided into two parts: a language of constraints for describing computations and a language of rules for metareasoning about concepts.

The language of constraints which we use can be described as a language of higher order functional constraints. Any constraints in it can be represented by a program. HoweveR, SOme constraints appear from declarations which are not programs, but still have functional semantics. An example of such a constraint is equation, for instance,

$$u = i * r$$

which can be interpreted as a source of the following three programs:

$$u := i * r \quad (11.1)$$
$$i := u/r \quad (11.2)$$
$$r := u/i \quad (11.3)$$

depending on what must be computed: u, i, or r.

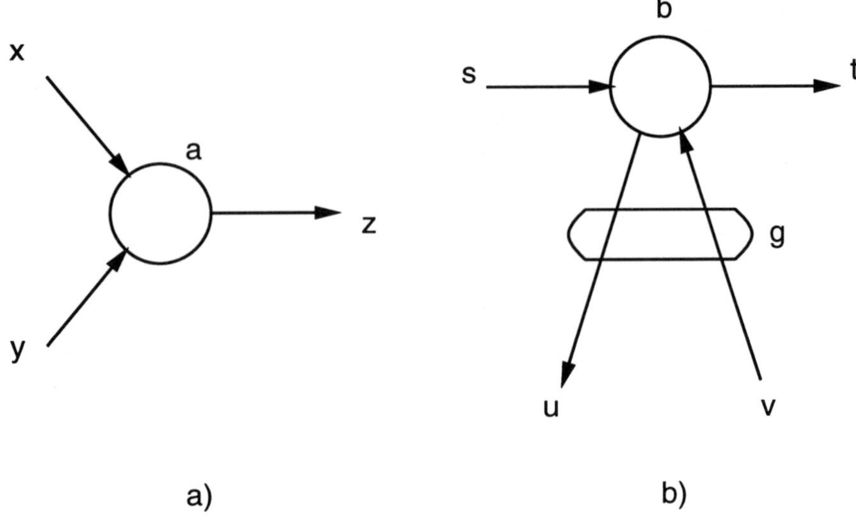

Figure 11.2. Functional constraints

To explain higher order functional constraints, we shall consider two programs a and b which represent the constraints shown in Figure 11.2.

The first program has input variables x, y, and an output variable z. The second program has an input variable s, an output variable t, and a procedural parameter g which is also an input of this program. Let us assume that this parameter can take values which have input u and output v. Figure 11.2 shows these programs as constraints between the variables x, y, z and s, t, u, v respectively. Arrows show possible directions of dataflow during the computations.

Observing the dataflow of the second program, we distinguish its input and output data (i.e. the values of the variables s and t) and the data passed between this program and the pogram f which is the value of its parameter g. The program b produces input of f (the value of the variable u) and gets back the value of the variable v which is the output of f. This happens every time the f is called, and that, in its turn depends on the computations performed by the program b. We must distinguish this dataflow, which occurs during the subcomputations performed by the program b, from the "ordinary" input and output of programs. This is reflected in markings of arcs binding the variables u, v which differ from the arcs binding other variables with programs. The role of the higher order variable g is binding the variables u and v as input and output of one and the same subcomputation. In general, a program can have more than one procedural parameter, as well as several input and output variables.

A higher order constraint network is a graph which has variables and constraints as its nodes. The constraints are functional or higher order dependencies represented by programs (or even by equations, etc.). Its arcs bind variables with constraints, as has been shown in Figure 11.2. It is remarkable, that this graph is a bipartitioned graph with the nodes divided into nonintersecting sets of variables and constraints (procedural parameters are not represented as nodes in this graph). An exmple of a higher order constraint network is shown in Figure 11.3. Such a network can be used as a program specification, provided a goal is given in the form of lists of input and output variables of the desired program. For instance, Figure 11.3 becomes a program specification as soon as realizations of its constraints are given and we say that x is the input and y is the output of the program desired.

It is quite easy to build a schema of the program required above. Two possible schemas are shown in Figure 11.4 *a,b* where we can see that the program implementing the constraint c has a subcomputation which is the application of the constraints a, b, d, and e in one case and the constraints d and e in another case. Program synthesis from specifications of this kind has been applied in several programming environments (Tyugu 1991a; Mints 1988).

The knowledge representation language described above is suitable for representing knowledge about the computability. But it is still too restricted to represent general knowledge required in universal software knowledge bases. We propose an extension of this language which is a metalanguage about constraint networks.

Let us have a knowledge base of concepts each of which is specified as a constraint network. Let a concept C have a speci-

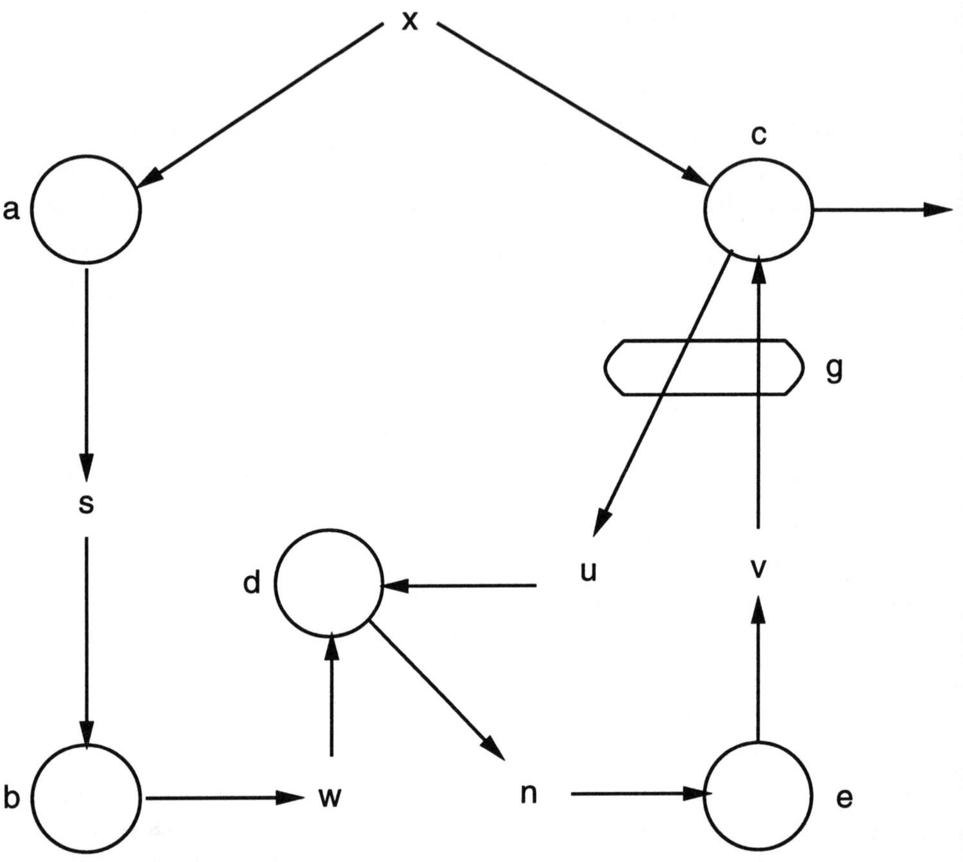

Figure 11.3. Higher-order functional constraint network

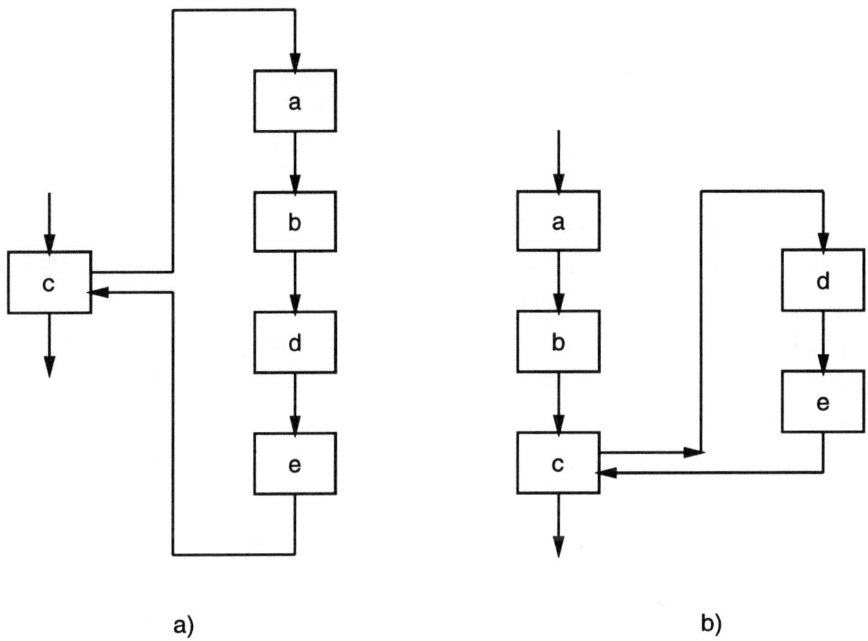

Figure 11.4. Synthesized program schemas

fication which contains variables $x, \ldots, y$. We shall consider this specification as a description of a relation between these variables, and shall introduce a predicate **C** for expressing this relation. Variables $x, \ldots, y$ in the atomic formula $\mathbf{C}(x, \ldots, y)$ can be unified with corresponding objects of any particular instance of the concept C. The predicates associated with concepts give us a possibility to represent theories about the concepts in the knowledge base as well as the situations described in terms of these concepts. In particular, if we restrict ourselves to Horn clauses, we get a knowledge system for metareasoning about concepts which can be efficiently implemented. This has been done in the NUT programming environment described in (Tyugu 1991a). Actually, the metareasoning component of the NUT system does more than described above. It also works with parts of constraint networks, not only with separate concepts.

3 REQUIREMENTS

A large knowledge base is intended for applications in the conditions where the scope of possible applications cannot be made precise in advance. This means that conformity, extendibility, reusability, etc. must be inherent features of such a knowledge base. Most of the existing knowledge bases have been designed for applications in some well-defined problem domain: in a particular field of CAD, management decision making, medicine, etc. This facilitates the design of a knowledge base considerably, enabling a designer to make reasonable commitments on the basis of information about the application domain.

From the other side, knowledge-based techniques of today are already powerful enough to enable one to build and to apply general purpose knowledge bases in knowledge areas such as basic geometry and physics (Tyugu 1988) as well as in general engineering fields such as machine parts in mechanical engineering of typical devices (filters, transformers, etc.) in electrical engineering. Our goal will be to make these knowledge bases applicable in various combinations and, ultimately, to unite them into a single knowledge body as is operational in human beings. Actually, the analogy with human reasoning and, in particular, with teaching people is much deeper here: conceptual design of a heterogeneous multidisciplinary knowledge base is principally not different from putting together a curriculum of studies for students. The differences appear at the stage of implementation. The students are considered to be universal in the sense that they must be able to absorb and apply any kind of knowledge systematically presented to them. The knowledge-based tools of today are still very restricted and we do not expect to be able to present knowledge in a natural language at the input of knowledge bases.

The following is a summary of requirements to a large heterogeneous knowledge base.

Support of knowledge acquisition is a requirement closely shared with the project CYC (Lenat 1990) where this requirement has been thoroughly investigated. This requirement means that all the knowledge available in a knowledge base must, first of all, be applicable for declarative specifications of new knowledge. This will give the cumulative effect when the knowledge is acquired. In the case of natural language input, this requirement necessitates associative reasoning. For formal knowledge representation languages, it requires a well-developed inheritance mechanism.

Transparency of knowledge for the user becomes a crucial feature of a knowledge base as soon as the latter grows out of easy comprehension. Analogy with large databases, and, in particular, understanding the reasons of introduction of data dictionaries in the thesaurus form is useful here. In the case of a large amount of knowledge, a user must discover what is available for the usage and how to use it. Invisibility leads to errors caused by misunderstanding of knowledge and to the necessity to solve the discovery task which can take up to 60% of the development time of procedural knowledge bases (Devanbu et al. 1991).

Modularity of knowledge is required for the following reasons. First of all, concentrating attention on a small number of comparatively small knowledge entities at each inference step improves the efficiency of an inference engine. Modularity enables one to represent various kinds of knowledge in a single integrated knowledge base by encapsulating different microtheories in separate knowledge modules (Tyugu 1991b).

Heterogeneity of knowledge is the feature that requires the ability to select a suitable form for representing each piece of knowledge. For instance, one can distinguish hard and soft knowledge, or in another dimension, shallow or deep knowledge. When shallow and soft knowledge can be well represented in the form of productions, this is not true for hard and deep knowledge. In engineering fields, for instance, the latter can exist in the form of precise mathematical models. A representation language for this kind of knowledge must be close enough to usual mathematical notations. The two knowledge representation formalisms presented in the previous section are very much intended to support the heterogeneity of knowledge.

Conformity means a possibility to use a knowledge base for different tasks and in various environments. One expects that a large knowledge base will be applicable whenever the knowledge in it becomes useful. In a more restricted context of procedural knowledge, conformity means interoperability, i.e. the ability of programs to communicate and work together even when having been written in different languages (Wileden et al. 1991).

Openness—extendability is the feature mentioned at the beginning of this section. Because a large knowledge base is not designed for any predefined application domain, it must be adaptable and extendable during its whole life-cycle.

Changeability is another feature which follows from the undefined scope of applications of a knowledge base at its development time. This feature is essentially different from openness and extendability, because it implies the nonmonotonic character of knowledge in a knowledge base. Nonmonotonicity is a strong requirement in knowledge handling and there are a number of logical models of it, see (Genesereth and Nilsson 1987).

Integrity is a feature which is known from database maintenance. In a logical framework it means preserving consistency after introducing any permissible change into knowledge. Practically, it can be only partially guaranteed by introducing integrity constraints which must be checked at every change of the knowledge base.

Size of a large heterogeneous knowledge base can be estimated only roughly, because there is no precise measure of knowledge. A possible measure of knowledge is the number of simple bindings (i.e. instances of symbolic names) in a knowledge base. A large knowledge base can, possibly, contain about the order of 10M bindings, as we shall see in Section 5.

Brittleness bottleneck is typically the problem which arises in connection with practical usage of knowledge-based techniques. In its essence, this means getting unsatisfactory responses to unexpected situations. When applying a program to solve a problem, one is always expected to know the precondition of the program, i.e. the predicate which determines whether the program could be applied to the problem. When a knowledge-based technique is being used, the preconditions are not so explicit. If knowledge is represented in the form of rules, one still can look at the conditions of the rules as preconditions. But, firstly, these conditions do not apply to the problem as a whole. Secondly, they may be expressed in imprecise terms the meaning of which remains hidden from a user. As a consequence, the user will not be satisfied with the responses he will get from a knowledge-based system. When the amount of knowledge increases and when the knowledge domain becomes wider, then the knowledge base appears to be unreliable to users—it will not satisfy their expectations, because they will be unable to get the inferences expected to be done automatically by the knowledge base. This is a brief review of requirements which must be taken into account in designing the architecture of large programming knowledge bases.

4 ARCHITECTURE

Requirements of modularity and diversity of knowledge representation cannot be satisfied in a flat and homogeneous knowledge base. Besides that, a knowledge representation formalism for internal use in a knowledge base must, first of all, satisfy the requirements of efficiency and need not be suitable for human understanding. This gives us the first architectural principle for building a large knowledge base:

1. The input language for knowledge representation should be different from the internal knowledge representation formalisms.

An input language can support a frame-based representation and it is highly preferable that its translation would be straightforward and simple. One way to achieve this is to use a calculus of inheritance for translating from an input language into a formal internal language which must be suitable for making inferences (Tyugu 1991b).

The second principle is concerned with the internal knowledge representation. Experience shows us that no universally efficient knowledge representation and handling technique exists. On the contrary, a number of very different methods have been developed for solving problems in various knowledge domains. Also considering human intelligence, one can distinguish basically different knowledge handling mechanisms that are associated with logical and intuitive ways of thinking. This gives us the following principle:

2. Several knowledge systems are needed in a large knowledge base.

Elaborating on the meaning of a knowledge system gives us the third principle:

3. Any knowledge system (KS) which is knowledge representation plus inference engine can be represented as a formal calculus with interpretation that adequately represents knowledge processing in this system.

This thesis cannot be proved formally without formalizing the adequacy requirement, which we will not do. However, looking at the numerous examples of knowledge systems, we can find good evidence in favour of this thesis. First of all, making inferences means using knowledge in a deductive way. The possibility of formalisation of a KS becomes obvious as soon as we loosen the requirement of adequacy: on a sufficiently low level, we can use the well-known formal systems—Turing machines or Post's systems for representing KS. This principle lies in the basis of inference engines of knowledge systems.

The following principle considers the stratification of knowledge:

4. Several levels of generality are required even in one and the same knowledge domain.

Even in one and the same knowledge domain one must be able to use very general knowledge as well as specific knowledge. The first kind of knowledge can be presented in the form of a metatheory about the domain knowledge, see Section 2.

In Figure 11.5, the operational part of a large knowledge base is shown. It contains all basic knowledge-paths for knowledge acquisition and application to external problems. It does not show the knowledge flow for self-referential reasoning and for explanations. At the knowledge acquisition stage, concepts are being formed by a translator (1). A concept has a layered structure and this structure is aalso preserved later, when a representation of a particular situation (a problem) is being constructed. Different layers need different inference engines for making inferences (3,4). The knowledge repository (5) can

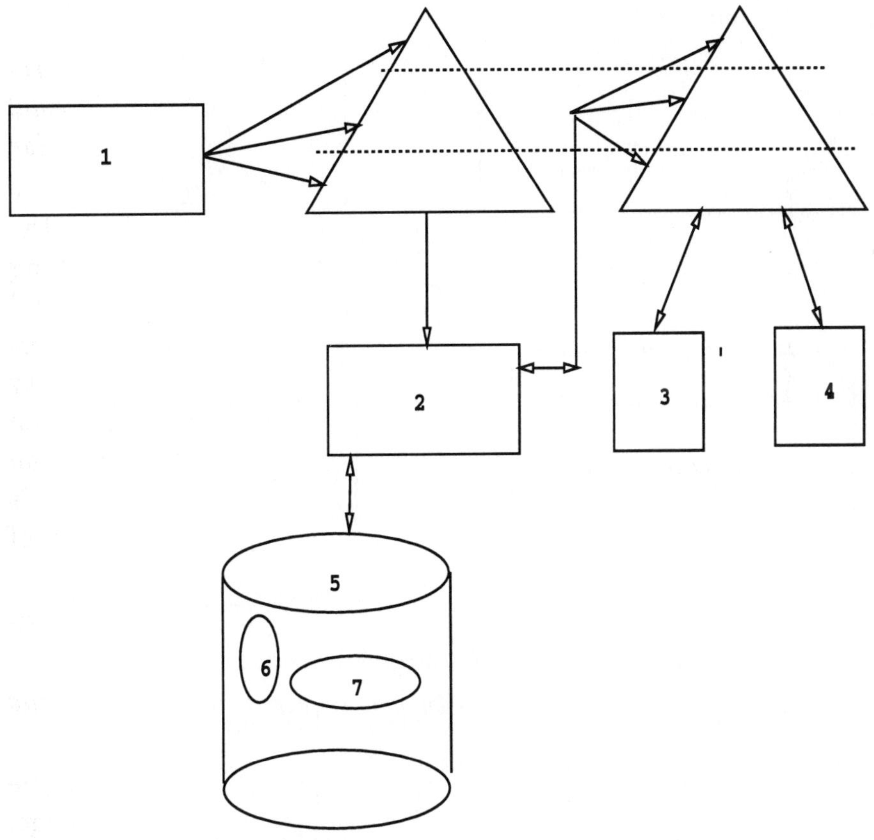

1 - language processor
2 - knowledge handler
3, 4 - inference engines
5 - knowledge repository
6 ,7 - knowledge packages

Figure 11.5. Operational part of a large knowledge base

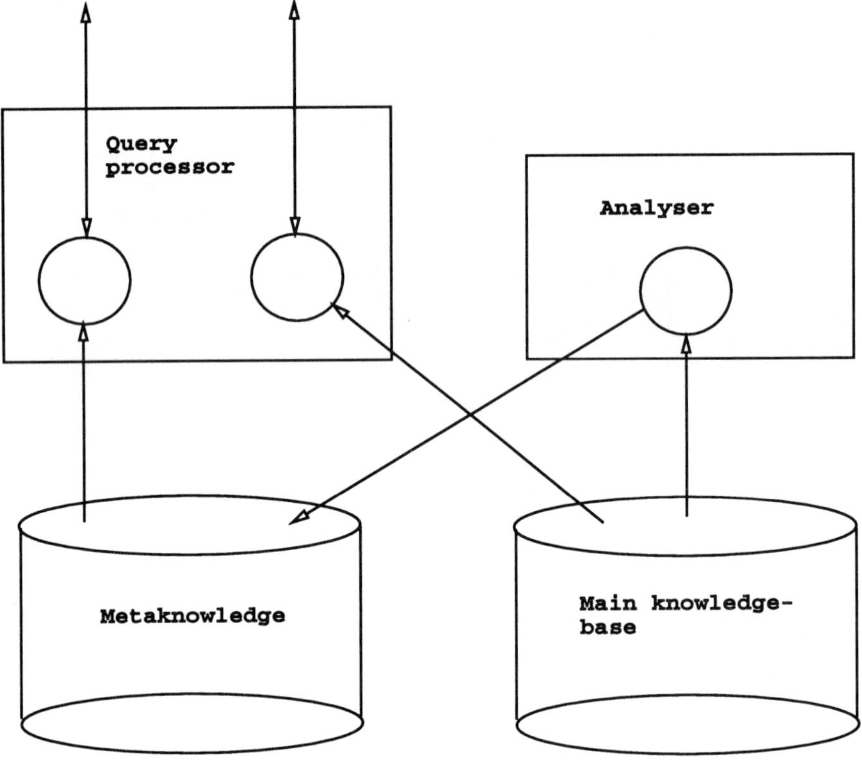

Figure 11.6. Referential part of a large knowledge base

be structured into knowledge packages (6,7) and can have self-referential (meta) knowledge about its own contents.

Figure 11.6 shows the referential part of a knowledge base. This part is intended for processing users' queries about the knowledge. It provides transparency of the knowledge in the knowledge base and supports both knowledge acquisition and its usage. There are two ways of processing users queries:

- direct analysis of the knowledge,
- answering on the basis of the metaknowledge.

In the first case, metaknowledge is not needed, but this method is applicable only to limited amounts of knowledge, i.e. not to the knowledge base as a whole. The second method is analogous to the usage of data dictionaries in data-bases. It requires the existence of metaknowledge about the contents of the knowledge base, which can be presented in the form of a thesaurus, inverted

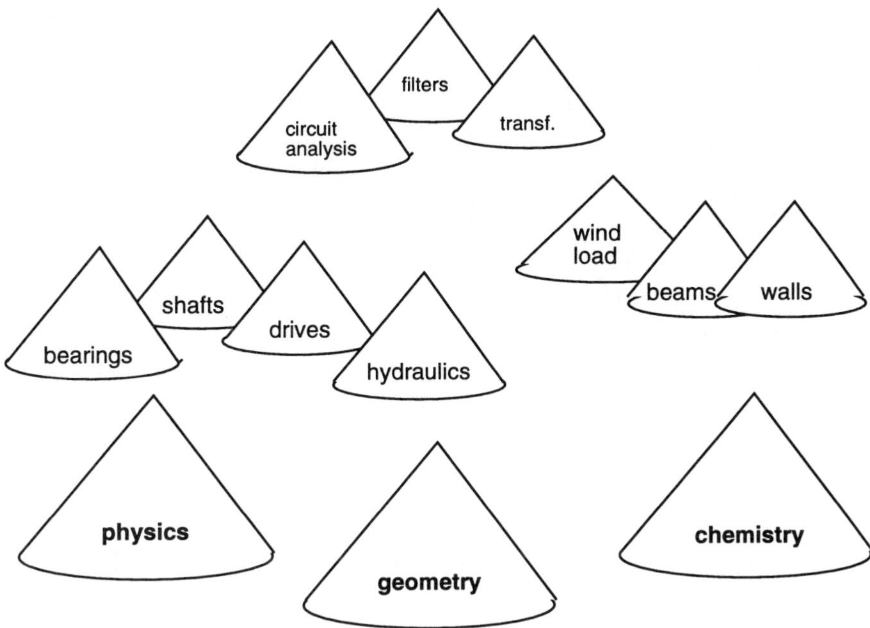

Figure 11.7. A "landscape" of engineering knowledge bases

files, or an object-oriented information system. The problems arising in the construction of information systems of this kind have been discussed in (Devanbu et al. 1991).

5 CONTENTS AND SIZE

This section is based on our experience gained in engineering applications of knowledge-based programming tools. Most of the applications belong to the electrical and mechanical engineering fields (Pahapill 1985), but there are also examples of successful development of knowledge bases in civil engineering design and other engineering fields. These experiments have given us the confidence that all kinds of engineering knowledge can already be presented in computer by means of knowledge tools available now.

Figure 11.7 shows a landscape formed by our development of engineering knowledge bases. It shows several clusters of knowledge corresponding to the electrical, mechanical, and civil engineering fields, as well as some general knowledge in physics, mathematics, and chemistry.

Once again, we shall use here the analogy between building up a knowledge base and teaching people. At a first glance, when the amount of stored knowledge is considered, it could seem that the more the better. But, even if we assume an unlimited capacity of the knowledge repository, we must take into account the costs of knowledge acquisition as well as maintenance costs. Therefore, it is reasonable to design our knowledge base for engineering applications similarly to the curriculum of a technical university, taking into account restrictions on the resources for teaching. In doing so, we assume that available knowledge handling tools are sufficiently powerful and can support all kinds of knowledge needed for applications. In other words, we feel free from technical restrictions in presenting a conceptual design of an engineering knowledge base in this section. However, we are restricted by the resources available for introducing the knowledge, for teaching a knowledge base.

We must remember that a number of different knowledge systems (representation forms and inference engines) for representing the knowledge we are going to use must be available. The choice of a knowledge system for representing a knowledge-module is not purely a technical question. For instance, nonspecific knowledge which, in general, is seldomly used by an engineer must be represented in a more transparent way than specific and well-understood knowledge. But, here we should not be concerned with the form of knowledge. We can still make a comment on the overall organization of the knowledge base. Knowledge is represented there in modules, each of which we can consider as a microtheory about some narrow part of the engineering field. These microtheories can be nested. But, in general, they constitute a metasemantic (conceptual) network, where these microtheories refer to one another in the same way as texts in a hypertext.

We classify engineering knowledge into three categories: nonspecific, general engineering knowledge, and specific engineering knowledge. Nonspecific knowledge corresponds to generally known knowledge in science which is usually taught in schools for technically oriented students. General engineering knowledge is mainly knowledge from handbooks and standards. Specific engineering knowledge is the most problem-oriented and supports particular designs or engineering solutions by composing different knowledge modules into a single body and by providing the design solutions and plans.

The following is a brief outline of a universal engineering knowledge base segmentation and this does not reflect in which form the knowledge is actually stored (organized) in the knowledge base.

Nonspecific knowledge:

- mathematics
    - geometry
    - symbolic computations
- physics
    - mechanics
    - electricity
    - heat and heat transfer
    - optics
- chemistry
    - ...

Part of this nonspecific knowledge can be represented only in a very shallow way today, as in general handbooks for engineers where only basic definitions and some formulas are given. This does not mean that mathematical knowledge could not be presented in a hard and deep form. The research towards formalizing mathematics has been going on for a long time, and special efforts have been directed towards automating mathematical reasoning (de Bruijn 1980).

General engineering knowledge:

- engineering graphics
- general standards
- general engineering handbooks (fits and tolerances, etc.)

Specific engineering knowledge which is divided into different engineering fields:

1. mechanical engineering

- machine parts
- devices and machines
- ...

2. electricity
    - basic laws
    - analysis of alternating current circuits
    - ...

3. civil engineering
    - ...

Looking at the contents above, one can see that a knowledge engineer is in a rather good position when building an engineering knowledge base. An essential part of the engineering knowledge is already presented in a well structured and formal form which is well suited for computerized processing. We have built knowledge modules which can be used separately as well as together. The architectural difference is that when a knowledge module is used independently, it must have an interactive user interface, whereas in the case of an integrated knowledge base the interface must, first of all, satisfy interoperability requirements.

We have estimated the amount of knowledge which must be put into a general engineering knowledge base. We suggest to measuring knowledge by the number of bindings. This is the number of arcs in a network representation or the number of instances of names in the case of symbolic representation of knowledge. The estimate is that a universal engineering knowledge base will contain at least 10 M bindings.

6 CONCLUDING REMARKS

As it has been stated at the beginning of the chapter, this work was provoked by the need for a sufficiently universal programming knowledge base which could be used as an aid in programming a great variety of problems. At present, this goal has been partially achieved—one can build a knowledge base which is an oracle for a restricted knowledge domain, and this has been done for many particular domains. However, we expect to get a significant breakthrough in problem-solving by uniting smaller and specialized knowledge bases into a large heterogeneous one. This task presents new requirements to the knowledge base which were reviewed in Section 3.

Some insights into the architecture of knowledge bases were obtained and contents of a large engineering knowledge base was outlined. This enabled us to estimate the amount of the knowledge needed, and an estimate for a general engineering knowledge base is 10 M bindings.

The amount of work needed for building a knowledge base can be estimated in two different ways: by guessing how much work is needed for introducing the given amount of knowledge or by summing up estimates for particular knowledge modules which constitute the contents of the knowledge base. The second is a direct way and, presumably, more precise. Our experience in introducing the knowledge modules gives us a rough estimate which shows that more than 140 man years are required for building a general purpose engineering knowledge base.

Finally, it seems that a realistic way to get a large programming knowledge base is by evolutionary development. This reminds us again how people are being taught. The crucial question in this case will be the ability of a knowledge base system to perceive the knowledge, i.e. to learn gradually.

Acknowledgments

I would like to thank my former colleagues Ahto Kalja and Jaak Pahapill from Tallinn for developing numerous microtheories for engineering knowledge-bases and Tiit Tiidemann for shaping a landscape of the engineering knowledge.

REFERENCES

de Bruijn, N. (1980). A survey of the project AUTOMATH. *Essays in combinatory logic, Lambda calculus and formalism* (Eds. J.P. Seldin and J.R. Hindley). Academic Press, 586–606.

Devanbu, P., Brachman, R. J., Selfridge P. G., Ballard B. W. (1991). LaSSIE: A Knowledge-Based Software Information System. *Comm*

ACM, v. 34, No. 5, 34–49.

Genesereth, M. R., Nilsson, N. J. (1987). *Logical Foundations of Artificial Intelligence.* Morgan Kaufman Publishers.

Lenat, D. (1990). CYC: Towards Programs with Common Sense. *Comm ACM, v. 33, No. 8*, 30–49.

Mints, G., Tyugu, E. (1988). The Programming System PRIZ. *Journal of Symbolic Computations, No. 5*, 359–375.

Pahapill, J. (1985). Programmpaket zur Modellierung der Hydromaschinen Systeme. *6. Fachtagung Hydraulik und Pneumatik*, Magdeburg. 609–617.

Tyugu, E. (1988). *Knowledge-Based Programming.* Addison Wesley Publishers Ltd (Turing Institute Press).

Tyugu, E. (1991a). Three New-Generation Software Environments. *Comm ACM, v. 34, No. 6*, 46–59.

Tyugu, E. (1991b). Modularity of Knowledge. *Machine Intelligence, 12.* (Eds. J.E. Hayes, D. Michie and E. Tyugu) Clarendon Press, Oxford, 3–16.

Wileden, J. C., Wolf, A. L., Rosenblatt, W. R., Tarr, P. L. (1991). Specification-Level Conformity. *Comm ACM, v. 34, No. 5*, 72–87.

EXPERIMENTAL MACHINE LEARNING

12

Learning Optimal Chess Strategies

M. Bain

Turing Institute

S. Muggleton

Oxford University Computing Laboratory

Abstract

Move-perfect databases for chess endgames can be constructed by full-width backup from won positions. Complete tabulations are available for certain endgames, such as King and Rook against King (KRK), which contain optimal depth-to-win information. Previous work has applied decision-tree learning to construct rules for the classification of positions from such databases as won or not won, but with limited success. The current work takes an Inductive Logic Programming approach, using methods of generalization in first-order logic and specialization by predicate invention. Results are given on learning rules from example black-to-move KRK positions which are won-for-white in a fixed number of moves.

1 MOTIVATION AND FIRST RESULTS

Could a machine learn to play a simple chess endgame *optimally* given only example positions and some facts about the geometry of the board ? Below is a chess program, in Prolog, machine-learned from examples, which could legitimately form part of a feasibility demonstration for such a project.

```
krk(0,Kfile,3,WRfile,1,Kfile,1) :-
        not(wrfile_threat(Kfile,WRfile)).
```

```
krk(0,c,WKrank,a,WRrank,a,BKrank) :-
        wrrank_safe(WKrank,WRrank,BKrank).

wrfile_threat(BKfile,WRfile) :- diff(BKfile,WRfile,1).

wrrank_safe(2,WRrank,1) :- lt(2,WRrank).
wrrank_safe(Krank,WRrank,Krank) :- diff(Krank,WRrank,2).
wrrank_safe(Krank,WRrank,Krank) :- diff(Krank,WRrank,3).
wrrank_safe(Krank,WRrank,Krank) :- diff(Krank,WRrank,4).
wrrank_safe(Krank,WRrank,Krank) :- diff(Krank,WRrank,5).
wrrank_safe(Krank,WRrank,Krank) :- diff(Krank,WRrank,6).
wrrank_safe(1,8,1).
```

This program is complete and correct for the canonical set of legal black-to-move (BTM) positions in the King and Rook against King (KRK) endgame which are won-for-white (WFW) at a depth of 0 moves (i.e. checkmate). The top-level predicate is **krk/7**, the first argument of which gives the depth of win in moves for a minimax-optimal strategy. The other six arguments specify positions by the file and rank (x and y) chessboard coordinates of, respectively, the White King, White Rook, and Black King. The two clauses for the top-level predicate are complete and correct in the sense that they cover all and only those positions in an exhaustive database (described in Section 2.1) which are won at depth 0. This solution, which is discussed in more detail in Section 4, calls the primitive predicates **lt/2** and **diff/3** which define the relations $<$ and symmetric difference for files and ranks. Apart from these background predicates, two machine-invented predicates which are here labelled **wrfile_threat/2** and **wrrank_safe/3** complete the definition [1].

Previous attempts to learn classification rules in chess endgame domains have typically used a much more powerful set of domain features to describe example positions than is the case in the current study. Quinlan (1983) has reported that in an application of ID3 to learning a classification for the "lost n-ply" relation in the King and Rook against King and Knight (KRKN) endgame the main obstacle to producing a complete and correct set of

[1] Variables and constants have been renamed to improve readability.

classification rules was the lack of suitable attributes. The approach taken in our work is to attempt to overcome this problem by relying on the learning system's ability to change its domain representation incrementally, as demanded by the learning task.

In the KRK endgame the maximum depth of win for BTM positions is 16 moves. Currently, we have Prolog programs learned by the Inductive Logic Programming system GCWS (described in Section 3) which are complete in the above sense for all depths of win, i.e. from 0 to 16. To our knowledge this is the first time an optimal strategy for a complete endgame has been learned automatically in this way from example positions and low-level background knowledge predicates. In this chapter we present results for complete and correct Prolog predicates for depths of win of 0 and 1 moves. We are continuing to work on correction by specialization of predicates covering the remaining depth levels, and expect to incorporate these results in subsequent reports on this work. For the present however we will describe only the results from learning the definitions of won at depths 0 and 1. The outline of the chapter is as follows. Section 2 contains descriptions of the KRK endgame database used. In Section 3 the learning approach is discussed. The definitions learned are given in Section 4 and the relation to other work in Section 5.

2 MATERIALS

In previous work (Shapiro 1987; Muggleton 1987; Muggleton *et al.* 1989; Bain 1991) chess endgame databases have been used as sources of training and testing examples for machine learning experiments. The current work employs an exhaustive database which is a complete tabulation for the KRK endgame. The entries of this database contain optimal depth-to-win values for all positions. These entries constititute examples of position classes at each depth of an optimal strategy. Our training and testing examples were randomly sampled from this database.

2.1 Retrograde analysis of endgame databases

The retrograde analysis method for generating chess endgame databases as described by Thompson (1986) employs reduction

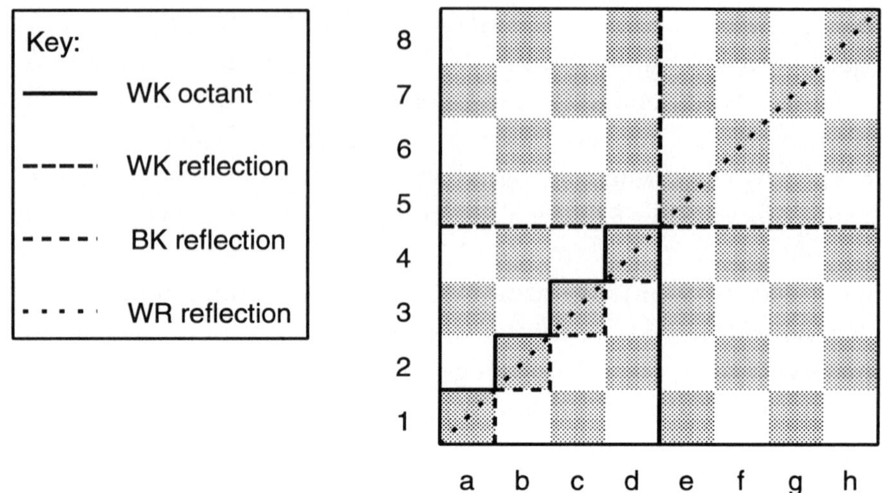

Figure 12.1. Canonical positions.
WK octant – the ten squares {a1,b1,c1,d1,b2,c2,d2,c3,d3,d4} are the canonical locations for the White King (WK).

WK reflection – reflection about the axes indicated places the WK in a canonical location:

> if the WK is above the central horizontal (rank 5 or greater), reflect the WK below (into the lower half of the board);
>
> if the WK is right of the central vertical (file e or greater), reflect the WK to the left of (into the left half of the board);
>
> if the WK is above the diagonal a1 to h8,
> reflect the WK below this diagonal (into the WK octant).

BK reflection – with the WK on squares a1, b2, c3 or d4 and the Black King (BK) above the diagonal a1 to h8, reflect the BK about this axis to place it below the diagonal.

WR reflection – with both WK and BK on the diagonal a1 to h8 and the White Rook (WR) above the diagonal, reflect the WR about this axis to place it below the diagonal.

Table 12.1. Depth of win, optimal play.

Depth is depth-of-win for white in moves with black-to-move; Number is number of positions. There are a total of 25 260 wins. Together with 2796 draws this gives a total of 28 056 positions used for positive and negative example sets.

Depth	Number	Depth	Number
0	27	9	1712
1	78	10	1985
2	246	11	2854
3	81	12	3597
4	198	13	4194
5	471	14	4553
6	592	15	2166
7	683	16	390
8	1433	Total	25260

of the space of positions by removing from consideration those positions equivalent to a canonical set by symmetry. Consequently, any legal position which could be encountered, for example in over-the-board play, must be translated to its canonical equivalent before its database value may be retrieved. The exact symmetries which may be exploited vary according to the pieces in the endgame. In the KRK database used to provide examples for the current work three types of symmetrical translation were applied. These are shown in Figure 12.1.

Only information on black-to-move (BTM) positions was extracted from the database. However this is sufficient to allow optimal play using only a legal move generator which is operated with 2-ply (one move) lookahead. Since every won position is tagged in the database with its minimax-optimal depth-of-win value, and the learned definitions contain the same values, this method holds also to allow the output of the learning-from-examples method to play optimally.

2.2 BTM WFW positions in the KRK database

By the removal of redundancy due to symmetries the total space of legal canonical positions in the KRK endgame is reduced from a potential 262 144 to 28 056. In the BTM database used in our experiments the number of legal positions won-for-white was 25 260. Each of these positions is tagged with its depth-of-win value. The number of positions in each depth-of-win class is given in Table 12.1.

3 METHOD

In the present work the goal was the induction of complete, correct, and concise theories in the KRK domain. Example positions for black-to-move (BTM) and win at depth D, where D is a number of moves, were extracted from an exhaustive database computed by the standard retrograde analysis method as discussed above. In this method, each entry in the database contains a position labelled by its minimax-optimal game-theoretic value, which in this case is its depth D. Such a database is taken to be a complete and correct definition of the endgame (although this has not yet been proved), containing as it does all legal positions for the given pieces together with their optimal depth-of-win labels. In logical terminology such a database may be thought of as an extensional definition of the endgame. Therefore the database is a complete and correct theory of KRK. Alternatively, the database is a relation $< P, D >$ where P is a canonical position and D is depth-of-win. However, despite reductions due to the removal of positions equivalent by symmetry, the database is too large to be called a concise theory. The goal was therefore the induction of a program which:

1. on input of a legal BTM position in KRK outputs the minimal depth-of-win for white;
2. on some measure was more concise than the database representation.

In this chapter the experimental tasks were restricted to positions with depth 0 or 1. Also, we did not apply any measure of relative conciseness [2]. A suitable candidate measure could be HP-compression as proposed in Muggleton *et al.* (1992).

---

[2]See discussion in Section 5.

<u>Algorithm GCWS</u>

<u>Input</u> : P_i, training, test, background

<u>if</u> training $= \emptyset$ <u>then</u>

$P_f = P_i$

<u>else</u>

$P = \text{Gen}(\text{training}, \text{background})$

$P' = P_i \cup P$

$\langle P'', \text{exceptions} \rangle = \text{Spec}(P', \text{test, background})$

$P_f = \text{GCWS}(P'', \text{exceptions, test, background})$

<u>Output</u> : P_f

Figure 12.2. GCWS algorithm schema.

We adopted an Inductive Logic Programming approach. Generalization steps were carried out using a version of Relative Least General Generalization (RLGG) as implemented in the GOLEM system (Muggleton and Feng 1992). This is shown in Figure 12.2 as the procedure call "Gen(training, backgound)". Correction of over-general clauses with respect to a target model was carried out using the Closed-World Specialization technique (Bain and Muggleton 1991). This is shown in Figure 12.2 as the procedure call "Spec(P', test, background)". The combined system is called GCWS and is as described in (Bain 1991) with a new extension which enables the automatic invention and introduction of non-negated exception predicates during the specialization process. The method can be viewed as implementing a form of automatic hierarchical problem decomposition.

An informal complexity constraint limiting the number of clauses used in any predicate definition to 7 ± 2 was applied when learning the depth 0 and 1 definitions. This is based on the hypothesized limit on human short term memory capacity of 7 ± 2 chunks. (A Prolog predicate is defined using one or more Horn clauses, where each clause contains a single positive literal with the same predicate symbol and arity.)

To avoid bias of the method by supplying the learning algorithm with a large number of predefined domain-specific background predicates, the background knowledge was restricted to contain only one specifically chess-oriented geometrical predicate, namely the symmetric difference between files and between ranks. This is defined as follows: for two file (resp. rank) values V_1 and V_2 the symmetric difference is $\mathrm{abs}(V_1 - V_2)$, where $V_1, V_2 \in \{1, .., 8\}$ and $\mathrm{abs}(X)$ is X if $X \geq 0$ and $0 - X$ otherwise. The other background predicate available was strictly-less-than ($<$) over files and over ranks.

These background predicates were selected as basic building blocks for the expression of piece relations in terms of the geometry of the chess board. In particular, they facilitate expression of the types of attack and counter-attack relationships which would seem to be essential for the expression of higher-level chess concepts such as capture, safety, check, etc. Thus they supply some of the raw materials for relevant predicate invention in chess domains.

Details of the experimental method are given in Figure 12.2 in the form of an algorithm schema. At each depth of win, the positive examples are all positions in the database which are won-for-white at that depth, and the negative examples are selected randomly from the remaining positions in the database.

Each depth of win was treated as a separate learning problem. The results for depth 0 are in Section 4.1 and for depth 1 in Section 4.2.

4 RESULTS

To recap on the Prolog representation used, the target predicate for the concept was **krk/7**. The first argument of this predicate indicates depth of win, in the range 0 – 16 (although only 0 and 1 were used in the present study). The remaining six arguments give the file and rank coordinates for, respectively, the White King, the White Rook, and the Black King. File arguments are in the range a – h while rank arguments are in the range 1 – 8. Note that these values are those used in the standard algebraic

```
krk(0,A,3,B,1,A,1) :- not(nonkrk1(A,B)).        % Clause 1
nonkrk1(A,B) :- diff(A,B,d1).

krk(0,c,A,a,B,a,C) :- krk2(A,B,C).              % Clause 2
krk2(2,A,1) :- lt(2,A).
krk2(A,B,A) :- diff(A,B,d2).
krk2(A,B,A) :- diff(A,B,d3).
krk2(A,B,A) :- diff(A,B,d4).
krk2(A,B,A) :- diff(A,B,d5).
krk2(A,B,A) :- diff(A,B,d6).
krk2(1,8,1).

% Key (labels for machine-invented and background predicates) :
%   nonkrk1(A,B) = "wrfile_threat(WKf,WRf)"
%   krk2(A,B,C) = "wrrank_safe(WKR,WRr,BKr)"
%   diff(A,B,d1) = "symmetric difference between A and B is 1"
%   lt(2,A) = "2 < A"
```

Figure 12.3. BTM WFW depth 0.

chess notation. For instance, the unit clause "krk(0, c, 1, a, 3, a, 1)" is read as "the black-to-move position WK:c1 WR:a3 BK:a1 is won-for-white at depth 0 (i.e. in 0 moves)". Recall that only legal positions are considered.

Background knowledge was restricted to the predicates **diff/3**, symmetric difference, and **lt/2**, strictly less than. Two argument types were used for file and rank arguments, as for the target predicate. A third type was used for symmetric difference arguments.

Throughout this section we illustrate each top-level clause in the induced definition with a figure containing diagrams of partial chessboards. The diagrams indicate position classes covered by the clause indicated. Dotted lines and arrows are used to show variations in placing the attacking White Rook.

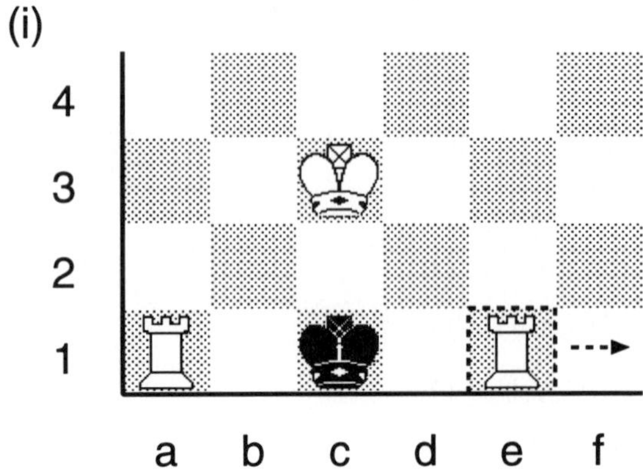

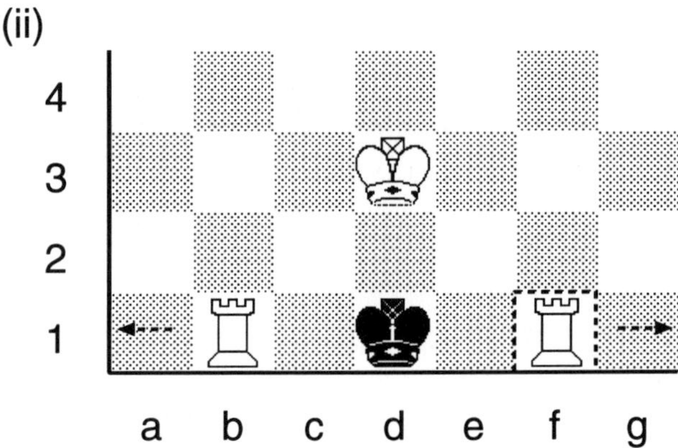

Figure 12.4. BTM WFW depth 0, clause 1.

4.1 BTM WFW depth 0

The induced Prolog definition for BTM WFW depth 0 is shown in Figure 12.3. Clause 1 of this definition is illustrated in Figure 12.4. As in clause 1, when the new predicate has only a single clause in its definition, it may be replaced with the clause body. In incremental learning this could however be undesirable — the invented predicate if retained might have further clauses added to its definition. For the time being we can interpret clause 1 as follows : any BTM KRK position with the kings on the same file, the white king on rank 3 and both the rook and black king on rank 1 is WFW depth 0 when the symmetric

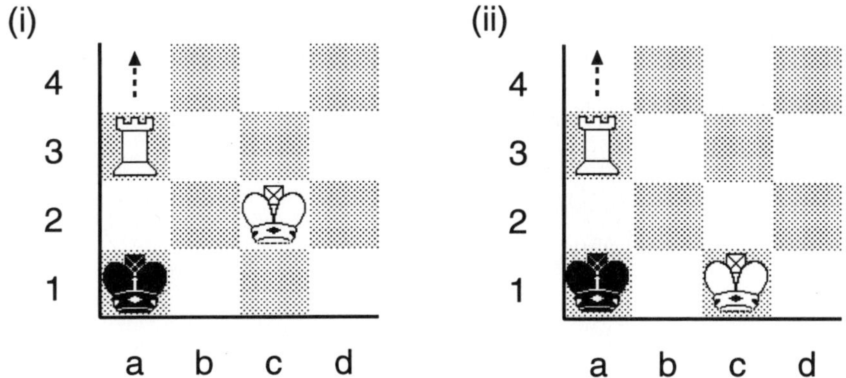

Figure 12.5. BTM WFW depth 0, clause 2.

difference between king and rook files is not 1. The machine-invented predicate **nonkrk1/2** clearly relates to the idea of rook safety when the black king is in check. It might be labelled **wr-file_threat/2**. This concept is apparent from the diagrams in Figure 12.4. Diagram (i) shows the placings of the White Rook on rank 1, essentially on file a and files e to h. Diagram (ii) illustrates the same pattern shifted one file to the right, with the White Rook on rank 1, files a, b and f to h.

The second clause is in a sense the dual of the first, this time for rank values, with the machine-invented predicate **krk2/3** being labelled **wrrank_safe/3**. This is shown in Figure 12.5. Diagram (i) illustrates the first clause of **krk2/3**, with the Kings not in opposition. Diagram (ii) covers the remaining clauses of **krk2/3** where the Kings are in opposition. In all cases the White Rook attacks the Black King on file a from the safety of rank 3 or above.

These two clauses cover all 27 positions won at depth 0. Although this is a small number of examples, the complexity of describing the concepts involved is clear from the diagrams of Figures 12.4 and 12.5.

4.2 BTM WFW depth 1

The induced Prolog definition for BTM WFW depth 1 is show in Figure 12.6. The top-level predicate **krk/7** has three clauses

```
krk(1,A,3,B,C,D,1) :-
      krk1(A,B,C,D), diff(A,B,d2), diff(A,D,d1).     % Clause 1

krk1(c,a,A,b) :- not(nonkrk12(A)).
krk1(c,e,A,d) :- not(nonkrk12(A)).
krk1(d,b,A,c) :- not(nonkrk12(A)).
krk1(d,f,A,e) :- not(nonkrk12(A)).

nonkrk12(1).
nonkrk12(2).

krk(1,c,2,A,B,a,C) :- krk2(A,B,C).     % Clause 2

krk2(A,4,3) :- not(nonkrk21(A)).
krk2(A,3,2) :- not(nonkrk22(A)).
krk2(A,4,1) :- not(nonkrk22(A)).
krk2(A,5,1) :- not(nonkrk22(A)).
krk2(A,6,1) :- not(nonkrk22(A)).
krk2(A,7,1) :- not(nonkrk22(A)).
krk2(A,8,1) :- not(nonkrk22(A)).

nonkrk21(a).
nonkrk21(b).

nonkrk22(a).

krk(1,c,1,A,3,a,2) :- not(nonkrk3(A)).     % Clause 3

nonkrk3(a).
nonkrk3(b).

%  Key (labels for machine-invented predicates) :
%    krk1(A,B,C,D) ="wrrank_safe(WKf,WRf,WRr,BKf)"
%    krk2(A,B,C) ="wrfile_safe(WRf,WRr,BKr)"
%    nonkrk12(A) ="ranklte2(WRr)"
%    nonkrk21(A) ="filelteb(WRf)"
%    nonkrk22(A)="cornerfile(WRf)"
%    nonkrk3(A) ="filelteb(WRf)"
```

Figure 12.6. BTM WFW depth 1.

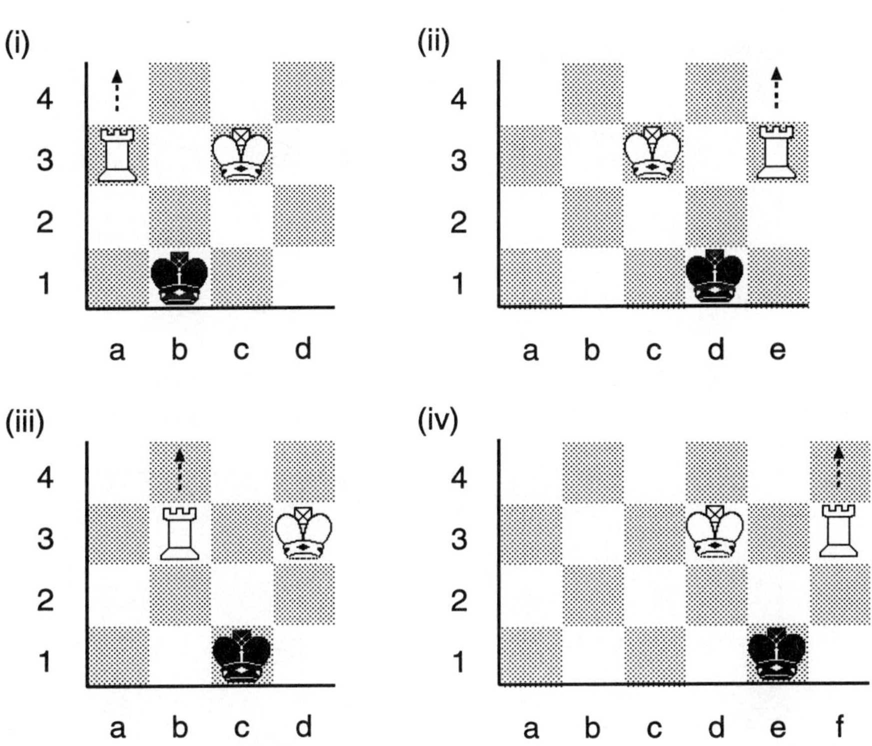

Figure 12.7. BTM WFW depth 1, clause 1.

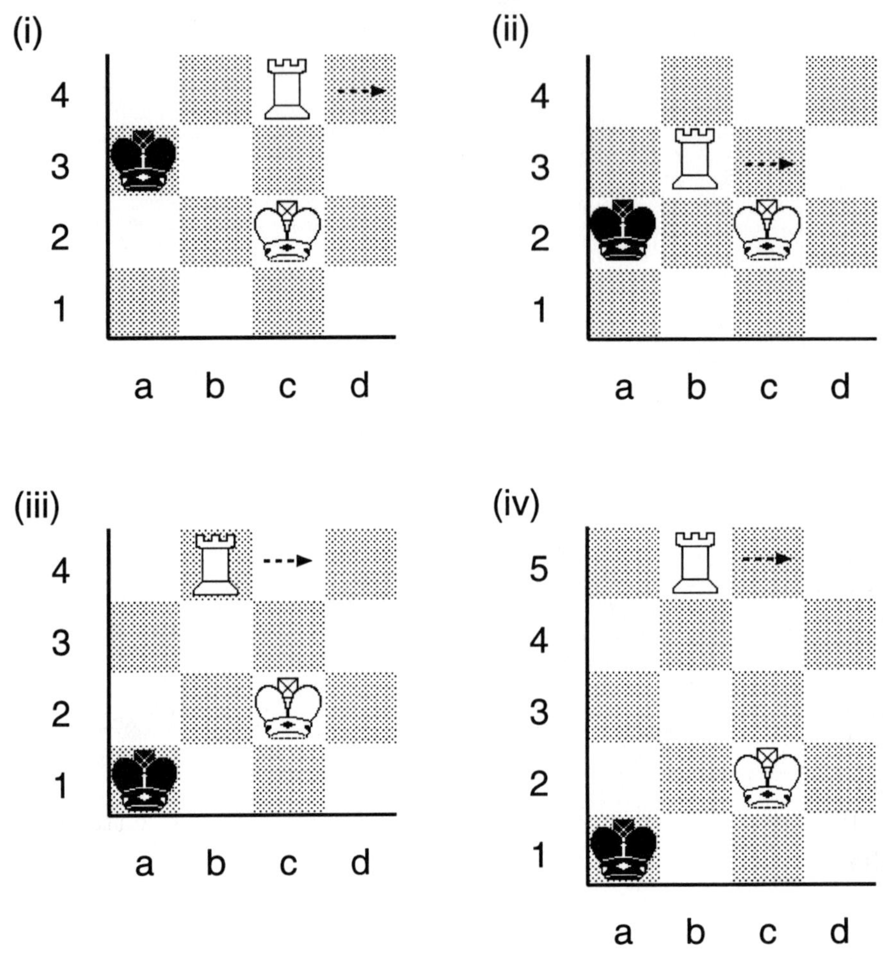

Figure 12.8. BTM WFW depth 1, clause 2.

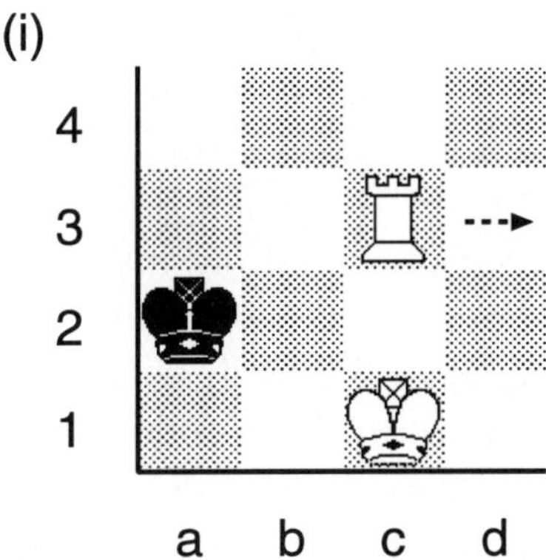

Figure 12.9. BTM WFW depth 1, clause 3.

in this definition. Clause 1 of this definition is illustrated in Figure 12.7. There are four diagrams in this figure, each of which show essentially the same pattern of attacking relationships between the pieces. Diagrams (i) – (iv) depict the patterns for the four clauses of the definition of the invented predicate **krk1/4**. This predicate might be labelled **wrrank_safe**, since it is principally a condition on the White Rook's rank safety. This is because the key attack relationships are defined by the top-level of clause 1. To see that the patterns do indeed cover checkmate at depth 1, refer to diagram (i) in Figure 12.7. The Black King is forced to move to c1, which is followed by the White Rook moving to a1. Black is in check with no further moves. Recall that this checkmate position is covered by the pattern of diagram (i) in Figure 12.4.

The second clause of the BTM WFW depth 1 definition has only the machine-invented predicate **krk2/3** in the clause body. The clause has the White King fixed on square c2, and a condition is required on the relation of the White Rook to the Black King. We might label this condition **wrfile_safe**, since the White Rook is restricting the movement of the Black King without directly attacking it or being attacked by it. The key patterns are shown in Figure 12.8. In diagram (i) the locations of the Kings are fixed with the White Rook free to occupy rank 4 on any file apart from a or b. This pattern corresponds to the first clause in the definition of **krk2/3**. Diagram (ii) presents an interesting configuration with the Kings in opposition. This pattern corresponds to the second clause of **krk2/3**, which covers one position where the White Rook is attacked by the Black King but defended by the White King. The remaining clauses in the definition cover positions falling into pattern classes closely related to that of diagram (i). For instance diagrams (iii) and (iv) correspond to clauses 3 and 4, with the Black King forced to move into opposition on file a followed by the White Rook moving to give checkmate on this file.

The third clause in the definition of the predicate **krk/7** covers positions fitting the pattern of Figure 12.9. Here the positions of both Kings are fixed, and the Black King is forced to move into square a1. From its safe position on rank 3 at file c or greater, the White Rook moves to a3 giving checkmate. The patterns shown account for all 78 of the canonical BTM positions which are WFW at depth 1.

5 DISCUSSION

In Section 3 we stated that a long-term goal of this work was to induce programs from database examples such that on some measure the programs were more concise than the examples. Although some work has been done on this front (Muggleton *et al.* 1992) we do not present any results in this chapter. However, the Prolog programs [3] from Section 4 are more compact on an informal lines of code criterion than their representation as ground examples. More importantly, these rules together with their accompanying chessboard diagrams have been verified as meaningful by chess expert and Prolog specialist Ivan Bratko.

Work on the inductive synthesis of knowledge employing the easy inverse trick (Michie 1986) pushed the nascent technology of decision-tree induction to its limits in the late 1970s and early 1980s (Quinlan 1983). These initial results in chess endgame domains led to a variety of successful applications. The landmark KARDIO system used a deep model of the heart to synthesize shallow rules for ECG interpretation (Bratko *et*

[3] Not including background predicates.

al. 1989). Among the rules which came out of this study were some previously undiscovered in over 200 years of cardiology. This route was also taken in an application to satellite fault diagnosis (Pearce 1988). Most recently, an ILP approach in the same domain allowed the learning of significant temporal relations not expressed in the earlier solution (Feng 1992).

The rôle of Machine Learning in previous work has focused on database compression. Even with decision tree induction this compression can be significant, as in the KARDIO work. Typically, however, in the chess endgame applications to date the bottleneck for decision-tree induction has been selection of an adequate set of attributes. For example, in experiments on the KPa7KR domain (Shapiro and Michie 1986) most of the effort was expended on hand-crafting the attributes which capture the necessary relational features of the won/not won predicate. The novelty of the present approach lies in the application of relational learning using RLGG as implemented in GOLEM coupled with specialization techniques based on predicate invention. This follows from earlier results with a similar approach in the simpler KRK illegality domain (Bain 1991).

The predicate invention methods by which induced clauses are specialized by introducing literals into the clause body are a special case of predicate invention within Muggleton's (Muggleton 1993) refinement lattice.

6 CONCLUDING REMARKS

We have presented results from learning optimal strategies in the KRK endgame. Optimality is achieved through the use of examples extracted from an exhaustive database for the endgame. A complete theory for black-to-move, won-for-white positions at all depths (0 to 16) has been learned automatically from ground instances of positions and only low-level background knowledge. The definitions for win at depths 0 and 1 have been fully specialized and were presented as a complete and correct definition of the target concept at these depths. They have been tested fully on the exhaustive example set and certificated by a human expert.

Acknowledgments

This work was supported by IED project 4/1/1320 on Temporal Databases and Planning and by the ESPRIT Basic Research Action ILP 6020. We thank Prof Donald Michie, Prof Ivan Bratko and the members of the Turing Institute ILP Group for their comments and suggestions regarding this work.

REFERENCES

Bain, M. (1991). Experiments in non-monotonic learning. In Birnbaum, L. and Collins, G., editors, *Proceedings of the Eighth International Workshop on Machine Learning*, pages 380–384, Morgan Kaufmann, San Mateo, CA.

Bain, M. and Muggleton, S. (1991). Non-monotonic learning. In Hayes, J. E., Michie, D., and Tyugu, E., editors, *Machine Intelligence, 12*, pages 105–119. Oxford University Press, Oxford.

Bratko, I., Mozetic, I., and Lavrac, N. (1989). *KARDIO: A study in deep and qualitative knowledge for expert systems.* MIT Press, Cambridge.

Feng, C. (1992). Inducing temporal fault diagnostic rules from a qualitative model. In Muggleton, S., editor, *Inductive Logic Programming*, pages 473–493, Academic Press, London.

Michie, D. (1986). Towards a knowledge accelerator. In Beal, D. F., editor, *Advances in Computer Chess*, volume 4, pages 1–8. Pergamon Press, Oxford.

Muggleton, S. (1993). Predicate invention and utility. *Journal of Experimental and Theoretical Artificial Intelligence* (to appear).

Muggleton, S., Srinivasan, A., and Bain, M. (1992). Compression, Significance and Accuracy. In Sleeman, D. and Edwards, P., editors, *ML92: Proceedings of the Ninth International Conference on Machine Learning*, pages 338–347, Morgan Kaufmann, San Mateo, CA.

Muggleton, S. (1987). An oracle-based approach to constructive induction. In *IJCAI-87*, pages 287–292, Morgan Kaufmann, Los Altos, CA.

Muggleton, S. and Feng, C. (1992). Efficient induction of logic programs. In Muggleton, S., editor, *Inductive Logic Programming*, pages 281–298, Academic Press, London.

Muggleton, S., Bain, M., Hayes-Michie, J., and Michie, D. (1989). An experimental comparison of human and machine learning formalisms. In Segre, A., editor, *Proceedings of the Sixth International Workshop on Machine Learning*, pages 113–118, Morgan Kaufmann, Los Altos, CA.

Pearce, D. (1988). The induction of fault diagnosis systems from qualitative models. In *AAAI-88: Proceedings of the 7th National Conference on Artificial Intelligence*, pages 353–357, Morgan Kaufmann, San Mateo, CA.

Quinlan, J. R. (1983). Learning efficient classification procedures and their application to chess end games. In Michalski, R., Carbonnel, J., and Mitchell, T., editors, *Machine Learning: An artificial intelligence approach*, pages 464–482. Tioga, Palo Alto, CA.

Shapiro, A. D. (1987). *Structured Induction in Expert Systems.* Turing Institute Press with Addison Wesley, Wokingham, UK.

Shapiro, A. D. and Michie, D. (1986). A self-commenting facility for inductively synthesised endgame expertise. In Beal, D., editor, *Advances in Computer Chess*, volume 4, pages 147–165. Pergamon, Oxford.

Thompson, K. (1986). Retrograde Analysis of Certain Endgames. *International Computer Chess Association Journal*, 8(3):131–139.

13

A Comparative Study of Classification Algorithms: Statistical, Machine Learning and Neural Network

R. D. King

R. Henery

Department of Statistics and Modelling Science,

University of Strathclyde, Glasgow.

C. Feng

Turing Institute, Glasgow.

A. Sutherland

Department of Statistics and Modelling Science,

University of Strathclyde, Glasgow.

Abstract

The aim of the StatLog project is to compare the performance of statistical, machine learning, and neural network algorithms, on large real world problems. This paper describes the completed work on classification in the StatLog project. Classification is here defined to be the problem, given a set of multivariate data with assigned classes, of estimating the probability from a set of attributes describing a new example sampled from the same source that it has a pre-defined class. We gathered together a representative collection of algorithms from statistics (Naive Bayes, K-nearest Neighbour, Kernel density, Linear discriminant, Quadratic discriminant, Logistic regression, Projection pursuit, Bayesian networks), machine learning (CART, C4.5, NewID, AC2, CAL5, CN2, ITrule – only propositional symbolic algorithms were considered), and neural networks (Backpropagation, Radial basis functions, Kohonen). We then applied these algorithms to eight large real world classification problems: four from image analysis, two from medicine, and one each from engineering and finance. Our results are still provisional, but we can draw a number of tentative conclusions about the applicability of particular algorithms to particular database types. For example: we found that K-nearest Neighbour can perform well on complex image analysis problems if the attributes are properly scaled, but it is very slow; machine learning algorithms are very fast and robust to non-Normal features of databases, but may be out-performed if particular distribution assumptions hold. We additionally found that many classification algorithms need to be extended to deal better with cost functions (problems where the classes have an ordered relationship are a special case of this).

1 INTRODUCTION

StatLog is an ESPRIT project with ten academic and industrial partners (Appendix A). Its aim is to evaluate the performance of Statistical, Machine Learning, and Neural Network Algorithms on large-scale, complex commercial and industrial problems. The problems are in the areas of classification, forecasting, control, and unsupervized learning. The objectives of the project are threefold:

1. to provide critical performance measurements, and criteria for measurement on available Learning Algorithms which improve confidence for full exploitation;
2. to indicate the nature and scope of the next-stage development which particular algorithms require to meet commercial performance expectations;
3. to indicate the most promising avenues of development for the commercially immature approaches.

This chapter describes the completed work on classification in the StatLog project. Classification is here defined to be the problem, given a set of multivariate data with assigned classes, of estimating the probability from a set of attributes describing a new example sampled from the same source that it has a pre-defined class (this problem is often known as discrimination in statistics, and supervised learning in machine learning - it fused with clustering which is also sometimes termed classification). We gathered together a representative collection of algorithms from statistics (Naive Bayes, K-nearest Neighbour, Kernel density, Linear discriminant, Quadratic discriminant, Logistic regression, Projection pursuit, Bayesian networks), machine learning (CART, C4.5, NewID, AC2, CAL5, CN2, Itrule – only propositional symbolic algorithms were considered), and neural networks (Backpropagation, Radial basis functions, Kohonen). We then applied these algorithms to eight large real world classification problems: four from image analysis, two from medicine, and one each from engineering and finance. This basic methodology can be thought of as a table: with the algorithms along one axis, the datasets along the other, and a performance measure at each position in the matrix. The objective measures of performance we use include processing time (for training and test data), and error rate (or cost if there is a cost function available). Subjective measures are much more difficult to use and we have done relatively little work on them, but they include: understandability of the decision rule, ease of use of the algorithm (particularly as perceived by a naive user), and robustness to required parameter input. Some results have already been presented by Henery and Taylor 1992, Sutherland *et al.* 1992. Tables of results obtained are at the end of this chapter.

1.1 Previous Comparative Studies

Several authors have recently compared the performance of neural algorithms to other machine learning methods such as ID3 (Quinlan 1986). Some of the tests indicated that neural algorithms worked better than other methods. Other tests have shown that neural algorithms performed worse.

Particular methods may do well in some domains, but not in others. For example, k-nearest neighbour methods usually do fairly well in handwritten character recognition, although backpropagation and/or radial basis function methods may be preferred for reasons of speed and memory.

Fisher and McKusick (1989), for example, compared ID3 with backpropagation (a neural net method) on two natural domains and found that backpropagation was a few percentage points more accurate. Shavlik *et al.* (1991) compared ID3 with backpropagation on five natural domains. The performance of both systems was rather similar, but on some datasets backpropagation worked better.

Weiss, Galen, and Tadepalli (1987) compared the PVM algorithm (that produces classification rules) with backpropagation. In this series of tests, the symbolic method performed better in three out of four domains. Weiss and Kapouleas (1989) showed that the CART system of Breiman *et al.* (1984) (similar to ID3) usually worked better than backpropagation. These comparative studies are further described in Weiss and Kulikowski (1991).

Quinlan (1990) compared the neural network approach as reported in Hinton with FOIL, a system that is capable of learning relational descriptions. Quinlan showed that FOIL can learn the given task as well as Hinton's neural network.

Difficulties in interpreting recent comparative studies arise from a number of problems, as the following extract from the StatLog Technical Annex makes clear:

- They have not always compared like with like; some methods are based on assumptions about the domain that can give the method an unfair advantage.
- Some learning methods are not complete, and require parameters to be 'tweaked' to tune the system to a particular domain. Sammut (1988) also reported that this tweaking was considerably important to achieve reasonable performance with some algorithms. Parameter tweaking, common, for instance, with neural net software, needs to be taken into account in the comparative evaluation.
- Some comparative studies use variant but not identical data sets and algorithms. A related problem is that some researchers preprocess their data sets in a manner that prevents direct comparison of results, for instance by treating unknown or unspecified values in some manner, or by partitioning a real-valued attribute into discrete attributes.
- When comparing different learning methods, the studies need to be carried out by experts both in the differ-

ent methods and in the problems tackled. Three studies were presented at the International Joint Conference on A.I. '89 that compared pattern recognition, neural networks and AI machine learning techniques (Fisher and McKusick 1989, Mooney *et al.* 1989, Weiss *et al.* 1989). Questions raised at this conference indicated that there was widespread concern that these comparisons were unfair. Two of the studies used decision-tree induction methods that were discarded by the applied statistics community in the early seventies. All three used a neural net method (back-propagation) that is several years out of date in a rapidly developing field. One study used a Bayesian classifier that would not have been applied by a trained Bayesian to the problems concerned. Because relative novices may well misuse techniques or apply outdated techniques, it is important that comparative trials be advised by experts in the areas concerned.

- Very frequently, studies use simulated data. These have the advantage of investigating the behaviour of algorithms under known conditions. For example, Cherkaoui and Cléroux (1991) investigated the performance of six procedures applied to data with a mixture of binary, nominal, ordinal, and continuous attributes.

2 TESTING METHODOLOGY

To ensure the fairness of comparison, a number of measures were taken in StatLog to reduce bias towards one category of algorithm or another.

- Firstly, all the data sets were collected at one centre for pre-processing. Each data set was issued to all testing sites at the same time with the same format and pre-processing. This was designed to eliminate the possible pre-knowledge the users have about the data sets.
- Part of the data was kept back from the testing sites in case the results were disputed. Missing values were replaced by a constant method.

- It was aimed for all algorithms to be tested by experts. For example, many of the statistical algorithms were tested at the Department of Statistics in Strathclyde University, and many machine learning algorithms were tested in The Turing Institute Ltd. Many results were validated by another naive partner. The validating sites were supplied with and used the *log* files from the testing sites, which kept records of the procedure followed.

3 STATLOG ALGORITHMS

This section describes a collection of algorithms that reflect the state of the art in statistical and logical learning. In the context of classification problems, they divide into three broad groups:

3.1 Statistical Algorithms

3.1.1 *Naive Bayes algorithm (Bayes)*

This algorithm directly applies the Bayes rule: $P(c|e) = P(e|c) \cdot P(c)/P(e)$, where c is the class and e is a given example. The aim is to obtain the most probable class given the data. This is c_i, if $P(e|c_i)P(c_i) > P(e|c_j)P(c_j)$ for all j $(i \neq j)$. Because the Bayes method requires complete and accurate probability data, for real problems it is not directly applicable. Some simplifying assumptions are made, i.e. all attributes are independent conditional on the classes, which entails $P(e|c) = P(a_1|c) \cdot P(a_2|c) \cdot ... \cdot P(a_n|c)$, where a_i $(i = 1, ..., n)$ are the n attributes. Despite the unrealistic nature of this assumption, it is found to perform well in many simple tasks in practice. In principle, it is possible to include prior information into the Bayesian analysis, but this is rarely done in reality. Naive Bayes is very simple to apply, it can cope with missing values and often it can produce reasonable results even when its assumptions are violated.

3.1.2 *K-nearest neighbour (K-N-N)*

This is a very simple algorithm, it assigns each new object to the class of the majority of its k nearest neighbours in attribute space (normally Euclidean).

3.1.3 *Kernel density estimation (ALLOC80)*

The program performs multigroup discriminant analysis; within each group the variability is modelled using a non-parametric density estimator, based on kernel functions. Suppose that we have to estimate the p–dimensional density function $f(x)$ of an unknown distribution. Information about f is given by n independent observations from this distribution, i.e. $Y_i = (Y_{i1}, \ldots, Y_{im}, \ldots, Y_{ip})$ with $i = 1, 2, \ldots, n$. Let $K^{(p)}(X; Y_i, \lambda)$ be a kernel function centred at Y_i and let λ denote the window width of the kernel. The estimate of $f(x)$ is given by

$$\hat{f}(x) = \frac{1}{n} \sum_{i=1}^{n} K^{(p)}(X; Y_i, \lambda)$$

Observations in the test data are then allocated to classes based on a calculation of the posterior odds by a standard Bayesian calculation. The smoothness of the kernel density estimate is determined in a data-based manner by a pseudo maximum likelihood method. The program can handle continuous as well as mixed (discrete) data. This program is computationally expensive, both in storage and CPU terms. The methods to choose the smoothing parameter automatically are not always successful, and the version in StatLog used is rather unwieldy . However, it has performed consistently well in the trials. It is expected to do better than standard methods where the data are highly non-Normal. For more bizarre datasets, such as two interlocking spirals, this method performs very well. In general, any situation where the boundaries between classes is not easily modelled by a straight line or quadratic may lend themselves well to this approach. The main difficulty, as with most nonparametric density estimators, is to ensure a good choice of smoothing parameter.

3.1.4 *Linear discriminants (Discrim)*

This is an implementation in Splus of Fisher's linear discriminant analysis (1936). The algorithm calculates a linear combination of the attribute values for each class and assigns a new observation to the class with the largest value. The algorithm is optimal when the data are multivariate normal with a common covariance matrix. The boundaries between classes are hyperplanes in attribute space. This algorithm like Quadra and LogReg were implemented in Splus/Fortran by the Department of Statistics and Modelling Science, University of Strathclyde, Glasgow, Scotland.

3.1.5 *Quadratic discriminants (Quadra)*

This a variant of the above linear algorithm for the unequal covariance case. The algorithm calculates a quadratic combination of the attribute values for each class and assigns a new observation to the class with the largest value. The boundaries between classes are conic sections in attribute space. The algorithm requires to estimate many more co-efficients than linear discriminants and so will only perform well when the training set is sufficiently large (and close to multivariate normal).

3.1.6 *Logistic regression (LogReg)*

Discriminant algorithms above cannot cope with combinations of continuous, categorical and qualitative attributes. So models of generalised linear models (GLIM) are proposed, which includes the logistic class (Cox 1966, Day and Kerridge 1967). In the logistic class of generalized linear models, the probability of an example e given the class of $c_e = k$ $(k = 1, ..., m)$ can be calculated from the relative probability $R(e|c_e = k)$ to a fixed class (the last one, say) which is a logistic function of the parametric linear combination of attributes:

$$P(e|c_e = k) = \frac{R(e|c_e = k)}{\sum\limits_{i=1}^{m} R(e|c_e = i)},$$

$$R(e|c_e = k) = \frac{P(e|c_e = k)}{P(e|c_e = m)} = e^{-(\alpha_{k1} a_1^e + ... + \alpha_{kn} a_n^e)},$$

where a_i^e $(i = 1, ..., n)$ are the attribute values of the example e and $R(e|c_e = m) = 1$. The coefficients α_{kj} $(k = 1, ..., m, j = 1, ..., n)$ are estimated for each class and it must maximize the total likelihood:

$p(\{e|e \text{ is in the sample}\})$. Assume that the sample is randomly chosen so the examples are independent:

$$p(\{e|e \text{ in sample}\}) = \prod_{\{e|c_e=1\}} p(e|c_e = 1) \prod_{\{e|c_e=2\}} p(e|c_e = 2) \ldots$$

$p(e|c_e = k)$ has close links with the binomial distribution, thus its arguments can be categorical.

The logistic regression method also produces a linear separation of classes although it appears to be different from other linear discriminants. It is identical, in theory, to Discrim for many restricted (e.g. normal or binomial) distributions with equal covariances. In fact, logistic regression may start the search for actual coefficients from the coefficients estimated by Discrim. So, the only differences between the two are in the way that the coefficients $\alpha_{kj}(k = 1, ..., m, j = 1, ..., n)$ (the parameters for the separation hyperplanes) are estimated. Fisher's linear discriminants optimize a quadratic cost function whereas logistic regression optimises on the total likelihood. There may well be occasions when a quadratic cost function is appropriate, in which case Fisher's linear discriminants are justified without appealing to the assumption of multivariate normality.

The logistic model can also be extended to include prior probability, but the number of parameters required in the function grows exponentially as the complexity of the model increases. In the training phase it is considerably more expensive computationally than linear discriminants, although in the testing phase the two methods are indistinguishable. While the model underlying logistic regression is more general, its success depends more critically on the correctness of the underlying assumptions.

3.1.7 *Projection pursuit (SMART)*

SMART (Smooth Multiple Additive Regression Technique) is a collection of FORTRAN subroutines written by Friedman. It is a generalization of projection pursuit regression PPR (Friedman and Stutzle 1981). The regression models take the form

$$E[Y_i|x_1, x_2, \ldots, x_p] = \overline{Y}_i + \sum_{m=1}^{M} \beta_{im} f_m(\sum_{j=1}^{p} \alpha_{jm} x_j)$$

with $\overline{Y}_i = EY_i$, $Ef_m = 0$, $Ef_m^2 = 1$ and $\sum_{j=1}^{p} \alpha_{jm}^2 = 1$. The coefficients β_{im}, α_{jm} and the functions f_m are parameters of the model and are estimated by least squares. The criterion

$$L_2 = \sum_{i=1}^{q} E[Y_i - \overline{Y}_i - \sum_{m=1}^{M} \beta_{im} f_m(\alpha_m^T x)]^2$$

is minimised wrt to the parameters β_{im}, $\alpha_m^T = (\alpha_{1m}, \ldots, \alpha_{pm})$ and the functions f_m.

Classification is closely related. The objective here is to minimise the misclassification risk

$$R = E[\min_{1 \leq j \leq q} \sum_{i=1}^{q} l_{ij} p(i|x_1, x_2, \ldots, x_p)]$$

where l_{ij} is the user specified loss for predicting $Y = c_j$ when its true value is c_i ($l_{ii} = 0$). The conditional probability is reformulated using a conditional expectation which is then modelled by the regression model above.

This algorithm would be expected to perform very well whenever a cost matrix is applicable, because it (unusually) uses the cost matrix in the **training** phase as well as in the classification stage. Although the training time is not competitive, the algorithm does generally produce good misclassification rates.

3.1.8 *Bayesian networks (CASTLE)*

CASTLE (CAusal STructures From Inductive Learning (Acid *et al.* 1991) is an implementation of the polytree algorithm defined by Pearl (1988). Causal networks are directed acyclic graphs (DAGs) in which the nodes represent propositions (or variables), the arcs signify the existence of direct causal dependencies between the linked propositions, and the strengths of these dependencies are quantified by conditional probabilities.

The structure of a causal network can be determined in the following way: each variable in the domain is identified with a node in the graph. We then draw arrows to each node X_i from a set of nodes $C(X_i)$ considered as direct causes of X_i. The strengths of these direct influences are quantified by assigning to each variable X_i a matrix $P(X_i|C(X_i))$ of conditional probabilities of the events $X_i = x_i$ given any combination of values of the parent set $C(X_i)$. The conjunction of these local probabilities defines a consistent global model, i.e., a joint probability distribution. Once the network is constructed it constitutes an efficient device to perform probabilistic inferences. The problem of building such a network remains. The structure and conditional probabilities necessary for characterizing the network could be provided either externally by experts or from direct empirical observations.

Under the Bayesian approach, the learning task in causal networks separates into two highly related subtasks, *structure learning*, that is, to identify the topology of the network, and *parameter learning*, the numerical parameters (conditional probabilities) for a given network topology.

CASTLE, currently being developed by members of the Department of Computer Science and Artificial Intelligence at the University of Granada, focuses on learning a particular kind of causal structure: polytrees (singly connected networks), networks where no more than one path exists between any two nodes. As a consequence, a polytree with n nodes has no more than $n-1$ links. It is in polytrees (and specially in trees) where the ability of networks to decompose and modularize the knowledge attains its ultimate realization. Polytrees do not contain loops, that is, undirected cycles in the underlying network (the network without the arrows or skeleton), and this fact allows a locally efficient propagation procedure (Pearl 1988).

3.2 Machine Learning

From the field of machine learning only propositional symbolic algorithms were considered by StatLog. No algorithms from the emerging field of Inductive Logic Programming (ILP) were included, nor were Genetic algorithms examined.

3.2.1 *CART*

CART, *Classification and Regression Tree*, is a binary decision tree algorithm. The acronym CART comes from *Classification And Regression Tree* (Breiman *et al.* 1984). CART is not really a single algorithm, but a collection of algorithms and analysis methods for classification and regression trees (or Discrimination And Forecasting Trees). The algorithm described in this section is the most commonly used version. CART is a binary decision tree algorithm A binary decision tree consists of nodes and each node has two branches. There is a single test (or decision) on each node, splitting the node into two subtrees. Depending on whether the result of a test is true or false, the tree will branch to left or right, at which this splitting process recursively continues. At each leaf node a decision is made on the class assignment.

The advantages of using a decision tree methodology are that it is non-parametric and produces classifications which can be easily understood. This latter feature has made them popular in machine learning.

The fundamental idea in CART's tree construction is to select each split so that the data in each of the descendant subsets are 'purer' than the data in the parent subset. The splitting evaluation function developed for CART is different from that used in the ID3 family of decision tree algorithms. Consider the case of a problem with two classes, and a node has 50 examples from each class, the node has maximum impurity. If a split could be found that divides the data into one subgroup of 40:5 and another of 10:45, then intuitively the impurity has been reduced. The impurity would be completely removed if a split could be found that produced sub-groups 50:0 and 0:50. In CART this intuitive idea of impurity is formalized in the *GINI* index for the current node c:

$$Gini(c) = 1 - \sum_j p_j^2$$

where p_j is the probability of class j in c. For each possible split the impurity of the subgroups is summed and the split with the minimum impurity chosen.

The GINI criteria is local: there is no guarantee that the overall tree is optimal. As a result, decision trees produced can potentially contain many nodes. This may result in the number of examples available to test on becoming very small for some branches. The original CART authors explored related impurity measures such as entropy.

In CART each split depends only on a single attribute. For ordered and numeric attributes, CART considers all possible splits in the sequence. For n values of the attribute, there are n splits. For categorical attributes CART examines all possible binary splits, which is similar to the attribute subsetting method used for C4.5. At each node CART searches through the attributes one by one. For each attribute it finds the best split. Then it compares the best single splits and selects the best attribute of the best splits.

CART uses a sophisticated form of pruning to try and avoid over-fitting the data. If no pruning is used then the trees generated by CART will be too specialized and biased towards the training data. CART uses cost-complexity pruning to decide the order of branches to prune, and cross-validation to decide on the pruning parameters.

Two versions of CART were investigated: the commercial version of CART (used by the University of Granada), and IN-DCART (a free version of CART supplied by Wray Buntine at NASA Ames, who also supplied the Naive Bayes program). IN-DCART differs from the standard version of CART (Breiman *et al.* 1984) by using a different (probably better) way of handling missing values, in a different default setting for pruning, not fully using costs, and in not implementing the regression part of CART.

3.2.2 *NewID*

NewID from the Turing Institute Ltd in Glasgow, Scotland, is a direct descendant of the decision tree algorithm ID3 (Quinlan 1986). It differs from the original ID3 in being designed to cope with continuous variables and noise. Its basic structure is very similar to that of CART. It differs from CART in using a different splitting criterion and a different pruning method. NewID uses entropy gain instead of GINI. The entropy of the example set at a node is $-\sum_j p_j \log p_j$, where p_j is the probability estimate of the jth class in that set. The entropy *gain* is the difference of entropy between the current set and the subsets created by the split. The attribute with the highest gain is selected, giving the most informative split at that node.

3.2.3 *C4.5*

C4.5 (Quinlan 1987) is also a direct descendant of the decision tree algorithm ID3 (Quinlan 1986). It also is designed to cope with continous variables and noise. It differs from NewID in using a slightly different splitting criterion and pruning method. C4.5 uses the entropy gain ratio as a splitting criteria. This takes the information provided by the attribute a into account. The attribute to split on should maximise the information gain relative to the information needed to determine the attribute value, i.e. the ratio:

$$gain\_ratio(c,a) = \frac{gain(c,a)}{I(a)} = \frac{I(c \wedge a) - I(c)}{I(a)}.$$

3.2.4 *AC2*

AC2 from Isoft in Paris is an extension of ID3 to deal with hierarchical data. It can learn structures from a predefined hierarchy of attributes. A hierarchy may be imposed to create trees that are more meaningful to the end-user: there is the additional advantage of savings in learning and testing time since certain tests may be ruled out by the hierarchical structure.

3.2.5 *Cal5*

Cal5 is a decision tree algorithm based on statistical methods and designed for continuous variables. It is contributed by the Institute of Automation, Berlin (Unger and Wysotzki 1981). Interestingly, it was developed by the Institute of Automation in East Berlin independently of the work on decision trees in the West. In Cal5 trees are constructed sequentially starting with one attribute and branching with other attributes recursively, if no sufficient discrimination of classes can be achieved. That is, if at a node no decision for a class c_i according to the above formula can be made, a branch formed with a new attribute is appended. If this attribute is continuous, a discretization, i.e. intervals corresponding to qualitative values, has to be used.

Let N be a certain non-leaf node in the tree construction process. At first the attribute with the best local discrimination measure at this node has to be determined. For that a method working without any knowledge about the result of the desired discretization is used. For continuous attributes the quotient

$$quotient(N) = \frac{A^2}{A^2 + D^2}$$

is a discrimination measure for a single attribute, where A is the standard deviation of examples in N from the centroid of the attribute value and D is the mean value of the square of distances between the classes. This measure has to be computed for each attribute. The attribute with the least value of $quotient(N)$ is chosen as the best one for splitting at this node. Note that at each current node N all available attributes $a_1, a_2, \ldots, a_n$ will be considered again. If a_i is selected and occurs already in the path to N, than the discretization procedure leads to a refinement of an already existing interval.

3.2.6 *CN2*

CN2 is a decision rule algorithm (Clark and Niblett 1988, Clark and Boswell 1991). It learns decision rules for each class in turn. Initially it starts with a 'universal rule': '**If** all conditions **Then** Current class'. This rule ought to cover at least one of the examples in the current class. Specializations of this rule are then repeatedly generated and explored until a rule has been found. This rule ought to cover examples of the 'right' mixture belonging to the current class and other classes. The mixture is usually determined by heuristics or is user-specified. Intuitively, as few as possible *negative* examples, i.e. examples in other classes, should be covered. Each specialization is obtained by adding a condition to the left-hand side of the rule, i.e. requiring one particular attribute to have value within a range. Similar to decision tree algorithms such algorithms are better suited to deal with logical attributes and classes.

CN2 is an extension of an earlier algorithm AQ (Michalski 1983) that can deal with noise in data. It can also accept continuous numeric values in attributes though not classes. The main technique for reducing error is use of Laplace's Law of Succession. If there are $m1, m2, ..., mk$ examples of classes $1, 2, .., k$, where the total no. of examples is n, then the probability that a new data item will fall into class i is:

$$(m_i + 1)/(n + k).$$

3.2.7 *ITrule*

ITrule (Goodman and Smyth 1989) from CalTech University produces rules of the form '**If** ... **Then** ... **with probability** ...'. This algorithm contains probabilistic inference in its evaluation function of rule candidates through the J-measure and varies from AQ and CN2 in its method of constructing rules by incorporating generalization as well (i.e. dropping conditions). The J measure is a product of conditional prior probabilities and the cross-entropy of class values conditional on the attributes values. ITrule cannot deal with continuous numeric values.

3.3 Neural Networks

3.3.1 *Back-propagation Multi-Layer Perceptron (Backprop)*

Back-propagation Multi-Layer Perceptron is a neural network algorithm (McClelland *et al.* 1986) which consists of a network of 'neurons' arranged in a number of layers, where each neuron is connected to every neuron in the adjacent layers. Each neuron sends a signal along the connections to the neurons in the layer above. The signals are multiplied by weights corresponding to each connection.

We used a version with three layers: an input layer, a hidden layer and an output layer. *strictly layered, 3-layer MLP*, which is the mapping

$$\begin{aligned} y_i^{(H)} &= f^{(H)}\left(\sum_j w_{ij}^{(HI)} y_j^{(I)}\right) \\ y_i^{(T)} &= f^{(T)}\left(\sum_j w_{ij}^{(TH)} y_j^{(H)}\right) \end{aligned}$$

from the inputs $y^{(I)}$ to the targets $y^{(T)}$, via the *hidden nodes* $y^{(H)}$. The parameters are the *weights* $w^{(HI)}$ and $w^{(TH)}$. The univariate functions $f^{(\cdot)}$ are usually each set to be logistic which varies smoothly from 0 at $-\infty$ to 1 at ∞.

This is also called a '2-layer' MLP by authors who prefer to count layers of weights rather than layers of nodes. It can be shown that, given enough hidden nodes, this can approximate any mapping to an arbitrary accuracy.

This mapping has a biological interpretation: Node (model neuron) i produces a strong output if its *activation potential* $\sum_j w_{ij} y_j$ is positive, and a weak output if it is negative. The activation is increased if node i is connected to an active node j via an *excitatory* synapse ($w_{ij} > 0$), and is decreased by active nodes connected via *inhibitory* synapses ($w_{ij} < 0$).

One input to each layer, say node 0, is traditionally assigned the constant value 1.0 so that w_{i0} provides a constant offset or *bias* for the activation of i. This device allows the *threshold* activation (due to the remaining nodes), at which $y_i = 0.5$, to be adjusted away from 0.

The code was written by R. Rohwer whilst at the Centre for Speech Technology Research, Edinburgh, Scotland.

3.3.2 *Radial basis functions (Radial)*

Radial basis function methods are closely related to the kernel estimator methods discussed under nonparametric discriminant analysis Poggio and Girosi, 1990). The radial basis function network mapping is given by

$$
\begin{aligned}
y_i^{(H)} &= f^{(H)}\left(\frac{\sqrt{\sum_j (y_j^{(I)} - c_{ij})^2}}{r_i}\right) \\
y_i^{(T)} &= \sum_j w_{ij}^{(TH)} y_j^{(H)}
\end{aligned}
$$

It has a linear output layer like an MLP with such an option, but the hidden layer is different. Hidden node i computes a function of the Euclidian distance of an input from its *centre* $c_{i\cdot}$, on a scale determined by its *radius* r_i. Usually the chosen function is the Gaussian.

As in the MLP, a bias node is introduced into the linear layer. Sometimes non-Euclidian distance measures are used. The loosely-defined region for which a radial basis function has a significant output is its *receptive field.* The functions computed at the hidden nodes are the 'radial basis functions' *per se.* They are 'radial' in that their receptive fields are spherically symmetric; there is an r_i for each centre i but not an r_{ij} for each centre and input coordinate. Such a generalization is often made, however, and if it is not, then it is desirable to prescale the input data to give it equal variance in each dimension. The code was supplied by Richard Rohwer then at the Department of Statistics and Modelling Science, Strathclyde University, Scotland.

3.3.3 *Kohonen net (Kohonen)*

Kohonen net is a self-organizing mapping algorithm (Kohonen 1989). Our implementation comes from J. Paul, Institut fuer Kybernetick und Systemtheorie, Am Hlsenbusch 54, W-4630 Bochum 1, Germany. It is capable of learning a mapping between an input space and an output space by establishing a topology-conserving map on a usually planar array of 'neuronal' nodes. A feature map is a data structure where the interrelationships of the data are captured in the spatial arrangement of the corresponding nodes. The map defines the ordering of the nodes allowing the algorithm to freely develop within this structure.

4 STATLOG DATASETS

The datasets studied by StatLog were all the large classification datasets of commercial and industrial interest that could be found.

4.1 Satellite Image (Satellite)

The database consists of the multi-spectral values of pixels in 3×3 overlapping neighbourhoods in a satellite image, and the classification associated with the central pixel in each neighbourhood. The aim is to predict this classification, given the multi-spectral values. In the sample database, the class of a pixel is coded as a number.

This sample database from LandSat Multi-Spectral Scanner image data was provided by: Ashwin Srinivasan, Department of Statistics and Modelling Science, University of Strathclyde. The original LandSat data for this database was generated from data purchased from NASA by the Australian Centre for Remote Sensing, and used for research at The Centre for Remote Sensing, University of New South Wales.

The sample database was generated taking a small section (82 rows and 100 columns) from the original data. The binary values were converted to their present ASCII form by Ashwin Srinivasan. Each line of data corresponds to a 3×3 square neighbourhood of pixels completely contained within the 82×100 sub-area. Each line contains the pixel values in the four spectral bands (converted to ASCII) of each of the nine pixels in the 3×3 neighbourhood and a number indicating the classification label of the central pixel. The number is a code for the following classes:

The data has 36 numerical attributes and six classes. The data was divided into a training set with 4435 examples and a test set with 2000 examples. A one-shot train-and-test was used to calculate the accuracy.

4.2 Handwritten Digits (Digits)

The purpose of the hand-written digit dataset is to classify 4×4 pixel images as the digits 0–9. Each member of a set of 18 000 handwritten digits was digitized onto a 16×16 pixel array with greylevels 0(white)-255(black). The pixel values were then averaged over 4×4 neighbourhoods to produce the 4×4 images. Each line of the dataset consists of the 16 pixel-values read out from left-to-right and top-to-bottom across the image, followed by the value of the digit appropriate to that image. The dataset has been divided into equal test and training sets with 9000 examples in each set. There are 900 examples of each digit in either set. The digits were gathered from postcodes on letters passing through the German Federal Post. A very small number of mistaken images have been allowed to appear, e.g. one of the ones is actually an '!' and one of the eights is actually a capital letter 'B'.

This dataset has already been studied by Kressel *et al* (1990), and Kressel (1991) of AEG, Ulm, who have published a comparative study of backpropagation and their own 'polynomial classifier'. They used the full 256 attribute version and achieved results of the order of 98% accuracy.

The data has 16 numerical attributes and 10 classes. The data was equally divided into a training set with 9000 examples and a test set with 9000 examples. A one-shot train-and-test was used to calculate the error rate.

4.3 Karhunen–Loeve Digits (KL)

The Karhunen–Loeve (kl) digits dataset is very closely related to the other handwritten digits dataset. Whereas the other dataset was produced by averaging over 4×4 neighbourhoods in the original 16×16 images, the kl dataset is produced by a linear transformation of the original 16×16 images. The eigenvectors of the covariance matrix of the original 16×16 images were computed. The scalar products of the top 40 eigenvectors with the original images were calculated. It was found that the original images could be reconstructed from these 40 eigenvectors with minimal loss of information. Therefore, the original 256 attributes had been compressed down to 40. It is these 40 scalar products which are the attributes of the kl dataset. In statistical terminology, the 256 attributes were replaced by the first 40 principal components.

The data has 40 numerical attributes and 10 classes. The data was equally divided into a training set with 9000 examples and a test set with 9000 examples. A one-shot train-and-test was used to calculate the error rate.

4.4 Vehicle Silhouettes (Vehicle)

The purpose of this dataset is to classify a given silhouette as one of four types of vehicle, using a set of features extracted from the silhouette. The vehicle may be viewed from one of many different angles.

This data was originally gathered at the Turing Institute in 1986-87 by J.P. Siebert (1987). It was partially financed by Barr and Stroud Ltd. The original purpose was to find a method of distinguishing 3D objects within a 2D image by application of an ensemble of shape feature extractors to the 2D silhouettes of the objects. Measures of shape features extracted from example silhouettes of objects to be discriminated were used to generate a classification rule tree by means of computer induction. This object recognition strategy was successfully used to discriminate between silhouettes of model cars, vans, and buses viewed from constrained elevation but all angles of rotation.

The features were extracted from the silhouettes by the HIPS (Hierarchical Image Processing System) extension BINATTS, which extracts a combination of scale independent features utilizing both classical moments based measures such as scaled variance, skewness, and kurtosis about the major/minor axes and heuristic measures such as hollows, circularity, rectangularity, and compactness. Four 'Corgi' model vehicles were used for the experiment: a double decker bus, Chevrolet van, Saab 9000, and an Opel Manta 400. This particular combination of vehicles was chosen with the expectation that the bus, van and either one of the cars would be readily distinguishable, but it would be more difficult to distinguish between the cars. The attributes were all real and the range for each attribute was different.

The data has 18 numerical attributes and four classes. There are 846 examples and nine-fold cross-validation was used to estimate the error rate.

4.5 Head Injury (Head)

The data set is a series of 1000 patients with severe head injury collected prospectively by neurosurgeons between 1968 and 1976. This head injury study was initiated in the Institute of Neurological Sciences, Glasgow. After four years two Netherlands centres (Rotterdam and Groningen) joined the study, and late data came also from Los Angeles.

The original purpose of the head injury study was to investigate the feasibility of predicting the degree of recovery which individual patients would attain, using data collected shortly after injury. Severely head injured patients require intensive and expensive treatment; even with such care almost half of them die and some survivors remain seriously disabled for life. Clinicians are concerned to recognize which patients have potential for recovery, so as to concentrate their endeavours on them. Outcome was categorized according to the Glasgow Outcome Scale, but the five categories described therein were reduced to three for the purpose of prediction. Titterington *et al* (1981) compared several discrimination procedures on this dataset. Our dataset differs by replacing all missing values with the class median. All attribute values are integers.

There are five numerical attributes and one binary (categorical), and there are three classes. There are 900 examples in the dataset. Nine-fold cross-validation was used to estimate the average cost.

This is one of the two datasets with a cost matrix. The matrix below gives the different cost of various possible misclassifications (d/v = dead or vegetative, sev = severe disability, and m/g = moderate disability or good recovery).

	d/v	*sev*	*m/g*
d/v	0	10	75
sev	10	0	90
m/g	750	100	0

4.6 Heart Disease (Heart)

This database comes from the Cleveland Clinic Foundation and was supplied by Robert Detrano, M.D., Ph.D. of the V.A. Medical Center, Long Beach, CA. It is part of the collection of databases at the University of California, Irvine collated by David Aha.

The purpose of the dataset is to predict the presence or absence of heart disease given the results of various medical tests carried out on a patient. This database contains 13 attributes, which have been extracted from a larger set of 75. The database originally contained 303 examples but six of these contained missing values and so were discarded leaving 297. 27 of these were left out as a validation set, leaving a final total of 270. There are two classes: presence and absence (of heart-disease). This is a reduction of the number of classes in the original dataset where there were four different degrees of heart-disease.

This data has been studied before, but without taking the cost matrix into account. In an unpublished study Detrano *et al.* got approximately a 77% correct classification accuracy with a logistic-regression-derived discriminant function. Aha and Kilber (1988) used instance-based prediction and got 77.0% accuracy with NTgrowth and 74.8% with C4. John Gennari got 78.9% with the CLASSIT conceptual clustering system.

There are eight numerical attributes and five binary/categorical ones, there are two classes. There are 270 examples in the dataset. Nine-fold cross-validation was used to estimate the average cost.

This is one of the two datasets with a cost matrix. The matrix below gives the different costs of various possible misclassifications. It was supplied by doctors in Leeds to Dr. C.C. Taylor, Statistics Dept, Leeds University.

	absent	*present*
absent	0	1
present	5	0

4.7 Credit Risk (Credit)

The purpose of this dataset is to evaluate various customers of the credit industry as good or bad credit risks. The dataset was supplied by Attar Software Ltd of Leigh, Lancashire. For commercial reasons the meaning of the attributes is secret. The original dataset had 39 attributes some of which were numerical and others categorical. Some of the categorical attributes had very large numbers of categories.

This original dataset presented structural problems for many of the StatLog algorithms (statistical, machine learning, and neural netwok). For example CART cannot deal with categorical attributes with large numbers of categories.

Therefore a new dataset with 16 attributes was produced by J. Mitchell of Strathclyde University. Many of the less informative attributes were thrown away, others were binarzsed and some of the numeric attributes were log-transformed to make them closer to normal. Missing values were replaced.

The ratio of good to bad customers in the dataset is almost 1:1. This is not representative of the population as a whole, where the ratio is more like 10:1. However, we assumed that the effect of this would be cancelled out by the different costs of misclassifying a good or bad customer; it is about ten times more costly to misclassify a bad customer as good compared to a good customer as bad.

The data has eight numerical and eight binary/categorical attributes, with two classes. The data was divided into a training set with 6230 examples and a test set with 2670 examples. A one-shot train-and-test was used to calculate the error rate.

4.8 Shuttle control (Shuttle)

The dataset was provided by Jason Catlett who was then at the Basser Department of Computer Science, University of Sydney, N.S.W., Australia. The data originated from NASA and concern the position of radiators within the Space Shuttle. The problem appears to be noise-free in the sense that arbitrarily small error rates are possible given sufficient data.

The data was divided into a train set and a test set with 43500 examples in the train set and 14500 in the test set. A one-shot train-and-test was used to calculate the accuracy. With samples of this size, it should be possible to obtain an accuracy of 99 – 99.9%. Approximately 80% of the data belong to class 1. At the other extreme, there are only six examples of class 6 in the learning set, so no rule could be constructed to have uniform accuracy for all classes.

The data has seven numerical attributes with seven classes. The data was divided into a training set with 43500 examples and 14500 examples test set. A one-shot train-and-test was used to calculate the error rate.

4.9 Characterization of Datasets

An important objective in StatLog is to investigate why certain algorithms do better on some datasets and other algorithms do better on different datasets Table 13.1. Appendix B describes a list of measures which will help to explain our findings. At present these measures are mostly very simple or statistically based, and none of these measures is really appropriate to decision trees as such.

There is a need for a measure which indicates when decision trees will do well. Bearing in mind the success of decision trees in the Tsetse Fly data described by Ripley (1992), it seems that some measure of multimodality might be useful in this connection.

Some algorithms have built–in measures which are given as part of the output. For example, CASTLE measures the Kullback–Leibler information in a dataset. Such measures are useful in establishing the validity of specific assumptions underlying the algorithm, and although they do not always suggest what to do if the assumptions do not hold, at least they give an indication of internal consistency.

5 RESULTS

The results fall naturally into three groups based on the databases. The first group contains the four image analysis datasets: satellite image, handwritten digits, Karhunen–Loeve digits, and vehicle recognition. The second group includes the two medical datasets (both involving cost matrices): head injury, heart disease. The last group includes the two datasets most difficult to characterize: credit risk, and shuttle.

5.1 Image Analysis

Within the group of image analysis datasets all the attributes are numerical. The attributes for satellite image, and handwritten digits come directly from the images with only minimum processing, i.e. are brightness levels. The attributes for the vehicle dataset were generated using an image analysis package.

5.1.1 *Satellite*

In the satellite dataset K-nearest Neighbour performs best, Table 13.2. Not surprisingly, radial basis functions and Alloc80 also do fairly well as these three algorithms are closely related. Their success suggests that all the attributes are equally scaled and equally important. There appears to be little to choose between any of the other algorithms, except that Naive Bayes does badly (and its close relative CASTLE does relatively badly also). This dataset has the highest correlation between attributes ($corr\_abs = 0.5977$). This may partly explain the failure of Naive Bayes (assumes attributes are conditionally independent), and CASTLE (confused if several attributes contain equal amounts of information). Note that only three linear discriminants are sufficient to separate all six class means ($fract\_3 = 0.9691$). This may be interpreted as evidence of seriation, with the three classes 'grey soil', 'damp grey soil' and 'very damp grey soil' forming a continuum. Equally, this result can be interpreted as indicating that the original four attributes may be successfully reduced to three with no loss of information. Here 'information' should be interpreted as mean square distance between classes, or equivalently, as the cross entropies of the normal distributions.

5.1.2 *Digits*

The results for the digits dataset and the KL-digits dataset are very similar so are treated together, Table 13.3 and Table 13.4. Most algorithms perform a few percent better on the KL-digits data. The Karhunen–Loeve version of digits is the closest to being normal. This could be predicted beforehand, as it is a linear transformation of the attributes that, by the Central Limit Theorem, would be closer to normal than the original. Because there are very many attributes in each linear combination, the KL-digits dataset is very close to normal ($skewness = 0.1802, kurtosis = 2.9200$) as against the exact normal values of ($skewness = 0, kurtosis = 3.0$).

In both Digits datasets dataset K-nearest Neighbour comes top and Radial basis functions and Alloc80 also do fairly well. These three algorithms are all closely related. Kohonen also does well in the Digits dataset (it has not yet been applied to KL-digits); Kohonen has some similarities with nearest neighbour type algorithms. The success of such algorithms suggests that the attributes are equally scaled and equally important. Quadratic discriminant also does well, coming second in both datasets. The KL version of digits appears to be well suited to quadratic discriminants: there is a substantial difference in variances ($SD\_ratio = 1.9657$), while at the same time the distributions are not too far from multivariate normality with kurtosis of order 3.

Backpropagation does quite well on the Digits dataset. This

might be expected from the literature (McClelland *et al.* 1986). Neural networks are widely considered to do well at character recognition.

5.1.3 *Vehicle*

The attributes for the vehicle dataset, unlike the other image analysis, were generated using image analysis tools and were not simply based on brightness levels. This suggests that the attributes are less likely to be equally scaled and equally important. This is confirmed by the lower performances of K-nearest Neighbour and Radial Basis functions Table 13.5. Alloc80 still does very well and appears to be more robust than the other two algorithms. The original Siebert (1987) paper showed machine learning performing better than K-nearest Neighbour. Quadratic discriminant does best. The high value of $fract\_2 = 0.8189$ might indicate that linear discrimination could be based on just two discriminants. This may relate to the fact that the two cars are not easily distinguishable, so might be treated as one (reducing dimensionality of the mean vectors to 3D). However, although the fraction of discriminating power for the third discriminant is low (1 minus 0.8189), it is still statistically significant, so cannot be discarded without a small loss of discrimination. Backpropagation also does well on this dataset.

5.2 Medical Datasets with Costs

Both medical datasets have cost matrices associated with them. In the results for both datasets the top ten algorithms (algorithms with the lowest costs) are all capable of utilizing costs in the testing phase. SMART performed best on the Head injury dataset, Table 13.6. It is the only algorithm that as standard can utilize costs directly in the training phase (we used in our results a modified version of Backpropagation that could utilise costs, but this is very experimental). Naive Bayes performed best on the heart dataset, Table 13.7. This may reflect the careful selection of attributes by the doctors. Logistic regression does very well and so do Linear and Quadratic discriminants.

It appears that in the head dataset a single linear discriminant is sufficient to discriminate between the classes (more precisely: a second linear discriminant does not improve discrimination). Therefore the head injury dataset is very close to linearity. This may also be observed from the value of $fract\_1 = 0.9787$, which is very close to unity. In turn, this suggests that the class values reflect some underlying continuum of severity, so this is not a true discrimination problem. Note the similarity with Fisher's original use of discrimination as a means of ordering populations. Perhaps this dataset would best be dealt with by a pure regression technique, either linear or logistic. If so, manova indicates that the middle group is slightly nearer to category 3 than 1, but not significantly nearer. It appears that there is not much difference between the covariance matrices for the three populations in the head dataset ($SD\_ratio = 1.1231$), so the procedure quadratic discrimination is not expected to do much better than linear discrimination.

In the heart dataset the leading correlation coefficient $cancor1 = 0.7384$, this is not very high (bear that in mind it is *correlation* that gives a measure of predictability). Therefore the discriminating power of the linear discriminant is only moderate.

5.3 Other Datasets

5.3.1 *Credit*

In the credit dataset most of the attributes are symbolic. Many of them are also irrelevant; it is possible to get 87% accuracy by using just one attribute (N.B. many algorithms do worse than this), Table 13.8. Most machine learning algorithms agree that this simple rule is indeed the best rule (NewID, AC2, CART, CN2), and this is also the conclusion of CASTLE. Due credit should be given to these procedures for finding a rule that is not only simple but efficient.

On the other hand, some algorithms (SMART and backprop) achieve about the same accuracy but the simple structure of the problem is completely masked. The statistical procedures (discrim, logdiscr and quadisc) also fail to find the simple rule: although there are procedures for dropping attributes that do not contribute usefully to the discrimination, these have not been implemented.

The poor performance of k-nearest neighbour on this dataset is thought to be due to the presence of symbolic attributes (which are difficult to scale): irrelevant attributes are also known to decrease its performance.

It is interesting to note that the characterization of this dataset is quite similar to that of the heart dataset. This would suggest that it would be an interesting experiment to see how well the algorithms do on the heart dataset without a cost matrix. If the dataset characterization is useful then algorithms that did well on credit should also do well on head without a cost matrix.

5.3.2 *Shuttle*

The shuttle dataset also departs widely from typical distribution assumptions. One important feature of the data is that there is very little 'noise', i.e. it is possible to get arbitrarily close to 100% accuracy Table 13.9. The attributes are numerical and appear to exhibit multimodality (we do not have a good statistical test to measure this).

In this dataset the data seem to consist of isolated islands or clusters of points, each of which is pure (belongs to only one class), with one class comprising several such islands. However, neighbouring islands may be very close and yet come from different populations. The boundaries of the islands seem to be parallel with the coordinate axes. If this picture is correct, and the present data do not contradict it, as it is possible to classify the combined dataset with 100% accuracy using a decision tree, then it is of interest to ask which of our algorithms are *guaranteed* to arrive at the correct classification given an arbitrarily large learning dataset. In the following, we ignore practical matters such as training times, storage requirements etc., and concentrate on the limiting behaviour for an infinitely large training set.

Procedures guaranteed to give the perfect rule for this dataset would seem to be: k-nearest neighbour, Bayes rule, CASTLE, backprop and Alloc80. Radial basis functions should also be capable of perfect accuracy, but some changes would be required in the particular implementation used in the project (to avoid singularities).

Decision trees will also find the perfect rule provided that the pruning parameter is properly set, but may not do so under all circumstances as it is occasionally necessary to override the splitting criterion (Gordon and Olshen 1978).

The statistical procedures Discrim, Quadisc and LogReg would not improve their accuracy much beyond that given in Table 13.9 (97.1% for Quadisc).

6 CONCLUSIONS

It is not easy to compare the performance of 19 algorithms tested on eight datasets, with three different criteria (accuracy/cost, time), especially as, at this stage in the project, some of the trials are not yet performed. However a number of tentative conclusions can be made.

The most important of these is: that there appears to be no one algorithm that is best for all types of dataset. What algorithm is best depends on features of the dataset. Much work needs to be done, both theoretical and empirical, to determine what features of datasets suit what types of algorithms.

If the distribution assumptions of Linear discriminant are met then this algorithm is provably optimal in terms of maximizing accuracy (not necesarily in terms of speed or human understandability. If the attributes are equally important and equally scaled then nearest neighbour algorithms can do very well; this appears to be true for some image analysis tasks where the attributes have not been overly transformed.

Machine Learning algorithms (symbolic propositional ones) can perform relatively well when the attributes are symbolic or far from assumptions of normality. They appear to be very robust, but may lose efficiency if certain distribution assumptions are met. There is a confusing profusion of Machine Learning algorithms, but they all seem to perform at about the same level.

With care, neural networks perform well on some problems. It is not clear how to characterize these problems. In terms of computational burden, and the level of expertise required, they are much more complex than, say, the machine learning procedures.

There were great differences in the time taken by the various algorithms. This may be important in some applications. The fastest algorithms were the simplest statistical ones Linear discriminant and Naive Bayes. The machine learning algorithms were also fast (AC2 is an exception, probably because of its complex interface and it is written in LISP). Backpropagation was very slow, but there are however some variations which make improvements to this. The nearest neighbour algorithms were extremely slow to classify new examples, however it is known (Hart 1968) that substantial time saving can be effected, at the expense of some slight loss of accuracy, by using a condensed version of the training data.

7 FURTHER WORK

Much more attention needs to be paid to the use of cost matrices in algorithms. Very few algorithms use costs in their learning phase, and many algorithms do not even use costs in their testing phase (this is easy to implement if a class probability estimate can be made). This use of costs has been virtually ignored by machine learning and neural network workers. Many problems are a combination of discrimination and regression, i.e. the classes are linearly ordered (e.g. Head injury). This problem can be considered as a special case of the use of cost matrices.

It can fairly be said that the performance of linear and quadratic discriminants was exactly as might be predicted on the basis of theory. Several practical problems remain however: (i) the problem of deleting attributes if they do not contribute usefully to the discrimination between classes; (ii) the desirability of transforming the data; and the possibility of including some quadratic terms in the linear discriminant as a compromise between pure linear and quadratic discrimination. Much work needs to be done in this area.

The performance of CASTLE should be related to how 'tree-like' the dataset is. A major criticism of CASTLE is that there is no internal measure that tells us how closely the empirical data are fitted by the chosen polytree. We recommend that any future implementation of CASTLE incorporates such a 'polytree' measure. Although it would seem that this measure could be based on the Kullback–Leibler information for the fitted polytree, it is not clear exactly how this should be done. The main reason for using CASTLE is that the polytree models the whole structure of the data, and no special role is given to the variable being predicted, viz. the class of the object. However instructive this may be, it is not the principal task in StatLog (which is to produce a classification procedure). So maybe there should be an option in CASTLE to produce a polytree which *classifies* rather than fits all the variables. To emphasize the point, it is easy to deflect the polytree algorithm by making it fit irrelevant bits of the tree (that are strongly related to each other but are irrelevant to classification).

In decision trees there are no indications in our results that any splitting criterion is best, but the case for using some kind of pruning is overwhelming, although, again, our results are too limited to say exactly how much pruning to use (it appears to be some function of the amount of noise in the data). Work needs to be done to relate the performance of a decision tree to some measures of complexity and pruning, specifically the average depth of the tree and the number of terminal nodes (leaves).

One major weakness of neural nets is the lack of diagnostic help. If something goes wrong, it is difficult to pinpoint the difficulty from the mass of inter-related weights and connectivities in the net. For example, in the prediction problems, the two neural nets performed rather badly. This is primarily due to their inability, as black boxes, to react to the several components of time series – trend, seasonality and correlated random effects. Any neural net predictor would need an architecture capable not only of incorporating these features but also of telling the operator when and how to adjust for changes in the features.

It may be possible to combine the best features of algorithms to produce a hybrid system than does better than any one single system.

Acknowledgments

This work has been supported by the Commission of the European Community under ESPRIT project no. 5170. We thank all the partners in StatLog, in particular Charles Taylor of Leeds University. We would also like to acknowledge the help of G. Nakhaeizadeh of Daimler-Benz, R. Molina of Granada University and J. Stender of Brainware, Germany. We also thank Pavel Brazdil of Porto University for work on software tools that were used in algorithm testing. A special acknowledgement is due to Ashwin Srinivasan of Strathclyde University, who contributed the satellite image database. The heart disease data came from the University of California in Irvine. It was given to us by David Aha of the Applied Physics Laboratory, John Hopkins University. Some algorithms came from Wray Buntine of the NASA Ames laboratory. Finally we would like to thank Donald Michie of the Turing Institute and Brian Ripley of Oxford University for initiating StatLog.

APPENDIX A – STATLOG KEY PERSONNEL: PARTNERS

Dr. Pavel B. Brazdil (Principal Investigator),
University of Porto,
Laboratory of AI and Computer Science (LIACC),
R. Campo Alegro 823,
4100 Porto,
Portugal.

Prof. E. von Goldammer (Principal Investigator),
Institut fuer Kybernetick und Systemtheorie,
Am Hlsenbusch 54,
W-4630 Bochum 1,
Germany.

Dr. R.J. Henery (Technical Director),
Department of Statistics and Modelling Science,
University of Strathclyde,
Glasgow G1 1XH,
UK.

Dr. R. Molina (Principal Investigator),

University of Granada,
Department of Computer Science and AI,
Facultad de Ciencias,
18071 Granada,
Spain.

Dr. G. Nakhaeizadeh (Project Director),
Daimler-Benz AG,
Forschungzentrum Ulm,
Ebeerhard-Finckh Str. 11,
D-7900 Ulm,
Germany.

Mr. H. Hendrix (Principal Investigator),
Isoft,
Chem de Moulon,
91190 Gif sur Yvette,
France.

Dr. S. Muggleton (Project Coordinator Algorithms),
Turing Institute,
Glasgow G1 2AD,
UK.

Dr. R. Rohwer,
Department of Computer Science and Applied Mathematics,
Aston University,
Birmingham B4 7ET,
UK.

Herr J. Stender (Project Coordinator Prediction and Control),
Brainware GmbH,
Gustav-Meye-Allee 25,
1000 Berlin,
Germany.

Dr. C.C. Taylor (Project Coordinator Data),
Department of Statistics,
School of Mathematics,
Leeds University,
Leeds LS2 9JT,
UK.

Prof. Wysotzki,
Fraunhofer-Gesellschaft IITB-EPO,

Kurstrasse 33,
O-1086 Berlin,
Germany.

Dr. Alejandro Moya (Project Officer),
Breydei 9/179,
45 Ave. d´Auderghem,
B-1049 Brussels,
Belgium.

APPENDIX B – MEASURES FOR DATASETS

Number of observations

This is the total number of observations in the whole dataset.

Number of attributes

The total number of attributes.

Number of classes

The total number of classes represented in the entire dataset.

Number of bin_cat

The total number of number of attributes that are binary or categorical.

Cost_matrix

If the dataset has a cost matrix (1 = yes).

Homogeneity of covariances

Homogeneity of covariances is the geometric mean ratio of standard deviations of the populations of individual classes to the standard deviations of the sample, and is tabulated as SD_ratio. The SD_ratio is strictly greater than unity if the covariances differ, and is equal to unity if and only if all individual covariances are equal to the covariances of the whole sample.

Mean absolute correlation coefficient

The correlations ρ_{ij} between all pairs of attributes indicate the dependence between the attributes. They are calculated for each class separately. The absolute values of these correlations are averaged over all pairs of attributes and over all populations to give the measure *corr_abs*, which is a measure of interdependence between attributes. If *corr_abs* is near unity, there is much redundant information in the attributes, and some procedures such as logistic discriminants may have technical problems associated with this. Also, CASTLE may be misled substantially by fitting relationships to the attributes instead of concentrating on relationship between the classes and the attributes.

Canonical discriminant correlations

Examples of n attributes from a sample are points in a n-dimensional space, where they form clusters of roughly elliptical shape. The sample points from one population of class form a cluster around its population mean. In general, if there are k populations, the k means lie in a $k-1$ dimensional subspace. On the other hand, it happens frequently that the populations form some kind of sequence so that the population means are strung out along some curve that lies in a $m--$dimensional space ($m < k-1$). For example, the simplest case occurs when $m = 1$ and the population means lie along a straight line.

Canonical discriminants are a way of systematically projecting the mean vectors in an optimal way to maximize the ratio of between-mean distances to within-cluster distances, successive discriminants being orthogonal to earlier discriminants. Thus the first canonical discriminant gives the best single linear combination of attributes that discriminates between the populations. The second canonical discriminant is the best single linear combination orthogonal to the first, and so on. The success of these discriminants is measured by the *canonical correlations.* If the first canonical correlation is close to unity, the k means lie along a straight line nearly. If the $q+1$th canonical correlation is near zero, the means lie in $q--$dimensional space.

Variation explained by first four canonical discriminants

The sum of the first q eigenvalues of the canonical discriminant matrix divided by the sum of all the eigenvalues represents the 'proportion of total variation' explained by the first q

canonical discriminants. We tabulate, as $fract\_q$, the values of $(\lambda_1 + ... + \lambda_q)/(\lambda_1 + \lambda_2 + \ldots + \lambda_p)$ for $q = 1, 2, 3, 4$. This gives a measure of collinearity of the class means. When the classes form an ordered sequence, for example soil types might be ordered by wetness, the class means typically lie along a curve in low dimensional space. The λs are the squares of the canonical correlations. The significance of the λs can be judged from the χ^2 statistics produced by 'manova'.

Univariate skewness and kurtosis

These are univariate measures of the non-Normality of the attributes when considered separately. The univariate skewness $\beta_1(i, j)$ is a measure of asymmetry of the distribution of the ith attribute in the jth class, essentially describing how the left and right tails differ. As a single measure of skewness for the whole data set, we quote the mean absolute value of $\beta_1(i, j)$, averaged over all attributes and over all classes. This gives the measure $skew\_abs$. For a normal population $skew\_abs$ should be zero: for uniform and exponential variables, the theoretical values of $skew\_abs$ are zero and 16 respectively.

To compare the thickness of the tails of the distributions compared to that of the Gaussian or normal, we use the kurtosis $\beta_2(i, j)$ of the ith attribute in the jth class. As an overall measure, we use the average of the univariate kurtosis $\beta_2(i, j)$, averaged over all attributes and populations. This gives the measure $kurtosis$. For a normal population, $kurtosis = 3$ exactly, and the corresponding figures for a uniform and an exponential are 1.8 and 9 respectively.

REFERENCES

Acid, S., de Campos, L., González, A., Molina, R. and Pérez de la Blanca (1991) CASTLE : A Tool for Bayesian Learning. In *Esprit Conference 1991.*

Aha, D. and Kibler, D. (1988) Detecting and Removing Noisy Instances From Concept Descriptions (Technical Report 88-12). Department of Information and Computer Science. University of California, Irvine, CA, 92717.

Breiman, L., Friedman, J.H., Olshen, R. and Stone, C. (1984) *Classification and Regression Trees.* Wadsworth and Brooks, Monterey, Ca.

Cherkaoui, O. and Cléroux, R. (1991) Comparative study of six classification methods for mixtures of variables. In *Computing Science and Statistics. Proceedings of the 23rd Symposium on the Interface, Fairfax, Virginia.* (ed. E. M. Keramidas), pp. 233–236.

Clark, P. and Niblett, T. (1988) The CN2 induction algorithm. *Machine Learning*, **3**(4), 261–283.

Clark, P. and Boswell, R. (1991) Rule induction with cn2: some recent improvements. In *EWSL '91: Machine Learning: Proceedings of the European Working Session on Learning*, pp. 151–163, Springer-Verlag, Berlin.

Cox, D.R. (1966) Some procedures associated with the logistic qualitative response curve. *Research Papers in Statistics: Frestschrift for J. Neyman, vol 45* (ed. F.N. David), Wiley, New York.

Day, N. and Kerridge, D. (1967) A general maximum likelihood discriminant. *Biometrics*, **23**, 313–323.

Fisher, D.H. and McKusick, K.B. (1989) An empirical comparison of ID3 and back-propagation and machine learning classification methods. In *International Joint Conference on Artificial Intelligence*, pp. 788–793. Morgan Kaufmann, Detroit.

Fisher, R.A. (1936) The use of multiple measurements in taxonomic problems. *Ann. Eugen.*, **7**, 179.

Friedman, J.H. and Stutzle, W. (1981) Projection pursuit regression. *J. Amer. Statist. Assoc.*, **76**, 817-823.

Gennari, J. *Models of Incremental Concept Formation* (to appear in the AI journal)

Goodman, R.M. and Symth, P. (1989) The induction of probabilistic rule sets – the itrule algorithm. In *Proceedings of the Sixth International Workshop on Machine Learning* (ed. B. Spatz), pp. 129–132, Morgan Kaufmann, San Mateo, CA.

Gordon, L. and Olshen, R.A. (1978) Asymptotically efficient solutions to the classification problem. *Annals of Statistics*, **6**, 515–544.

Hart, P.M. (1968) The condensed nearest neighbour rule. *IEEE Trans. Comput.*, **24**, 515–516.

Henery, R.J. and Taylor, C.C. (1992) StatLog: An evaluation of machine learning and statistical algorithms. *Compstat 1992*, Neuchatel.

Kohonen, T. (1989) *Self-Organisation and Associative Memory.* Springer-Verlag.

Kressel U., Franke J. and Schuermann J. (1990) Polynomial Classifier versus Multilayer Perceptron, *DAGM.*

Kressel U., (1991) The impact of the learning set size in handwritten digit recognition. In *ICANN 91, Helsinki.*

McClelland, J.L., Rumelhart, D.E. and Hinton, G.E. (1986) *Parallel Distributed Processing: explorations in the microstructure of cognition. Volumes I, II and III.* MIT Press, Cambridge, MA.

Michalski, R.S. (1983) A theory and methodology of inductive learning. *Artificial Intelligence*, **20**(2) 111–161.

Mooney, R., Shavlik, J., Towell, G. and Gove, A. (1989) An experimental comparison of symbolic and connectionist learning algorithms. In *International Joint Conference on Artificial Intelligence,* pp. 775–780 Morgan Kaufmann.

Pearl, J. (1988) *Networks of Belief: Probabilistic Reasoning in Intelligent Systems.* Morgan Kaufman.

Poggio, T. and Girosi, F. (1990) Networks for approximation and learning. *Proceedings of the IEEE*, **78**(9), 1481–1497.

Quinlan, J.R. (1986) Induction of Decision Trees. *Machine Learning,* **1**, 81–106.

Quinlan, J.R. (1987) Simplifying decision trees. *International Journal of Man–Machine studies*, **27**(3), 221–234.

Quinlan, J.R. (1990) Learning logical definitions from relations. *Machine Learning,* **5**, 239–266.

Ripley, B.D. (1992) Statistical aspects of neural networks *SemStat*, Sanbjerg, Denmark.

Sammut, C. (1988) Experimental results from an evaluation of algorithms that learn to control dynamic systems. In *Proceedings of the Fifth International Conference on Machine Learning*, pp. 437–443, Morgan Kaufmann

Shavlik, J.W., Mooney, R.J. and Towell, G. (1991) Symbolic and neural learning algorithms: an experimental comparison. *Machine Learning*, **6**, 2, 111–143.

Siebert, J.P. (1987) Vehicle recognition using rule based methods.

Turing Institute Technical Report TIRM-87-018.

Sutherland, A., Henery, R., Molina, R., Taylor, C. and King, R. (1992) Statistical Methods in Learning In *Conference on Information Processing and Management of Uncertainty. (IPMU '92)*, Palma de Mallorca.

StatLog Technical Annexe (1990) *Esprit project no 5170.*

StatLog Report on phase II (1992) *Esprit project no 5170.*

Titterington, D.M., Murray, G.D., Murray, L.S., Spiegelhalter, D.J., Skene, A.M.,Habbema, J.D.F. and Gelpke, G.J. (1981) Comparison of Discrimination Techniques Applied to a Complex Data Set of Head Injured Patients (with discussion). *J. Royal Statist. Soc.*, **144**, 145–175.

Unger, S. and Wysotzki, F. (1981) *Lernfaehige Klassifizierungssystems, Academieverlag*, Berlin.

Weiss, S.M., Galen, R.S. and Tadepalli P.V. (1987) Optimizing the predictive value of diagnostic decision rules. In *Proceedings AAAI-87: Sixth National Conference on Artificial Intelligence*, pp. 521–526

Weiss, S.M. and Kapouleas, I. (1989) An empirical comparison of pattern recognition, neural nets and machine learning classification methods. In *IJCAI 89: Proceedings of the Eleventh International Joint Conference on Artificial Intelligence*, pp. 781–787

Weiss, S.M. and Kapouleas, I. (1991) *Computer Systems that Learn.* Morgan Kaufmann, San Mateo.

Table 13.1. The characterization of the eight datasets: Satellite, Digits, KL-Digits, Vehicle, Head, Heart, Credit, and Shuttle (see Appendix A for details). The measures for KL-digits are based on the training examples (except no. of examples). For Shuttle a reduced dataset was used with all members of classes 2,3,6,7 and only 2000 examples each from classes 1,4,5. Formally speaking, the skewness and kurtosis figures for classes 2 and 7 are undefined as there are variables here with attributes whose values are constant (attribute 4 for class 2 and attribute 1 for class 7).

	Satellite	Digits	KL-digits	Vehicle	Head	Heart	Credit	Shuttle
N_examples	6435	18000	18000	846	900	270	8900	58000
N_attributes	36	16	40	18	6	13	16	9
N_classes	6	10	10	4	3	2	2	7
N_bin_cat	0	0	0	0	1	5	8	0
cost_matrix	0	0	0	0	1	1	0	0
SD_ratio	1.2970	1.5673	1.9657	1.5392	1.1231	1.0612	1.0273	1.6067
corr_abs	0.5977	0.2119	0.1093	0.4828	0.1217	0.1236	0.0825	0.3558
cancor_1	0.9366	0.8929	0.9207	0.8420	0.7176	0.7384	0.7618	0.9668
cancor_2	0.9332	0.8902	0.9056	0.8189	0.1057	0.0000	0.0000	0.6968
cancor_3	0.7890	0.7855	0.8440	0.3605	0.0000	NA	NA	0.2172
cancor_4	0.2385	0.6982	0.7761	0.0000	NA	NA	NA	0.1458
fract_1	0.3586	0.2031	0.1720	0.4696	0.9787	1.0000	1.0000	0.6252
fract_2	0.7146	0.4049	0.3385	0.9139	1.0000	1.0000	1.0000	0.9499
fract_3	0.9691	0.5621	0.4830	1.0000	1.0000	NA	NA	0.9814
fract_4	0.9923	0.6862	0.6053	1.0000	NA	NA	NA	0.9957
skew_abs	0.7316	0.8562	0.1802	0.8282	1.0071	0.9560	1.2082	4.4371
kurtosis	4.1737	5.1256	2.9200	5.1800	5.0408	3.6494	4.4046	160.3108

Table 13.2. Satellite image results – '@' times based on transputer; '#' no time given for classifying training examples; '!' uses new version of Cal5, so real time should be higher; examples were cycled 40 times and 1600 nodes are used in this Kohonen feature map.

Algorithm	*Source*	*Accuracy(%)*		*Time(sec.)*	
		Train	Test	Train	Test
k-N-N	leeds	91.1	90.6	2105	944
Radial	strath	88.9	87.9	723	74
Alloc80	leeds	96.4	86.8	63840	28757
INDCART	strath	98.9	86.3	2109	9
CART	granada	NA	86.3	348	14
Backprop	strath	88.8	86.1	54371	39
NewID	turing	93.3	85.0	296	53
C4.5	turing	95.7	84.9	449	11
CN2	daimler	98.6	84.8	1718	16
Quadra	strath	89.4	84.7	276	93
Cal5!	fraunh	87.8	84.6	1345	13
AC2	isoft	NA	84.3	8244#	17403
SMART	leeds	87.7	84.1	83068	20
LogReg	strath	88.1	83.1	4414	41
Kohonen@	luebeck	NA	82.1	12627	129
Discrim	strath	85.1	82.9	68	12
CASTLE	granada	81.4	80.6	NA	NA
Bayes	strath	71.3	69.3	56	12

Table 13.3. Digits results – '@' times based on transputer

Algorithm	*Source*	*Accuracy(%)*		*Time(sec.)*	
		Train	Test	Train	Test
k-N-N	leeds	98.4	95.3	2231	2039
Quadra	strath	94.8	94.6	194	152
Alloc80	leeds	93.4	93.2	3250	134370
Kohonen@	luebeck	NA	92.5	67176	2075
Backprop	strath	92.8	92.0	28910	110
Radial	strath	92.0	91.7	1150	250
LogReg	strath	92.1	91.4	5110	138
SMART	leeds	90.4	89.6	51435	33
Discrim	strath	89.0	88.6	65	30
CN2	turing	99.9	86.6	2229	78
NewID	turing	91.9	85.5	516	80
C4.5	turing	95.9	85.1	543	39
INDCART	strath	98.9	84.6	3615	51
AC2	isoft	NA	84.5	32965	22384
CASTLE	granada	82.5	82.1	4341	4090
CART	granada	84.1	81.9	291	40
ITrule	brainwr	NA	77.8	8283	NA
Bayes	strath	78.0	76.7	104	62
Cal5	fraunh	78.5	71.5	570	55

Table 13.4. KL digits results

Algorithm	*Source*	*Accuracy(%)*		*Time(sec.)*	
		Train	Test	Train	Test
k-N-N	leeds	100.0	98.0	0	13881
Alloc80	leeds	100.0	97.6	48106	48188
Quadra	strath	98.4	97.5	1990	1648
Backprop	strath	95.9	95.1	129840	240
LogReg	strath	96.8	94.9	3538	1713
Radial	strath	95.2	94.5	2280	580
SMART	leeds	95.7	94.3	389448	58
Discrim	strath	93.0	92.5	141	54
C4.5	daimler	NA	82.2	1434	35
CASTLE	granada	87.4	86.5	49162	45403
NewID	isoft	100.0	83.8	785	109
AC2	isoft	100.0	83.2	27382	24791
INDCART	granada	99.7	83.2	3550	53
CN2	turing	96.4	82.0	9183	103
ITrule	brainwr	NA	78.4	NA	8175
Bayes	isoft	79.5	77.7	141	76
Cal5	fraunh	75.2	66.9	3049	64.0

Table 13.5. Vehicle data results – '@'times based on transputer; '*' indicates that the time includes training and testing

Algorithm	*Source*	*Accuracy(%)*		*Time(sec.)*	
		Train	Test	Train	Test
Quadra	strath	91.5	85.0	251	29
Alloc80	leeds	100.0	82.7	30	10
LogReg	strath	83.3	80.9	758	8
Backprop	strath	83.2	79.3	14411	4
Discrim	strath	79.8	78.4	16	3
SMART	leeds	93.8	78.3	3017	1
C4.5	turing	93.5	73.4	153	1
k-N-N	leeds	100.0	72.5	164	23
CART	granada	NA	71.6	29	1
AC2	isoft	NA	70.3	595	23
NewID	turing	97.0	70.2	18	1
INDCART	strath	95.3	70.2	85	1
Radial	strath	90.2	69.3	1736	12
CN2	turing	98.2	68.6	100	1
ITrule	brainwr	NA	67.6	985*	NA
Kohonen@	luebeck	88.5	66.0	5962	50
Cal5	fraunh	70.3	64.9	41	1
CASTLE	granada	49.5	45.5	23	3
Bayes	strath	48.1	44.2	4	1

Table 13.6. Head injury results – '*' indicates that the time includes both training and testing; '!' present version does not utilize costs fully.

Algorithm	*Source*	*Avg. Cost*		*Time(sec.)*	
		Train	Test	Train	Test
LogReg	strath	16.6	18.0	736	7
Discrim	strath	19.8	19.9	28	3
Quadra	strath	17.8	20.1	253	32
CASTLE	granada	18.9	20.9	30	3
CART	granada	19.8	20.4	20	1
Backprop	strath	18.2	21.5	656	32
SMART	leeds	13.6	21.8	420	4
Bayes	strath	23.6	25.0	2	1
INDCART!	strath	21.9	29.3	56	1
k-N-N	leeds	9.2	35.3	9	11
ITrule	brainwr	NA	37.6	7*	NA
Cal5	fraunh	32.3	38.4	5	1
Alloc80	leeds	45.3	46.1	322	276
NewID	isoft	18.9	53.6	16	2
Radial	strath	53.4	63.1	17	5
C4.5	daimler	59.8	82.0	49	1

Table 13.7. Heart disease results – '*' indicates that the time includes both training and testing; '!' present version does not utilize costs fully; '@' times based on transputer.

Algorithm	*Source*	*Avg. Cost*		*Time(sec.)*	
		Train	Test	Train	Test
Bayes	strath	0.351	0.374	6	3
Discrim	strath	0.315	0.393	14	3
LogReg	strath	0.271	0.396	128	7
Alloc80	leeds	0.394	0.407	31	5
Quadra	strath	0.274	0.422	60	16
CASTLE	granada	0.374	0.441	16	3
CART	granada	0.463	0.452	7	1
k-N-N	leeds	0	0.478	0	1
SMART	leeds	0.264	0.478	725	1
ITrule	brainwr	NA	0.515	5*	NA
Cal5	fraunh	0.517	0.559	8*	NA
Backprop	strath	0.381	0.574	128	13
INDCART!	strath	0.261	0.630	8	1
Kohonen@	luebeck	0.429	0.693	227.1	1.9
AC2	isoft	0	0.744	250*	NA
CN2	turing	0.206	0.767	25	5
C4.5	turing	0.439	0.781	34	1
Radial	strath	0.303	0.781	26	4
NewID	isoft	0	0.844	12*	NA

Table 13.8. Credit data results – '!' These algorithms may have classified the test set exactly the same way (perhaps based only on attribute 14; '*' indicates times based on training and testing; '@' times based on transputer.)

Algorithm	*Source*	*Accuracy(%)*		*Time(sec.)*	
		Train	Test	Train	Test
INDCART	isoft	92.0	92.0	206	193
SMART	leeds	89.5	89.1	2151	5
CASTLE	granada	88.3	88.1	81	33
Cal5	fraunh	89.5	87.4	76	12
CART!	granada	87.2	87.0	19	19
NewID!	isoft	100	87.0	380	4
Discrim!	strath	87.2	87.0	71	16
AC2	isoft	100	87.0	7970	410
Radial	strath	87.5	87.0	837	54
LogReg	strath	87.3	86.9	251	30
Backprop	strath	88.2	86.9	28819	19
CN2	daimler	100	86.7	2309	13
Quadra	strath	86.3	86.0	78	20
Bayes	isoft	85.0	84.4	44	8
ITrule	strath	83.9	83.3	773*	NA
Alloc80	leeds	97.7	83.0	24	738
Kohonen@	luebeck	83.3	81.0	30704	71
k-N-N	leeds	100.0	80.6	0	1851

Table 13.9. Shuttle control data results – '!' indicates that only a sample of the training data could be used (C4.5 32760; AC2 4351; Bayes, INDCART 32625; Discrim, Quadra, LogReg 20000); '*' indicates times based on training and testing.

Algorithm	*Source*	*Accuracy(%)*		*Time(sec.)*	
		Train	Test	Train	Test
NewID	daimler	100	99.99	6180*	NA
CN2	daimler	100	99.97	11160*	NA
C4.5!	turing	99.90	99.96	11131	11
SS1 0.81 INDCART!	strath	99.96	99.92	1152	16
AC2!	isoft	100.0	99.68	4493	3397
Cal5	fraunh	NA	99.60	552	18
k-N-N	leeds	99.61	99.56	65270	21698
SMART	leeds	99.39	99.41	110010	93
Alloc80	leeds	99.05	99.17	55215	18333
CASTLE	granada	96.34	96.23	819	263
LogReg	strath	96.06	96.17	6946	106
Bayes!	strath	95.42	95.45	1030	22
Discrim!	strath	95.02	95.17	508	102
Backprop	strath	95.1	95.1	28800	75
Quadra!	strath	93.65	93.28	709	177

LEARNING CONTROL

14

Recent Progress with BOXES

C. Sammut

School of Computer Science and Engineering
University of New South Wales
Sydney, Australia

Abstract

The BOXES algorithm of Michie and Chambers (1968) has proved to be an effective and flexible method for learning to control dynamic systems. The algorithm, in its original form has been used as a benchmark for many experiments in control tasks such as pole balancing. Recent work in our laboratory has shown that the BOXES algorithm can be improved to yield very good learning rates. We describe experiments on a variety of update functions and discuss their robustness. We also develop the notion of freezing of BOXES, suggested by Michie and implemented by Bain (1990).

We have also been concerned with synthesising a readable account of the control strategy employed by a set of boxes. Some preliminary work has begun in combining decision tree learning algorithms with BOXES. Using this method, we regard BOXES as the acquirer of sub-cognitive skills and the decision tree induction as a means of introspecting on the learned strategy to generate understandable control rules.

1 POLE BALANCING

The BOXES algorithm addresses the problem of learning to control a pole and cart system by trial and error. The physical plant consists of a cart which can run on a track of fixed length. A pole is hinged to the cart such that it can only swing in one

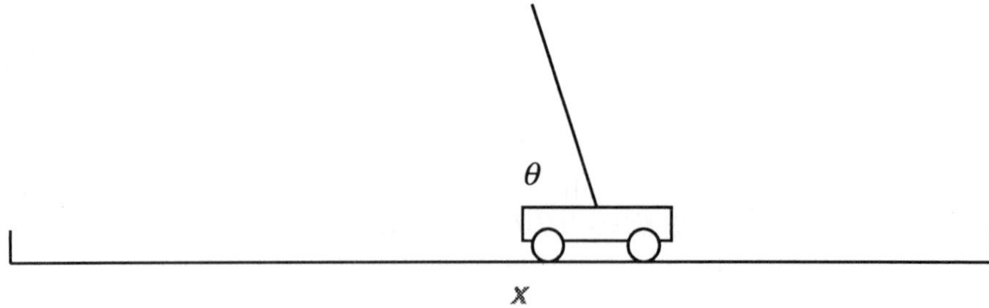

Figure 14.1. The pole and cart

dimension. As usually stated, the task only allows bang-bang control. That is, only a force of fixed magnitude can be applied to push the cart to the left or right. The task of the learner is to construct a control strategy that will keep the pole from falling over and the cart from hitting the ends of the track. It is important to note that the problem is to avoid failure rather than to reach a specified target value. A pole and cart system is depicted in Figure 14.1.

The state of the pole and cart system can be fully determined by the variables: x (the position of the cart), $\dot{x}$ (the velocity of the cart), θ (the angle of the pole), $\dot{\theta}$ (the angular velocity of the pole). That is, the system can be represented by a four dimensional state space.

The most significant problem to be overcome in designing a learning algorithm for this task is that of credit assignment. The control system may make an incorrect choice as to whether to push the cart left or right. However, the consequences of that incorrect choice may not be noticed for some time, when the system finally fails. Many actions may have been taken between the incorrect choice and failure. So how can the learner decide which of those actions was truly incorrect? This is the main task of the BOXES algorithm.

2 BOXES

The algorithm derives its name from the way in which it partitions the state space. The state space is partitioned into regions (boxes) by dividing the range of each dimension into intervals. Thus, the entire space is tessellated by four dimensional boxes.

The state of the system determines a box. Each box contains an action setting indicating that when the system enters the box, the action to be performed is given by the setting of the box.

We regard each box as an independent learning element. The task of a box is to learn which action setting is appropriate for that region of the state space. In order to accomplish learning, each box contains statistics on its performance. These are:

- How many times each action has been performed (the action's *usage*).
- The sum of the lengths of time the system has survived after taking a particular action (the action's *lifetime*).

Each sum is weighted by a number less than one which places a discount on earlier experience. This will be represented by, DK, in the algorithms described below.

The BOXES algorithm proceeds by making an initial random selection of setting for the boxes. A trial is performed by using the current box settings to control the system. If the controller fails, then the actions settings in each box are reviewed and possibly changed. These new settings are then used for a new trial. This process repeats until the system can be kept under control for a predetermined time which signifies success. Clearly, the critical operation in the algorithm is the choice of action setting.

2.1 Deciding which action to take

After a failure, a decision is made in each box, whether to set the action of that box to be 'push right' or 'push left'. This decision must trade off exploration versus exploitation. That is, should the setting be chosen such that the most successful action, so far, is chosen or should the learner switch actions if the other action has not been tried very frequently and thus there is little information about its likelihood of success?

To make the choice between actions, BOXES determines the relative merits of pushing left or right and then applies the following rule:

$$\text{if } value_L > value_R \text{ then}$$

choose left
if $value_L < value_R$ then
choose right
if $value_L = value_R$ then
make random choice

The calculation of the value of an action must include the trade-off described above. In the following subsections we will describe a succession of formulas that have been used in this calculation. We begin with the original Michie and Chambers formula and then proceed to describe, in historical order, improvements that have yielded faster learning rates.

2.1.1 *The Michie and Chambers algorithm*

The original method for calculating an actionUs merit in a box is as follows. To find the value for 'push left' we use the left action's lifetime (LL) and usage (LU) statistics.

$$value_L = \frac{LL + K \times target}{LL + LU}$$

where

$$target = C_0 + C_1 \times merit$$

and

$$merit = \frac{GL}{GU}$$

K, $C0$ and $C1$ are experimentally determined parameters. GL is the global life time (the sum of the lengths of time the system has survived after taking any action) and GU is the global usage (the number of actions taken). Like the local lifetimes and usages, the global statistics are also subject to decay.

This formula was devised to provide for trade-off between exploration and exploitation. The target was introduced to tie the level of exploration to the overall performance of the system. The learner becomes more conservative in its exploration as the overall performance improves. The Michie and Chambers algorithm was a milestone in learning to control dynamic systems. However, the number of trials required to learn to control the pole and cart is quite high. A number of variations of the original algorithm have lead to improved learning rates. We described these variations and follow these with comparisons of performance.

2.1.2 *Cribb's local merit*

James Cribb (Cribb, 1989) introduced a variation in which each box used a local merit function. Cribb argued that the level of exploration within a box should be tied to the performance of the box rather than the whole system. Thus, he devised the following local merit, replacing the global merit in the Michie and Chambers formula: The local merit is the larger of $\frac{LL}{LU}$ and $\frac{RL}{RU}$. Local merit was found to halve the number of trials to learn to control the pole and cart.

2.1.3 *Variations on local merit*

Cribb's version of BOXES can be simplified further by making the trade-off between exploration and exploitation explicit in the control structure of the algorithm rather than hiding it inside the update formula.

```
if an action has not been tested
        choose that action
else if LL/LU > RL/RU
        if LU/RU < C0 + C1 × LL/LU
                choose left
        else
                choose right
else
        if RU/LU < C0 + C1 × LL/RU
                choose right
        else
                choose left
```

C_0 and C_1 are experimentally determined parameters.

In this variation of the BOXES algorithm, running averages of the lifetimes of actions are compared. Assuming that the left average is greater than the right, the default action is to push left. However, before taking that action, we ensure that the ratio of usages, $\frac{LU}{RU}$, does not exceed a target value.

The intuition behind this formula is that the trade-off of exploration and exploitation is related to the ratio of usages of the actions. If the ratio favours the left action this suggests that the left action has been used in preference to the right. If the merit of the box is not sufficiently large, then the program should try the right action.

This variation learns more quickly than Cribb's version as well as being easier to understand.

2.1.4 *Law's algorithm*

Law (1991) replaced the test $\frac{LU}{RU} < C_0 + C_1 \times \frac{LL}{LU}$ by $\frac{\left(\frac{LL}{LU}\right)}{\left(\frac{RL}{RU}\right)} > \frac{LU}{RU}$.

If the ratio of average lifetimes exceeds the ratio of usages then the action represented by the numerator in the ratios should be favoured. Thus the selection algorithm in BOXES can be simplified to:

```
if an action has not been tested
        choose that action
else if LL/LU^2 > RL/RU^2
        choose left
else if LL/LU^2 < RL/RU^2
        choose right
else
        choose an action at random
```

2.1.5 *Variation on Law's algorithm*

The exponent in Law's variation can be seen as an exploration factor. Let us rewrite the algorithm above as:

```
if an action has not been tested
        choose that action
else if LL/LU^k > RL/RU^k
        choose left
else if LL/LU^k < RL/RU^k
```

```
        choose right
    else
        choose an action at random
```

As K approaches one, the level of exploration falls to zero. The higher the value of K, the greater the level of exploration. This variation is the most successful version of BOXES to date. The following section describes comparisons between some of the variations.

3 COMPARISON

Three variations of BOXES are compared: Law and Sammut, Cribb and Sammut and the original Michie and Chambers algorithm. Learning experiments were repeated 20 times for each variant. The system is started in random states for each trial. Table 14.1 shows the number of trials required to learn to control the system for each of the 20 experiments. The average number of trials is shown at the bottom of the table. To study the consistency of the results, logs were calculated. The standard deviations of the logs are also shown at the bottom of the table.

One of the points of comparison is that the Michie and Chambers algorithm can vary considerably in learning rates, whereas the other two variants are more consistent. The Law and Sammut variant has a learning rate comparable to Selfridge, Sutton and Barto (1985).

4 PROPERTIES OF THE LAW AND SAMMUT ALGORITHM

The problem of most learning systems for this domain is that their properties are not well understood. In particular, the values of parameters can only be determined experimentally. This section describes some of the properties of the Law and Sammut variant, determined by experiment.

Figure 14.2 shows a plot of the average number of trials against K values for a series of learning experiments. The average was obtained over five learning runs. The graph shows that the algorithm is stable over the range 1.4 to 1.8. For the standard pole balancing problem, 1.7 was found to be the best.

Table 14.1. Comparison of number trials for variations on BOXES

Law & Sammut	Cribb & Sammut	Michie & Chambers	
79	152	115	
120	80	60	
94	131	140	
69	171	158	
45	90	211	
89	209	747	
80	224	285	
73	105	279	
68	116	2207	
81	166	2586	
81	230	484	
90	138	455	
43	130	387	
43	143	873	
103	85	189	
48	146	505	
81	61	249	
89	112	228	
68	122	392	
65	61	581	
75	134	557	Average
0.29	0.38	0.92	Std. Dev. of log

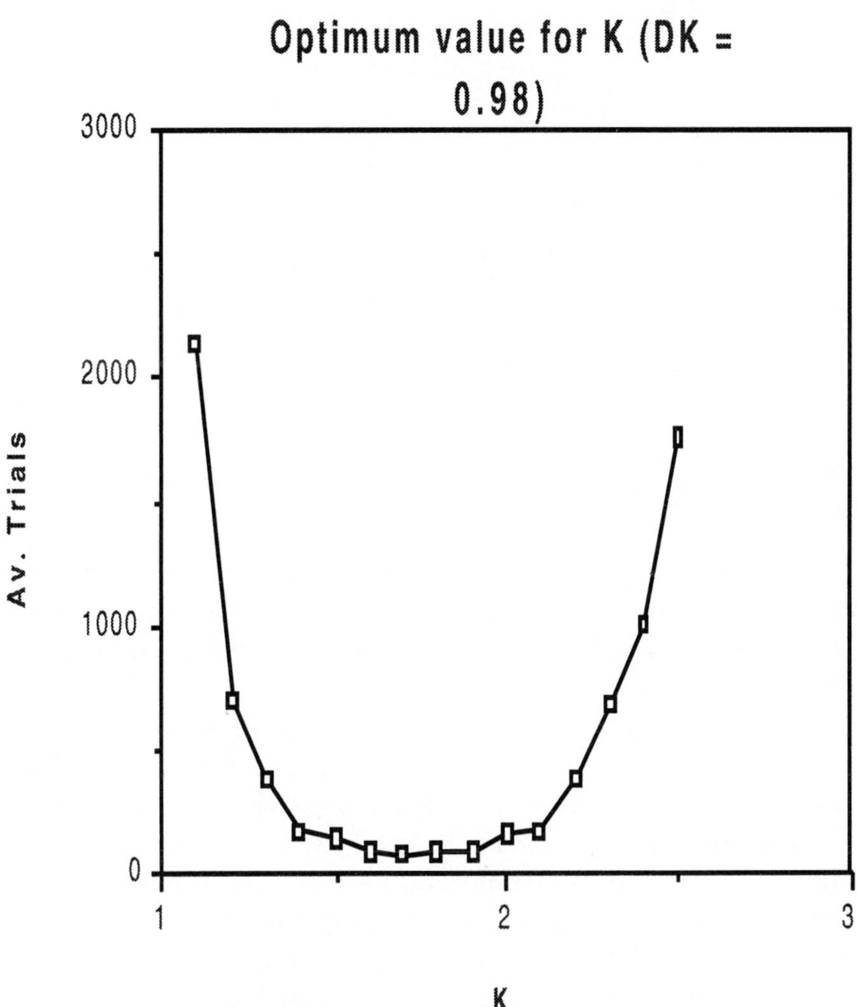

Figure 14.2. Determining the value of K

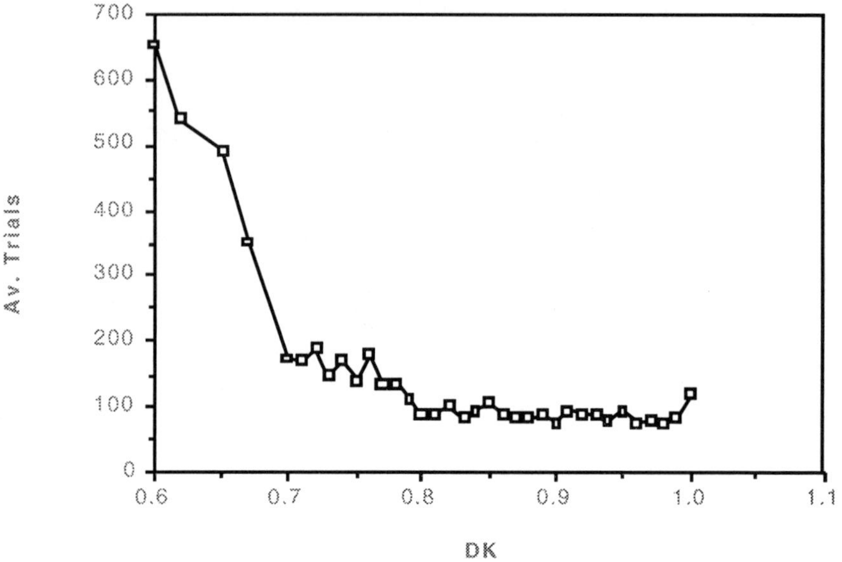

Figure 14.3. Determining the value of DK

The DK value in the previous experiments was set at 0.98. While exhaustive experiments have not been completed, Figure 14.3 shows the results of experiments to determine an appropriate value for DK.

Once workable values for parameters have been found, it is reasonable to ask what are these settings sensitive to. One possibility is that the value of K depends on the number of boxes in the state space partition. Throughout these experiments we have used the standard pole and cart simulation of Anderson (1987). This uses 162 boxes. The original Michie and Chambers set up used 225 boxes. Determination of K was redone using 225 boxes. The results are shown in Figure 14.4. As can be seen, the different number of boxes had little effect. While this is not conclusive evidence, it suggests that K does not depend heavily on the number of boxes used in the state space partition.

The final property tested was that of sensitivity to the problem. Often, the parameters in reinforcement learning algorithms are dependent on the learning task. To find out if this was the

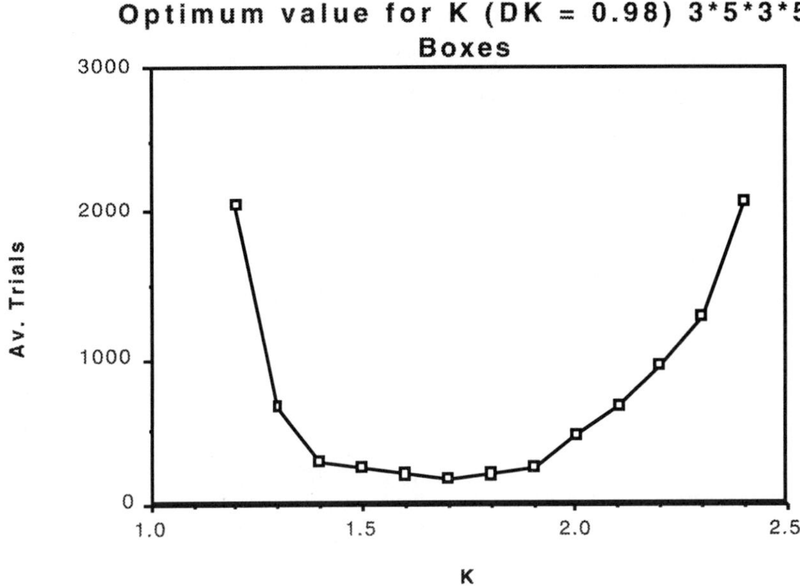

Figure 14.4. Sensitivity to the number of boxes

case with the Law and Sammut algorithm, the problem was varied as follows. Rather than using an equal force to push left and right, asymmetric pushing was used. That is, only half the force was used in a right push as in a left push. This makes the system less easily controlled. Therefore, we expect the learning times to increase. We wish to observe how the learning rate degrades and if parameters must be changed to find the best learning rate. Figure 14.5 shows the determination of K for the asymmetric pushing task. As can be seen, the best value for K still lies within roughly the same range. However, the best value is 1.4. Thus, K is slightly sensitive to the problem.

Table 14.2 shows a comparison of the three variants from Table 14.1 for the asymmetric pushing problem. We also show the Law and Sammut algorithm for $K = 1.4$ and $K = 1.7$. It is interesting to note that, proportionally, the Cribb and Sammut variant does not degrade as much with constant parameter setting as the Law and Sammut variant.

Table 14.2. Comparison on asymmetric pushing

Law (K=1.4)	Law (K=1.7)	Cribb & Sammut	Michie & Chambers	
134	545	43	1382	
562	445	168	487	
120	911	314	1360	
224	123	917	1195	
132	383	2789	3145	
394	101	780	431	
413	1977	253	1768	
83	125	1726	1916	
273	1044	376	709	
821	735	236	816	
249	1155	607	297	
262	611	319	2008	
611	661	214	833	
150	308	265	1643	
278	1179	94	565	
547	227	1125	565	
73	439	157	574	
305	107	97	6493	
669	284	493	5657	
532	517	360	712	
342	594	567	1628	Average

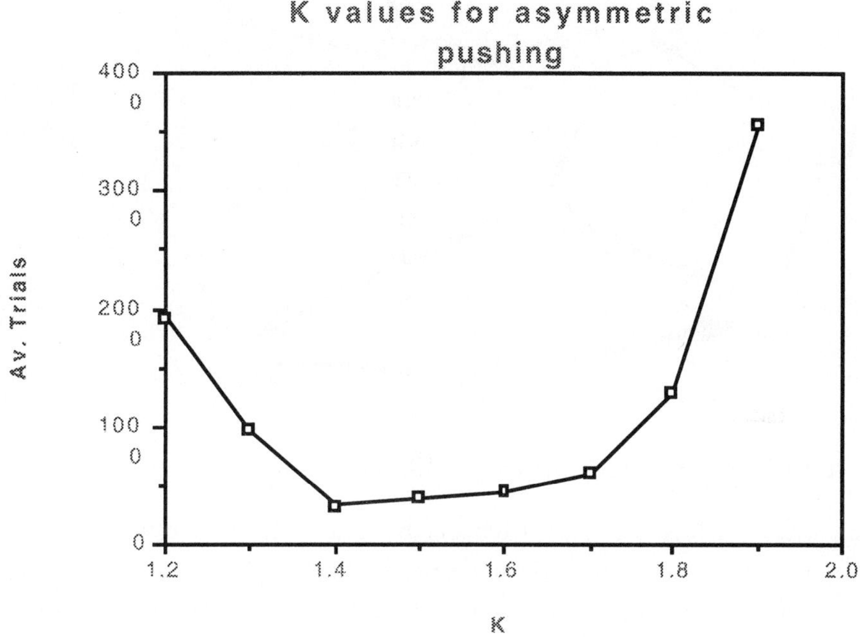

Figure 14.5. Sensitivity to the problem

5 RELIABLE CONTROLLERS

Sammut and Cribb (1990) noted that a program that learns to control the pole and cart in a single learning run does not necessarily learn how to control the pole and cart in general. That is, the program has not learned how to control the system, no matter what the starting state is. Instead, it has learned to control the system from one particular start state. Figure 14.6 illustrates why this may be so.

The dynamic behaviour of BOXES can be characterised by a graph in which the nodes represent boxes and the edges represent transitions from one box to another when a left or right push is performed. Strictly, the edges should have labels corresponding to TleftU and TrightU transitions. It should be noted that the transitions are non-deterministic since the same action in the same box does not guarantee a transition along the same edge. Thus the control of the pole and cart system can be viewed as a Markov process.

The goal of the learning algorithm is to keep the system from

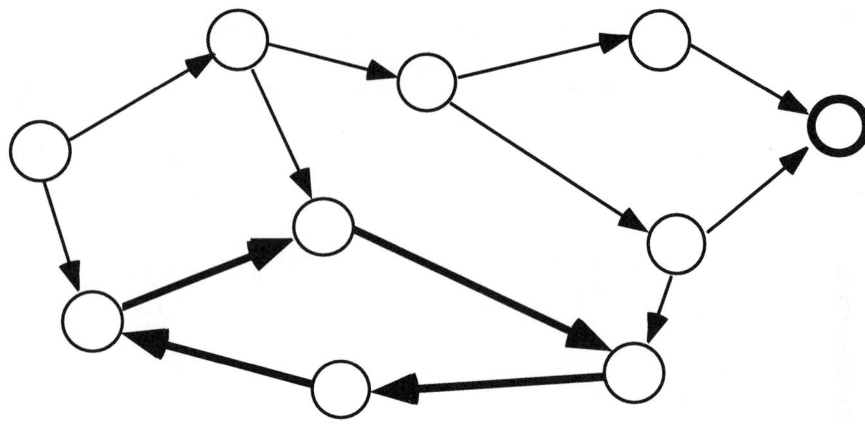

Figure 14.6. State transitions in BOXES

entering a fail state, indicated by the heavy circle. In other words, if the learning algorithm can find a closed set (i.e. a set of nodes which the system never leaves, indicated by the heavy arrows) then its task will be accomplished. If the learning algorithm is able to find such a set quickly from a particular starting state, then it may never explore many regions of the state space. Thus, when the system is restarted in a state which has not be explored, the learned control strategy may fail. Indeed, it is very likely to fail. Sammut and Cribb (1990) described a solution to this problem where the system was allowed to learn to control the pole and cart a number of times and the results of each of these runs were pooled into one control strategy.

5.1 Voting and incremental freezing

The results from different learning runs are pooled by a system of voting. Corresponding boxes in successive learning runs contribute votes for the correct action in that box. A χ^2 test is used to determine when a vote is significant, so when a vote passes the χ^2 test, the action for that box is frozen. Sammut and Cribb report that when 20 to 30 learning runs are collected and then voting is applied, the result is a reliable controller that can control the pole and cart system from any recoverable random starting position without requiring further learning.

Bain (1990) reports on a method called incremental freez-

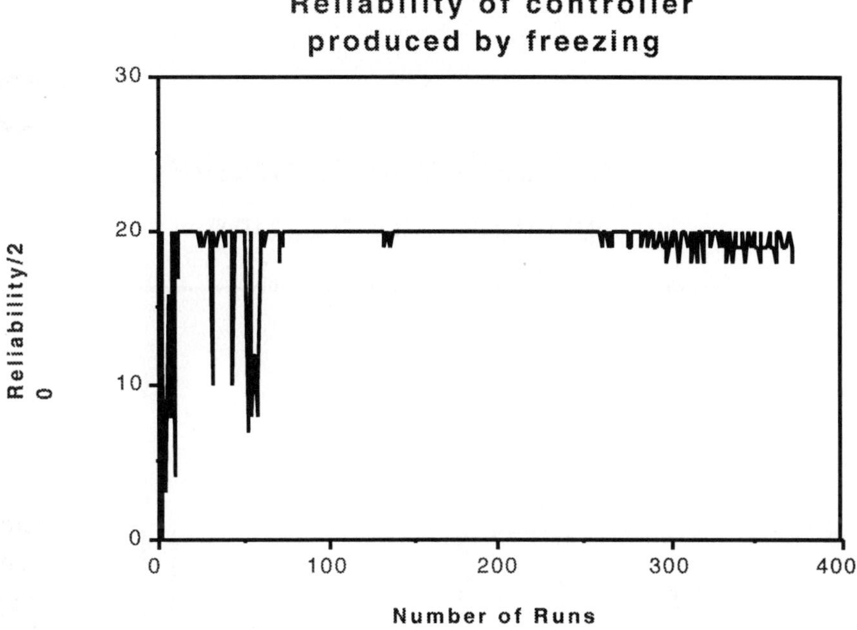

Figure 14.7. First attempt at incremental freezing

ing, suggested by Donald Michie, where votes are collected as learning runs are completed and freezing occurs as soon as the χ^2 is passed. We conducted similar experiments with previous versions of the BOXES algorithm which confirmed the efficacy of this method. However, when incremental freezing was attempted with the Law and Sammut variant the results were not encouraging.

Figure 14.7 shows a plot of the reliability of the controller obtained by incremental freezing. The reliability of the controller is tested by running it 20 times on new pole balancing tasks which all start in random states. If the system successfully controls all 20 runs, then it is deemed reliable. Despite some erratic behaviour, controllers produced beyond about 70 runs are mostly reliable. When this experiment was repeated with a different random number sequence the results were as shown in Figure 14.8. Thus, the latest version of BOXES causes incremental freezing some problems. It is not yet clear why this is the case.

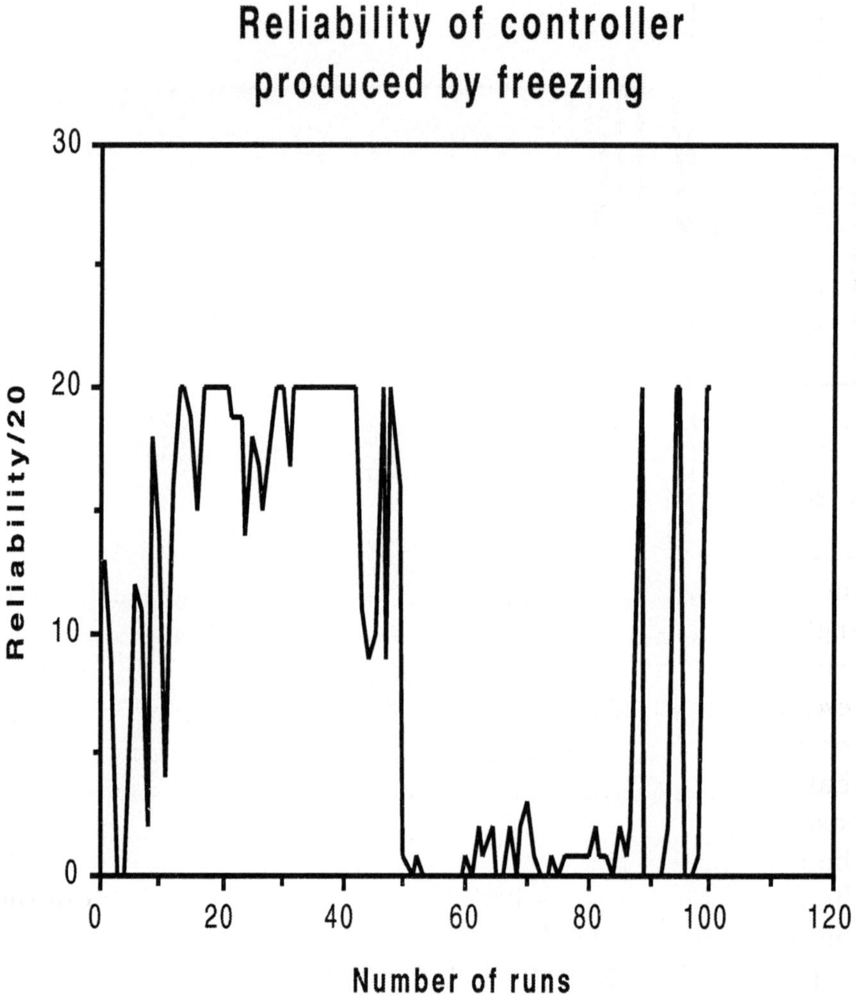

Figure 14.8. Second attempt at incremental freezing

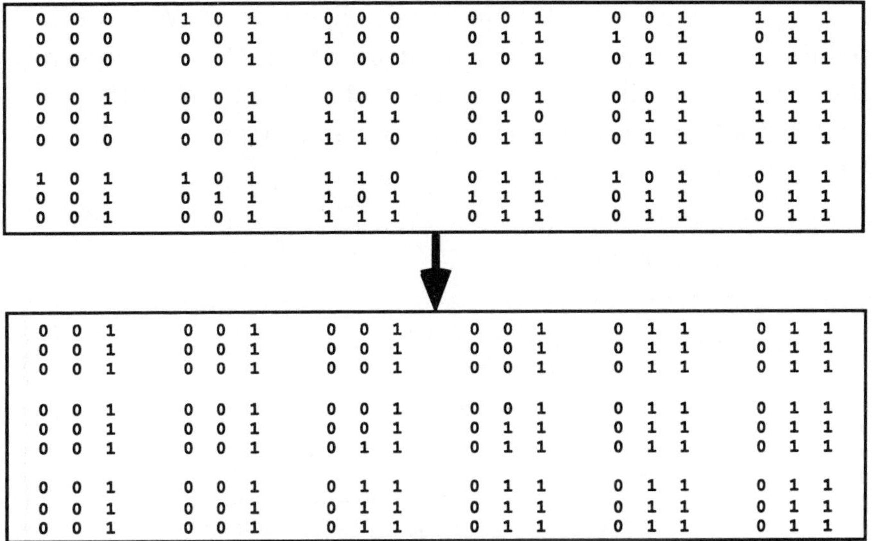

Figure 14.9. Cleaning up BOXES

6 DISCOVERING PATTERNS IN BOXES

Sammut and Cribb (1990) also reported on progress in making the results of BOXES more understandable. The upper rectangle in Figure 14.9 shows the settings of all the boxes after learning. A '0' represents a 'push left' action and a '1' represents a 'push right'. The set of boxes can be stored in the computer as a four dimensional decision array and it this array that has been reproduced in the figure. The display groups all boxes in the region where the pole is leaning to the far left in the left hand side major column. All boxes in the region where the cart is placed at the far left of the track are grouped in the top major row.

It is very desirable that the output of a learning program should be readable. In that way, humans can learn something as well as the program. Unfortunately, the top display is not very informative. However, Sammut and Cribb noted that there is noise in the display. That is, some boxes have settings that are inconsistent with their neighbours. In quite a number of cases, the setting of a box is not critical and can be either 0 or 1. Therefore, Sammut and Cribb experimented with coercing boxes to conform to a regular pattern. For example, the lower rectangle in Figure 14.9 shows a cleaned up set of boxes which is reliable. More importantly, the regularity in the boxes permits compression of this representation to the point where the boxes can be read as a simple rule. This was first noted by Makarovic. His rule is shown below.

```
theta_dot = -inf..-0.87
        push left
theta_dot = -0.87..0.87
        theta = -0.2..-0.017
                push left
        theta = -0.017..0.017
                x_dot = -inf..-0.5
                         push left
                x_dot = -0.5..0.5
                         x = -2.4..0
                                 push left
                         x = -2.4..0
                                 push right
                x_dot = 0.5..inf
                         push right
        theta = 0.017..0.2
                push right
theta_dot = 0.87..inf
        push right
```

6.1 Combining induction and reinforcement learning

In recent experiments, we have tried to improve the method of discovering patterns in boxes by using decision tree induction. The method used is as follows.

- Each box contributes one example to an ID3-like algorithm.
- The description of a box gives the attributes and the left/right decisions are the class values.
- After running the ID3 algorithm, the decision tree is pruned top-down, breadth-first.
- Each non-leaf node in the tree is replaced by a leaf node whose class value is the majority class of the unpruned

node.

- If the boxes defined by the decision tree preserve 20/20 performance then pruning of the node is made permanent.
- Otherwise, we try pruning a sub-tree.

20/20 performance refers to the reliability test described earlier. The decision tree that results from this method is shown below.

```
theta_dot = -inf..-0.87
        push left
theta_dot = -0.87..0.87
        theta = -0.2..-0.017
                push left
        theta = -0.017..0
                x = -2.4..-1
                        push left
                x = -1..2.4
                        push right
        theta = 0..0.017
                x = -2.4..-1
                        push left
                x = -1..1
                        x_dot = -inf..-0.5
                                push left
                        x_dot = -0.5..inf
                                push right
                x = 1..2.4
                        push right
        theta = 0.017..0.2
                push right
theta_dot = 0.87..inf
        push right
```

This tree generates a set of boxes which is regular as can be seen in Figure 14.10. The great advantage of readable control rules is that they can instruct humans about the nature of control. Sammut and Michie (1991) describe the way in which they

```
0 0 1    0 0 1    0 0 1    0 0 1    0 1 1    0 1 1
0 0 1    0 0 1    0 0 1    0 0 1    0 1 1    0 1 1
0 0 1    0 0 1    0 0 1    0 0 1    0 1 1    0 1 1

0 0 1    0 0 1    0 1 1    0 0 1    0 1 1    0 1 1
0 0 1    0 0 1    0 1 1    0 1 1    0 1 1    0 1 1
0 0 1    0 0 1    0 1 1    0 1 1    0 1 1    0 1 1

0 0 1    0 0 1    0 1 1    0 1 1    0 1 1    0 1 1
0 0 1    0 0 1    0 1 1    0 1 1    0 1 1    0 1 1
0 0 1    0 0 1    0 1 1    0 1 1    0 1 1    0 1 1
```

Figure 14.10. Pattern of decision tree

were able to transpose knowledge of pole balancing to controlling the attitude of an orbiting spacecraft.

7 CONTINUING WORK

Work on improving the readability of BOXES output continues. We have found that the order of pruning the decision tree influences the outcome. To avoid this problem, it may be necessary to re-learn the boxes when a pruned decision tree imposes new patterns on the boxes. That is, the boxes specified by the tree are fixed, but all other boxes are subject to further training. It is also important that intervals can be re-defined, that is, that adjacent nodes in the tree can be merged. This may be easily achieved by the adoption of more sophisticated decision tree induction programs.

REFERENCES

Anderson, C. W. (1987). *Strategy Learning with Multilayer Connectionist Representations.* Technical Report No. TR87-509.3. GTE LAboratories Incorporated, Waltham MA.

Bain, M. (1990). Machine-learned rule-based control. In M. Grimble, S. McGhee, & P. Mowforth (Eds.), *Knowledge-base Systems in Industrial Control.* Peter Peregrinus.

Cribb, J. (1989). *Comparison and Analysis of Algorithms for Reinforcement Learning.* Honours Thesis, Department of Computer Science, University of New South Wales.

Law, J. K. C. (1992). *Adaptive Rule-based Control.* Master of Cognitive Science Thesis, School of Computer Science and Engineering, University of New South Wales.

Michie, D. & Chambers, R. A. (1968). Boxes: An Experiment in Adaptive Control. In E. Dale. and D. Michie (Eds.), *Machine Intelligence 2*. Edinburgh: Oliver and Boyd.

Selfridge, O. G., Sutton, R. S. & Barto, A. G. (1985). Training and Tracking in Robotics. In *Proceedings of the Ninth International Conference on Artificial Intelligence* (pp. 670-672). Los Altos: Morgan Kaufmann.

Sammut, C. & Cribb, J. (1990). Is Learning Rate a Good Performance Criterion of Learning? In *Proceedings of the Seventh International Machine Learning Conference*, Austin, Texas: Morgan Kaufmann.

Sammut, C. & Michie, D. (1991). Controlling a 'Black-Box' Simulation of a Spacecraft. *AI Magazine*, *12*(1), 56–63.

15

Building Symbolic Representations of Intuitive Real-time Skills from Performance Data

D. Michie and R. Camacho

The Turing Institute,
Glasgow, UK

Abstract

Real-time control skills are ordinarily tacit — their possessors cannot explicitly communicate them. But given sufficient sampling of a trained expert's input–output behaviour, machine learning programs have been found capable of constructing rules which, when run as programs, deliver behaviours similar to those of the original exemplars. These 'clones' are in effect symbolic representations of subcognitive behaviours.

After validation on simple pole-balancing tasks, the principles have been successfully generalized in flight-simulator experiments, both by Sammut and others at UNSW, and by Camacho at the Turing Institute and Oxford. A flight plan switches control through a sequence of logically concurrent sets of reactive behaviours. Each set can be thought of as a committee of subpilots who are respectively specialized for rudder, elevators, rollers, thrust, etc. The chairman (the flight plan) knows only the mission sequence, and how to recognize the onset of each stage.

This treatment is essentially that of the 'blackboard model', augmented by machine learning to extract subpilot behaviours (seventy-two behaviours in Camacho's auto-pilot for a simulated F-16 combat plane). A 'clean-up' effect, first noted in the pole-balancing phase of this enquiry, results in auto-pilots which fly the F-16 under tighter control than the human from whom the behavioural records were sampled.

1 INTRODUCTION

The labels 'strong AI' and 'weak AI' have sometimes been used to differentiate two schools. Criteria are summarized in Table 15.1.

Table 15.1. Criteria of strong and weak AI

	Strong	Weak
Feasibility of goals	Human-level intelligence will be achieved in machines within foreseeable time.	Human-level intelligence will be implemented only in some unimaginable future, or perhaps never.
Forms of implementation	All thought can be mechanized as sequential logical reasoning from axiomatic descriptions of the world. The 'physical symbol system hypothesis': all agents, including intelligent, are best implemented symbolically.	Most thought is intuitive, not introspectable, non-logical, associative, approximate and 'fuzzy': best modelled by brain-like ultra-parallel networks.
Personnel	Vintage AI professionals, e.g. Turing, Simon, Newell, McCarthy, Feigenbaum, Nilsson, and their followers.	Members of other professions, particularly in linguistics, neurobiology, physics, and philosophy.

The taxonomy in Table 15.1 lays emphasis on the 'physical symbol system hypothesis' of Newell and Simon (1976). Their intended interpretation restricts symbol systems to those which can transparently support communication with human users.

Thus the lists of numerical weights in which neural nets express themselves constitute 'symbols' of a sort, but not in the sense intended by the above authors. This restriction has persuaded some practitioners that the physical symbol system hypothesis excludes intuitive processes from AI's domain of discourse. Such separatism is unsafe, since much knowledge-based thought seems irredeemably intuitive and sub-articulate (for a recent commentary see French, 1990). For its subcognitive processes, there is no direct evidence that the brain employs a symbolic regime. Hence those who accept subarticulate expertise as a proper AI concern may wonder whether for this purpose they should abandon symbolic representations as untrue to nature. The present chapter advocates a different position, namely that a conceptually transparent symbolic style offers a way of improving on nature. By representing intuitive processes symbolically, inductive inference can do something which is both non-brainlike and also highly useful, catering to the client who says: 'My in-house experts may be 'intuitive'. But I want an expert system to formulate its reasons more explicitly than that.'

2 KNOWLEDGE AND THOUGHT

In industrial knowledge systems the implementer has to distinguish between thought as something to be communicated and thought as problem solving. Choice of representation remains a developer's option. In implementing intuition, he or she may decide that it is something over which to draw a veil. The veil may be woven of neural nets, or of hand-crafted spaghetti-code, or of something else. But suppose that the developer has to supply the customer also with means to draw the veil aside, for purposes of interrogation about goals, plans, evidence, justification, and the like. At the price of being less true to nature, he or she might then be better off not to have veiled it in the first place. Like cognitive and brain scientists, knowledge engineers also study the structure of expertise. Unlike cognitive and brain scientists, they do this (or should do) for the purpose not of emulating but of transcending the brain's limitations. First among these is the relative inarticulacy of what both cognitive scientists and knowledge engineers call 'procedural knowledge', thus distinguishing it from 'declarative'.

2.1 Declarative knowledge

It is characteristic of the retrieval and use of declarative knowledge that it is ordinarily done in conscious awareness. From a wealth of neurobiological observations concerning the effects of brain lesions on memory, L. R. Squire (1987, chapter 11) distinguishes declarative memory from procedural as 'memory that is directly accessible to conscious recollection'. By contrast, the hall-mark of a highly trained expert brain is that it does much of its work intuitively. 'Dialogue elicitation' of rules for building expert systems may therefore be frustrated whenever a given expertise involves strategies stored in procedural memory. Inaccessibility to consciousness of even parts of a targeted expertise can then cause serious problems for large knowledge engineering projects, such as Japan's ambitious 'Fifth Generation' (Michie, 1988). Differentiation of the two forms is thus desirable.

Declarative knowledge comprises whatever lends itself to logical formulation: goals, descriptions, constraints, possibilities, hypotheses. The declarative category also includes facts. When these relate directly or indirectly to events in the agent's own experience, their place of storage is referred to as 'episodic' memory. Another subdivision of declarative knowledge is held to reside in 'semantic' memory, which Squire defines as follows:

> Semantic memory refers to knowledge of the world. This system represents organised information such as facts, concepts, and vocabulary. The content of semantic memory is explicitly known and available for recall. Unlike episodic memory, however, semantic memory has no necessary temporal landmarks. It does not refer to particular events in a person's past. A simple illustration of this difference is that one may recall the difference between episodic and semantic memory, or one may recall the encounter when the difference was first explained.

A school founded by John McCarthy (1959) aims to extend formal logic to serve as a vehicle for mechanizing declarative knowledge (see a recent collection edited by Ginsberg, 1987).

We will say little further about the project, beyond expressing respect for such work. Its philosophical importance is matched only by its difficulty. Our theme is closer to the name and nature of expert systems. These are not so much to do with giving computers knowledge of the world, as with equipping them with useful know-how. In face of the difficulties which confront the McCarthy project, there is something to be said for separately studying the mechanization of procedural knowledge and only later integrating the two levels.

2.2 Nature of procedural knowledge

In Anderson's (1990) text on cognition, skilled procedures are pictured as arising in part by derivation from pre-existing mental descriptions. No direct evidence is offered. Knowledge engineers concerned with real-time skills have been led by practical experience in a rather different direction. The empirical picture is one of inductive compilation from sensorimotor data gathered in the course of trial and error. In this picture the role of higher-level knowledge is not to participate directly, but to steer the learning process, setting and adjusting the frame within which skill-bearing rules are constructed.

The final phase of skill-learning, described by Anderson and others as 'automatization', does not ordinarily support introspective report by the expert performer, hence the 'knowledge-acquisition bottleneck' of applied AI. Procedural knowledge, as we have seen, limits itself to the 'how to' of skilled tasks, whether physical as in making a chair, or more abstract as in prediction of sterling rates against the dollar or the diagnosis of acute abdominal pain. A common synonym for such knowledge is know-how, and its manifestation in observable behaviour is called 'skill'. One difficulty is that observed task-performance does not necessarily reveal whether a given expert's behaviour really exemplifies a skill in the procedural sense or whether he or she is using declarative-semantic memory to form action-plans on the fly. Squire's earlier-cited definition supplies a test, namely the ability to give a verbal account of the way in which each decision was made, possible only for declarative memory. A second criterion is the frequency of the recognize–act cycle: this may simply be too fast for 'what-if' inferential planning to be feasible.

For those concerned to recover procedural rules, as in building expert systems, lack of verbal access (on which Anderson also remarks) is a problem. Yet there is widespread faith among knowledge engineers that special methods of 'dialogue elicitation' can be found which will permit the construction of rule-based systems on the scale of such inductively built systems as the GASOIL (Slocombe *et al.*, 1986) and BMT programs of Table 15.2.

Is rule induction from expert-supplied data nevertheless in some sense a second-best option for building systems on the BMT scale? On the contrary. Experts can rapidly and effectively communicate their skills (as in the BMT case) solely via illustrative responses to selected cases. Does he or she thereby omit something indispensable? Certainly the practitioner's explicit and communicable awareness is basic to expertise in some task domains. But other domains, which lack this property, can be found not only among a rather wide variety of industrial tasks, but even in such purely 'mental' forms of expertise as playing a strong game of checkers (see below).

As a paradigm of procedural knowledge, Feigenbaum and McCorduck (1983, p.55) give the example of tying one's shoes. It is interesting that once this skill has reached the stage known as automatization it can continue unaffected by destruction of the individual's brain mechanisms for acquiring and handling important forms of declarative knowledge. Damasio describes a patient named Boswell. The following summary is from Patricia Smith Churchland (personal communication).

> In addition to losing the hippocampal structures, he has massive damage to frontal cortex. He can identify a house, or a car, but he cannot identify his house or his car; he cannot remember that he was married, that he has children, and so forth. He seems to have no retrograde episodic memory, as well as no anterograde episodic, ... Boswell can still play a fine game of checkers, though when asked he says it is bingo. He cannot learn new faces and does not remember 'pre-morbid' faces such as that of his wife and his children... Boswell can play checkers, tie his shoes, carry on a conversation, etc.

Table 15.2. Of the world's three largest expert systems the two latest (GASOIL and BMT) were not constructed from rules obtained in dialogue fashion, but by automated induction from expert-supplied data. In each case the induction engineer trained the system in the desired skill in the style that the master of a craft trains an apprentice, by a structured sequence of selected examples. Rates of code production are typically in excess of 100 lines of installed Fortran, C, Pascal, etc., per programmer day. The methodology allows validation to be placed on a user-transparent basis (Michie 1989), and maintenance costs are in many cases trivialized. Tabulation is from Slocombe *et al.* (1986) with 1990 data on BMT added. The BMT program is described on p.10 of Pragmatica, vol. 1 (ed. J.E. Hayes Michie), Glasgow, UK: Turing Institute Press.

	APPLICATION	NO. OF RULES	DEVELOP. MAN-YRS	MAINTENANCE MAN-YRS/YR	INDUCTIVE TOOLS
MYCIN	medical diagnosis	400	100	N/A	N/A
XCON	VAX computer configuration	8,000	180	30	N/A
GASOIL	hydrocarbon separation system configuration	2,800	1	0.1	ExpertEase and Extran 7
BMT	configuration of fire-protection equipment in buildings	>30,000	9	2.0	1st Class and RuleMaster

Of considerable interest is the survival of Boswell's checkers skills. Evidently what we shall later term 'fast' skills are not the only ones for which procedural knowledge may dominate over declarative. In contrast to chess skill, checkers was already known not to lend itself to the planning approach and to be essentially 'intuitive'. When A.L. Samuel was engaged in his classic studies of machine learning using the game of checkers, he had numerous sessions with leading checkers masters directed towards dialogue acquisition of their rules and principles. Samuel reported (personal communication) that he had never had such frustrating experiences in his life. In terms of relationship to what the masters actually did, the verbal material which he elicited contained almost nothing which he could use or interpret. In similar vein, Feigenbaum and McCorduck (loc. cit., p.82) describe this type of expert response in the following terms: 'That's true, but if you see enough patient/rocks/chip-designs/instrument readings, you see that it is not true after all.' They conclude 'At this point, knowledge threatens to become ten thousand special cases.'

The message from clinical studies is that skilled performance of even sophisticated tasks can still be manifested, and learned, when the brain is so damaged that knowledge of new happenings cannot be retained and previously stored facts and relations (declarative-semantic memory) are seriously disrupted. Another circumstance under which the mediation of declarative memory is at least equally disabled can be observed in the normal brain by imposing a sufficiently restrictive constraint on the time available for the recognize–act cycle, as in touch-typing. This skill does not depend on the storage and retrieval of declarative knowledge, and can be acquired and executed in its virtually complete absence. Recall that when copy-typing at speed the typist does not need to understand the words as he or she reads them. Indeed, after a speed test little or nothing of the text's content can be recalled. Moreover, educated onlookers are surprised, although they should not be, by the outcome of a request to the typist (supposing that he or she has been using a typewriter with unlabelled keys) to label the keyboard correctly with the proper alphanumeric symbols. Lacking a declarative model, the touch-typist is ordinarily unable to do so (see, for example, Posner, 1973), other than by deliberately typing a symbol and observing where the finger went!

Simon (in press) has recently re-emphasized that simple recognition of a familiar object takes at least 500 milliseconds. Operations involving reference to a semantic model of the task domain require retrieval from long-term memory of relatively complex knowledge-structures and an associated apparatus for inferring, storing, and utilizing intermediate results. Such elaborate transactions are to be found only in the 'slow lane'. Here seconds, minutes, or even hours are required to incubate a decision. The bare bones of an explicit rationale for a slow-lane decision, when it comes, can usually be elicited from the expert by verbal report. Not so in the fast lane, to which the present discussion is confined. 'Fast' skills cannot be accessed by 'dialogue elicitation' methods. How then are expert systems to be built for these skills? A solution is to record behavioural traces from the expert subject. Inductive inference then reconstructs from recorded decision-data rule-based models of the brain's hidden strategies. As reported in this review, machine execution of data-derived models has been found to generate performance exceeding in reliability the trained subject's own.

2.3 Postulates of skill acquisition

Experimental work which will now be described was animated by a point of view about brains, summarized below as a list of postulates. Declarative knowledge is abbreviated to 'D' and procedural to 'P'. P designates only procedural knowledge which has already reached the automatized stage.

1. human agents are able verbally to report their own D;
2. human agents cannot verbally report their P;
3. D can be augmented by being told, and also by deduction;
4. P is built by learning, whether by imitation or by trial and error;
5. P can be executed independent of D, but not vice versa;
6. decision-taking via P is fast relative to use of D;
7. sufficiently fast control skills depend on P alone;

8. even for some slow skills P is sufficient for expert performance;
9. rule-induction can extract an explicit form of P from behavioural traces.

Experiments on dynamical control have yielded illustrations of the listed postulates, culminating in a test of 9 above, namely induction of rules from silent brains. But a comment is first requisite on the undoubted existence of expert systems (EXCON was mentioned earlier) whose rule-bases have, with whatever difficulty, been constructed by dialogue acquisition.

Many observers have noted that experts seek to escape from the requirement of rule-formulation (which they find uncongenial) by supplying 'rules' of such low-level form that they constitute no more than concocted sample cases, i.e. specimen decision-data. The phenomenon has been described by Sterling and Shapiro (1986) in their description of the construction of a credit evaluation expert system. The finance specialists continually gravitated towards concrete instances rather than general rules. This has indeed been a universal finding in knowledge engineering, in line with the known facts concerning procedural memory and its mode of access.

But what if knowledge engineers in search of improvements on raw formulations were consciously or unconsciously to apply their own powers of inductive inference to such sample cases? They could then themselves create the kind of high-level rule structures that they had hoped to elicit. The result would of course be testimony more to their own powers of inductive generalization than evidence that experts can introspect their own rules. In a recent aerospace application two knowledge engineers were able, by deliberately exploiting this style of 'rule-conjecture and test', to construct a rule-based solution with no more than a black-box simulator of the task domain to provide corrective feed-back. No set of rules pre-existed, either in an expert's brain or anywhere else.

3 AN EXPERIMENT IN RULE-BASED CONTROL

The role of the systems developer postulated above requires only a reactive oracle. This source need not be an expert. Indeed, it need not be human. As will be described it could be a simulator on which the developers can play 'what-if' games with their latest conjectured rules (what if we modify the rules like this? ... what would result from that adjustment? ... etc.). In an R & D contract for a US space consortium Sammut and Michie (1991) were given access to just such an interactive oracle.

When building a controller for a physical process, traditional control theory requires a mathematical model to predict the behaviour of the process. Many processes are either too complicated to model accurately or insufficient information is available about the process environment. Space-craft attitude control is an example of the latter. The client was interested in the development by machine learning of a rule-structured controller. A check was desirable as to whether dynamical control tasks can be satisfactorily handled by production rules at all, whether these are captured by learning algorithms or developed in some other way.

If the attitude of a satellite in low Earth orbit is to be kept stable by means of thrusters, the control system must interact with many unknowns. For example, although very thin, the Earth's atmosphere can extend many hundreds of kilometers into space. At different times, the solar wind can cause the atmosphere's density to change, thus altering the drag and aerodynamic torques on the vehicle. These are factors which earthbound designers cannot predict and even after three decades of space flight, attitude control is still a major problem.

The client required a trial of rule-based control, using a computer simulation of an orbiting space-craft under 'black box' conditions. By this is meant that knowledge of the simulation's structure and parameters was unavailable to the developers and hence to the controller. Constraints and assumptions included minimal human supervision. Only one ground station was to be used for control. The ground crew therefore have only a 16-minute window in each 90-minute orbit during which they can communicate with the space-craft. A premium was thus placed on the controller's aptness for generating intelligible reports.

The client's 'black box' simulated three-axis rigid body attitude control with three non-linear coupled second order differential equations, and was supplied as Fortran object code. The use of pseudo-random generators introduced various time-varying disturbances, not only concerned with aerodynamic effects of solar wind variations and of atmospheric density and altitude changes, but also effects of propellant expenditure, payload redistribution, solar array articulation, extension and retraction of the gravity gradient boom and the motion of robotic and other on-board manufacturing appliances. Due to such unpredictabilities and to the possibility of a failure while out of communication with the ground, interest in a rule-based back-up controller centred on robustness, simplicity, and conceptual transparency.

The BOXES adaptive rule-based control algorithm (Michie and Chambers, 1968; Chambers and Michie, 1969) was recently the subject of new work by Sammut (1988) who also reviewed trials of other algorithms for learning rule-based solutions to the 'pole and cart' problem. A rigid pole is hinged to a cart which is free to move along a track of fixed length. The learning system attempts to keep the pole balanced, and the cart within the limits of the track, by applying to the cart a force of constant magnitude but variable sign, either right or left ('bang-bang' control). The pole and cart system is characterized by four state variables which make up a four-dimensional space. By dividing each dimension into intervals, the state space is filled by four-dimensional 'boxes'. With each box (i.e. local region of state-space, or 'situation' in the terminology of situation-action rules) is associated a setting which indicates that for any point within the given box the cart should be pushed either to the left or to the right. Essentially this representation was tested on the client's simulated spacecraft.

3.1 The black box

The task was to drive the system from its initial state to the specified final state and maintain that state. Included in the black box was a fourth order Runge–Kutta numerical algorithm which integrated the dynamics of the equations of motion. The time step had a fixed value of 10 seconds. The black box kept track of time and randomly injected various time-dependent disturbances as earlier described.

The state variables:
  Attitudes: yaw (x), roll (y), pitch (z)
  Body rates: ω_x, ω_y, ω_z.
Initial values of the state variables:
  $x = y = z = 10$ deg
  $\omega_x = \omega_y = \omega_z = 0.025$ deg/sec

The desired state:
  $x = y = z = 0 \pm 3$ deg
  $\omega_x = \omega_y = \omega_z = 0.005$ deg/sec

Failure conditions:
  x or y or z exceeds $\pm$ 30 deg
  ω_x or ω_y or ω_z exceeds $\pm$ 0.05 deg/sec

A flag is turned on if any of these go out of bounds.
Available control inputs:
  Torque: T_x, T_y, T_z.

Torque was applied by the firing of thrusters which were aligned to the body axes. Although other attitude control devices (momentum exchange systems) will be used on the satellite in addition to thrusters, this work only addressed the use of thrusters. The following are minimum and maximum torques which can be applied by the thrusters:

$T_x(\text{Min}) = T_y(\text{Min}) = T_z(\text{Min}) = 0$ ft-lbf
$T_x(\text{Max}) = \pm$ 0.5 ft/lbf; $T_y(\text{Max}) = T_z(\text{Max}) = \pm$ 1.5 ft/lbf.

3.2 The rules

The first trial was made by directly adapting a set of BOXES-derived rules from the pole-and-cart domain to a sequential logic suggested by hand-derived rules due to Makarovic (1987, 1991). In each recognize–act cycle rule-matching follows a certain priority order, cycling through the state variables until an action is selected. For each in turn the rule first checks that the first derivative does not exceed certain bounds. If it does, then a force is applied to oppose it. If it does not, then with respect to the same variable check its magnitude. If it exceeds given bounds then a force is applied accordingly.

In the case of the pole and cart, there was a clear priority to the order in which dimensions were checked. It was critical that the angular velocity and the angle of the pole were considered before the cart variables, since neglect of the pole leads to failure much more rapidly than neglecting to keep the cart away from the ends of the track. If this principle is applicable to the case of the space-craft then it is necessary to determine which of the state variables changes most rapidly. This was done, yielding rules expressible in 'if-then-else' form, thus:

if ω_z < -0.002 then	apply a T_z of +1.5
else if ω_z > 0.002 then	apply a T_z of -1.5
else if z < -2 then	apply a T_z of +1.5
else if z > 2 then	apply a T_z of -1.5
else if ω_y < -0.002 then	apply a T_y of +1.5
. . . and so on ...	

Note the use of 'bang-bang' control, i.e. the torquers were set either fully positive or fully negative just as in the pole-balancing experiments. With a space vehicle there are three dimensions, not one, to which a control motor (torquer) can apply a positive or negative thrust, corresponding to the yaw, roll, and pitch dimensions of rotation respectively. The thresholds for the variables were determined by choosing an arbitrary value slightly within the bounds given for the desired values of the variables.

This control strategy proved to be successful but slow, requiring 8700 seconds to bring the vehicle within desired bounds, and it also consumed 11.2 units of propellant. The question arose whether the control of each dimension could be decoupled. The cited rule only allows one thruster to be fired at any one time. If

Table 15.3. A decision array for control of the yaw dimension.

Yaw too-positive	$T_x/4$	0	$-T_x/4$	$-T_x/2$	$-T_x$
Yaw OK	$T_x/2$	0	0	0	$-T_x/2$
Yaw too-negative	T_x	$T_x/2$	$T_x/4$	0	$-T_x/4$
	Yaw-rate too-neg.	Yaw-rate negative	Yaw-rate OK	Yaw-rate positive	Yaw-rate too-pos.

each axis of the craft were considered separately then all three thrusters could be fired simultaneously. This modification resulted in rules which brought the vehicle under control very quickly, requiring only 4090 seconds. But propellant consumption, although improved, was still too high, using 7.68 units before the vehicle became stable. Therefore a partial retreat was made from pure 'bang-bang', with a view to replacing it with finer control of the thrusters.

The resulting strategy is best understood by a decision array. For example, yaw control can be displayed as in Table 15.3 and the resulting performance as in Figure 15.1. Each of the 15 boxes corresponds to one control rule. Thus the box in the top left hand corner states that if the yaw is positive (i.e. above the bounds on the desirable yaw) and the yaw rate ω_x is well below the bounds of desirability then apply a quarter of the full torque in the positive direction. Thresholds were set for angles at ± 2 deg and for angular velocities they were ± 0.002 and ± 0.003. The decision arrays for roll and pitch dimensions were of the same form. The resulting control behaviour was highly satisfactory. The pitch dimension was the slowest of the three to be brought within the desirability zone.

The client's engineers stated that both in speed of recovery and in propellant expenditure results were close to calculated

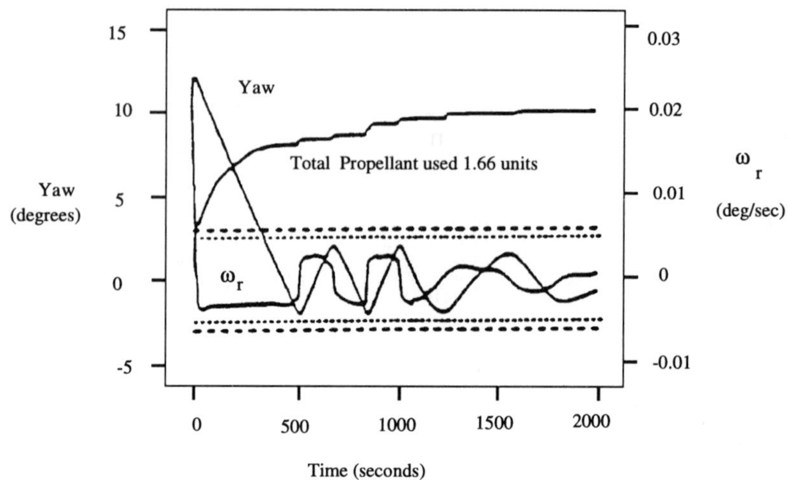

Figure 15.1. Plot over time of vehicle's yaw behaviour (see text)

optima. Since however it appeared that the satellite had greater inertia in the z-axis (pitch) than in the other two the thrust of the z-torquer was increased. This brought the vehicle under control in 5290 seconds, somewhat more slowly than the previous controller. But it only required 1.7 units of propellant, a substantial saving. Also calculations and simulations by the client's engineers made the result appear slightly better than optimal. This doubtless arose from minor approximations and/or distributional assumptions made in their numerical work. Time did not permit the point to be elucidated. But the broad conclusion was seen as extremely encouraging. An industrial-strength problem had shown that the simplicity, robustness, and conceptual transparency of rule-based control does not have to be purchased at the cost of significant degradation of performance.

4 EXPERIMENTS WITH SKILL-GRAFTING

Supported by the freedom interactively to test each conjectured modification on the simulator, Sammut and Michie found their own powers of inductive conjecture adequate. But tasks of higher complexity, such as remote control of pilotless aircraft, demand a less primitive approach. Present ideas are oriented towards the industry's use of interactive simulators for training pilots. A simulator-trained performer cannot tell you his or her strategy, but can demonstrate it. What is demonstrated can be automatically recorded. What is recorded can be inductively analysed by computer. With psychology-trained colleagues, Michael Bain, Jean Hayes-Michie, and Chris Robertson, one of us (D.M.) engaged in an investigation into the use of the rule-induction algorithm C4.5 (see Quinlan, 1987) to uncover effective control rules from such behavioural records. Experimental subjects were trained on an interactive simulation of a task illustrated in Figure 15.2. Control was exercised through a joystick of a pole-and-cart simulation which refreshed the screen approximately 20 times per second. New results together with earlier findings with this experimental system (Chambers and Michie 1969) lead to conclusions as follows (details are available in Michie, Bain, and Hayes-Michie 1990).

4.1 Conclusions from pole-balancing

First conclusion: role of problem representation. Chambers and Michie used two regimes of training, identical except for the graphical animation seen by the subject. In one variant the picture was as shown. In the other the subject saw only a display of four separate horizontal lines, along each of which a pointer wandered to and fro. The subjects in this second variant were kept in ignorance of the nature of the simulated physical system. Unknown to them, the pointers actually represented the current status of four state variables, namely position of cart, velocity of cart, angle of pole, and angular velocity of pole. Our hypothesis was that when the system is run fast, leaving only time for use and up-dating of procedural memory, then there will be no difference in the learning curves of subjects using the two different representations. Although not explicitly reported in their paper, an indication of this was observed by Chambers and Michie. In recent work a rate was additionally used sufficiently slow for subjects to report the task as having a major 'planning' component. This slow-trained group learned more slowly, at least in the initial stages. In the new work trials have not yet been made of the lines-and-pointers representation.

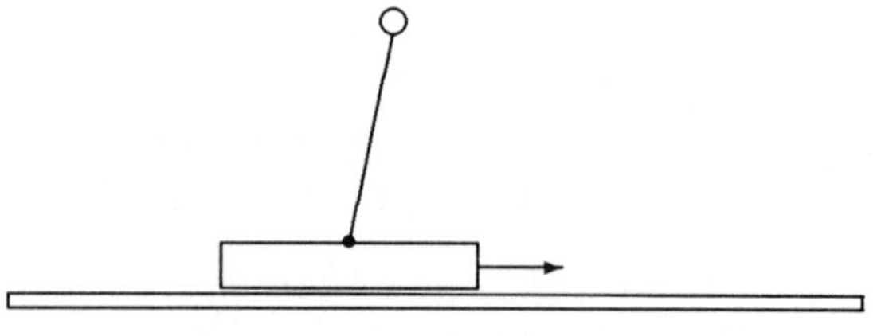

Arrow indicates current direction of motor

Figure 15.2. Diagram of the pole and cart task

Second conclusion: induction of rules from behaviour. Machine learning by imitation of a trained human was first shown for the inverted pendulum by Donaldson (1960) and partially reproduced under bang-bang conditions by Widrow and Smith (1964). Our concern was to test the ability of modern induction algorithms to extract from the behavioural record the kinds of rules believed to accumulate in procedural memory during skill-learning. Results have been positive. A task was investigated where the object was to cross the centre of the track as often as possible in an allotted time-span without dropping the pole or crashing the cart. When induction-extracted rules were installed in the computer as an 'auto-pilot', performance on the task was similar to that of the trained human who had generated the original behavioural trace, but more dependable, as described below.

Third conclusion: the clean-up effect. Rules induced from a behavioural record can be assessed in two different ways. Predictive mode tests the ability of a rule-set correctly to predict other behaviour sampled from the same source. Performance mode tests the ability of the rules to substitute for the human source in executing the skilled task.

Induced rule-sets performed satisfactorily in the second mode while consistently showing high prediction error, often exceeding 20 per cent. One of the team, Mr. Michael Bain, pointed out that when watching a machine-generated rule-set's performance on the screen one is struck by an appearance of superhuman precision and stability. A trained human skill, although

Table 15.4. Clean-up effect shown by induced control rules over a 5-minute test period. x = position, θ = angle: 'dot' denotes first derivatives. These results are typical, and have been many times confirmed in test runs with the same, and with other, subjects.

	x	$\dot{x}$	θ	$\dot{\theta}$
Trained human (ranges)	2.79	4.85	0.562	5.021
Induced rule (ranges)	0.46	1.83	0.134	2.276
Range differences	2.33	3.02	0.428	2.745
'Clean-up'	83%	62%	76%	55%

controlled by an equally precise and stable set of production rules, is obliged to execute via an error-prone sensorimotor system. Inconsistency and moments of inattention would then be stripped away by the averaging effect implicit in inductive generalization, thus restoring to the experimenters a cleaned-up version of the original production rules. When tested in predictive mode, such a rule-set can do no better than the cumulative sum of human perceptual and execution errors allow. But in performance mode one would expect a super-reliable stereotype of the behaviour of the human exemplar. Direct confirmation of this idea was obtained by calculating the magnitude of the pole and cart's excursions during a control session along each of the four dimensions of the state space. Observed ranges tabulated in Table 15.4 were obtained from a behavioural trace recorded from Mr. Bain's own trained performance.

The findings suggest that 'skill-grafting' from behavioural traces may be possible for more demanding tasks, such as those encountered in aircraft flight control. The key idea is that if we could look inside the head of the ground-based pilot of a remotely controlled aircraft, or of the on-board pilot of a difficult vehicle such as a helicopter, we might see a neural encoding of a fully sufficient skill, but degraded in real-time execution by sensorimotor delays and errors. Recovery of a logically equivalent rule structure and its transplantation to an error-free device (i.e. to a control computer) then offers a source of enhanced and more reliable performance. In advanced rotorcraft control there is a current need for libraries of individual autopilot manouevres ('circle at 50 feet', 'fly slowly sideways for one minute', etc.) which the pilot could activate in difficult weather or other conditions, so as to free his attention for some main task in hand, visual search of water surface, target acquisition, etc.

4.2 Learning to fly

Sammut and colleagues have recently been able to reproduce the 'skill-grafting' phenomenon in the complex task of flying a simulated aircraft (Sammut, Hurst, Kedzier, and Michie, 1992). Using a flight simulator developed by Silicon Graphics, three subjects trained themselves by repeatedly piloting a simulated Cessna through the successive stages of a defined flight plan, consisting of the following manouevres:

1. Take off and fly to an altitude of 2000 feet.
2. Level out and fly to a distance of 32 000 feet from the starting point.
3. Turn right to a compass heading of approximately 330°.
4. At a North/South distance of 42 000 feet, turn left to head back towards the runway.
5. Line up on the runway.
6. Descend to the runway, keeping in line.
7. Land on the runway.

Taking 'events' as being signalled by the occurrence of control actions, then up to 1000 events were recorded per flight. Each of three trained subjects performed 30 flights, so that the complete data comprised about 90 000 events. For each event the control action was recorded, together with values of state variables measured at a moment selected 1–3 seconds earlier. The 'offset' makes approximate allowance for the pilot's delay in responding to complex stimuli. To give a rough impression of the data, the following are names of recorded variables:

boolean variables: *on-ground, g-limit, wing-stall*;
integer variables: *twist, elevation, azimuth, roll-speed, elevation-speed, azimuth-speed, air-speed, climb-speed, fuel, thrust, flaps*;
real variables: *E/W distance, altitude, N/S distance, rollers, elevator.*

The simulation program was modified to log the subjects' actions during flight. Log files from trained subjects were used to create the input to an inductive rule-learning program. The learning program was Quinlan's (1987) C4.5. Its output took the form of separate decision trees for each of the four different control actions, further sub-divided into the seven stages listed above. For example, to quote from the original paper,

> The critical rule at take-off is the elevator rule:
>
> elevation $>$ 4: level-pitch
> elevation $\leq$ 4
>   airspeed $\leq$ 0: level-pitch
>   airspeed $>$ 0: pitch-up-5
>
> This states that as thrust is applied and the elevation is level, pull back on the stick until the elevation increases to 4°. Because of the delay, the final elevation usually reaches 11° which is close to the values usually obtained by the pilot. 'pitch-up-5' indicates a large elevator action, whereas 'pitch-up-1' would indicate a gentle elevator action. The other significant control at this stage is flaps:
>
> elevation $\leq$ 6: full-flaps
> elevation $>$ 6: no-flaps

> Once the aircraft has reached an elevation angle of 6°, the flaps are raised.

The 28 decision trees were automatically converted to C-code routines, arranged as a suite of seven flight control modules, each responsible for all aspects of a given stage. A new module was invoked as soon as a pre-programmed precondition was satisfied for the onset of the next stage. Within each module, four sets of if–then rules separately supervized the four separate control actions.

An autopilot was generated in this fashion from each of the trained subjects. Tests were made by running the simulator in autopilot mode, substituting as autopilot code one or another of the three inductively synthesized program suites. The entire flight plan was executed with conspicuous competence, but with individual mannerisms characteristic of the flying styles of the individual human data source. Indications of the 'clean-up effect' (see earlier) were also evident, particularly during the approach stage.

4.3 Learning to fly straight

What is the significance of the foregoing experiment? Primarily that a suitable decomposition of the problem allows the skill-grafting methodology to be scaled up. Inductive skill-grafting evidently is not just applicable to pole-balancing but also to more complex domains such as flight control.

The same workers also reported indications of the 'clean-up' effect earlier found in the pole-balancing experiments, but these indications were of a preliminary nature only. We now report a more detailed examination of this phenomenon independently conducted by Camacho (1992) using a more challenging flight control task. He used a computer simulation (ACM public-domain software down-loaded onto a Sun Sparcstation 2) of the F-16 combat aircraft. Using Quinlan's C4.5 (see Quinlan, 1987) decision-tree induction package Camacho not only found that clean-up was operating, but was also able to show that in his experimental context it played a very large, almost dominating, role.

Camacho followed a similar methodology to that of Sammut *et al.*, details being as follows.

Flight plan stages:

1. Take off.
2. Climb to 1500 feet.
3. Reduce climbing angle and thrust attaining level flight at 2 kilo-feet.
4. Fly parallel to the runway's long axis for a distance of 200 kilo-feet.
5. Turn left 270°.
6. Turn right to line up with the runway.
7. As soon as distance to runway is less than 70 kilo-feet, start descent to runway keeping in line.
8. Land on the runway.

Variables sampled.

real: magnitude of airspeed (knots) (Geoparallel system)
real: y coordinate of airspeed (knots) (Geoparallel system)
integer: x position (ft) (Geoparallel system)
integer: y position (ft) (Geoparallel system)
integer: altitude (ft)
real: climb rate (ft/h)
real: g-force vector in acft system (only z coordinate)
real: roll rate (rad/sec)
real: pitch rate (rad/sec)
real: yaw rate (rad/sec)
real: heading (rad) /* Euler angles for acft */
real: pitch (rad) /* Euler angles for acft */
real: roll (rad) /* Euler angles for acft */
real: angle of attack (rad)
real: angle of sideslip (rad)
real: elevators setting (radl)
real: ailerons setting (rad)
real: rudder setting (rad)
real: elevator trim setting (NOT used)

real: flaps setting (rad)
real: speedBrake setting (rad)
integer: throttle
boolean: gear handle
boolean: brakes
boolean: afterBurner

Control commands used were: elevators, rollers, rudder, flaps, speed brake angle, throttle, gear handle, brakes and after burner The last three are boolean valued. Thus for each of the flight plan's eight stages nine separate decision trees were synthesized.

From each of typically twenty missions, successive 'state-vectors' were sampled and written to file, making about 213 thousand 'state-records' in all. As a post-processing operation, between 1100 and 1600 'events' were then machine-selected from each of these, making about 25 thousand 'events-records'. As in Sammut *et al.*, only those state vectors were selected which precede by a fixed interval in the file the subsequent record of a control action. The set of events so constructed formed the 'training set' for inductive synthesis of a complete autopilot of the form: flight plan plus 72 decision trees.

The earlier-mentioned clean-up effect became evident when autopilots synthesized according to the above formulation were substituted for human control. The magnitude of this gain in steadiness of control can be appreciated by a study of Figures 15.3 and 15.4, which relate only to one dimension, namely control of horizontal deviation from flight plan during the first four stages (straight-line flight on a constant bearing).

This has so far been the main result of an investigation still in its early stages. It should be emphasized that on the other criteria there are local stages of the total mission where improvement is needed. In particular, probably because the human pilot himself (R. C.) has not yet adequately mastered stage eight, the autopilot induced from these records has not either. Self-training, as well as autotraining, is currently continuing. Since the foregoing was written, both human and clone have become able routinely to land the simulated F–16 without mishap.

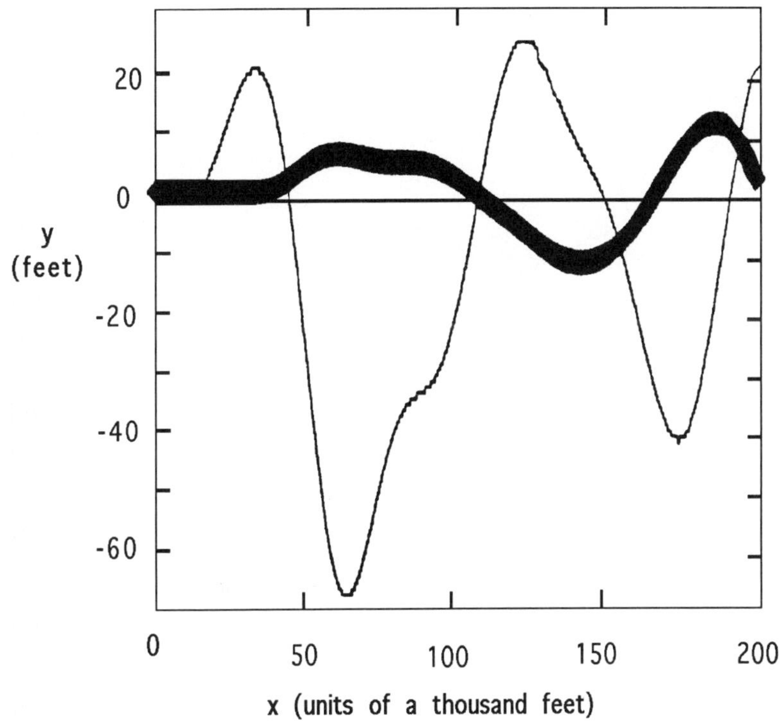

Figure 15.3. The 'clean-up' effect. Plotted lines show distances travelled in the horizontal plane from take-off by a human pilot (thin line) and the autopilot (heavy line) using the ACM flight simulator of the F-16 combat plane (see text): the y axis represents deviations in the horizontal plane from straight flight.

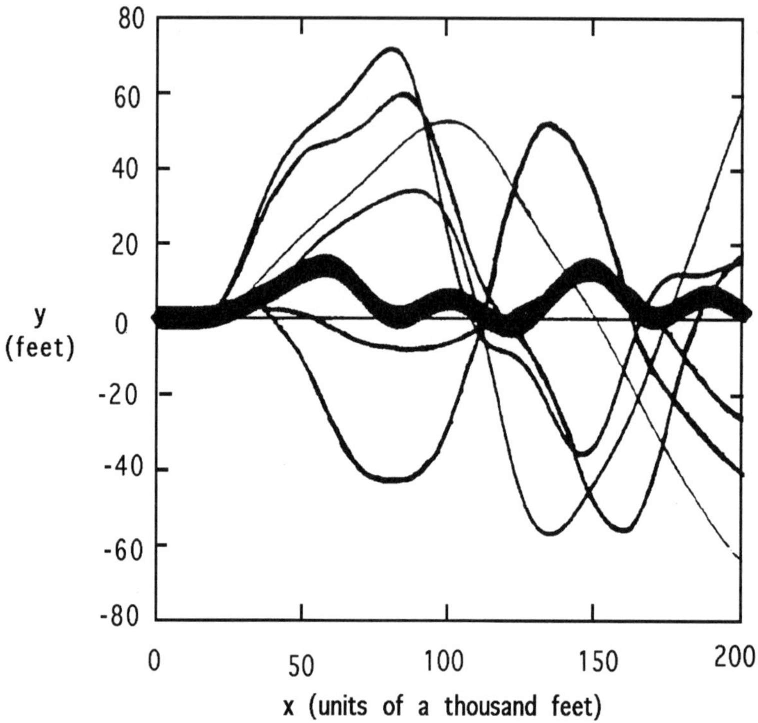

Figure 15.4. Further measurements of the 'clean-up' effect, see previous figure. The six thin-line plots represent the first six missions of a total of 20 flown by the human pilot (R. Camacho) to form the training set of about 25 000 events. The general appearance of the remaining 14 was very similar when plotted in the same way (see text): x represents distance travelled in the horizontal plane from take-off; the y axis represents deviations in the horizontal plane from straight flight.

4.4 A blackboard-like model for coordination among agents

In both of the flight control implementations reviewed above, there is a two-level hierarchy of control: a high level 'chairman' (the flight plan) and a set of low-level 'agents' (decision trees). So far the only role played by the chairman is to monitor the stage of flight and switch the subset of active agents according to context (stage of flight). Each low-level agent has a very specialized task of deciding upon one control in one given stage. All active low-level agents have the same view of the situation (inspect all the state variables) including access to the decision values of their peer agents. Despite the two-level design of the current controller there is no supervized coordination of the low-level agents. How then is the work done?

There is a strong similarity between the community of low-level agents and the AI paradigm of the blackboard (see Nii, 1986, for review). Each agent behaves like a blackboard's 'knowledge source' responsible for a specialized problem solving activity (decide one of the controls in one particular stage of flight). The blackboard (shared memory) role is played by the variables of the aircraft depicting the overall situation. Since all variables are visible to every agent the agents have the same view of the situation and, most importantly, have information about their peer agents by watching their corresponding decision values in the 'blackboard'. This latter facility is responsible for the coordination among low-level agents. As an example, if the rudder agent decides to change its value (moving the physical rudder) the change will be noted on the 'blackboard' and the roller agent, seeing it, may compensate the banking effect of the physical movement of the rudder. For achieving this coordination effect the decision values of the other agents must also be used as attributes during the learning phase. The chairman represents explicit knowledge that is easily articulated and therefore can be hand-crafted. The specialist agents, on the other hand, implement low-level real-time control skills that, in a human, are not performed at a conscious level and therefore cannot be articulated. To create this kind of knowledge, each agent is separately derived by inductive learning from recorded human performance of the skilled task — a step which is an extension of previous blackboard models. The current implementation of the low-level agents may be effective if some small variations to 'normal conditions' appear (mild wind). But if the wind is abnormally strong (serious exception to normal flying conditions) then an understanding of the situation is needed and possibly a reformulation of some current goal, like make a slight change in the bearing to accommodate the wind component in the final velocity. Therefore to improve the skills of the chairman and to incorporate planning capabilities, a deep model (possibly qualitative, as suggested by Sammut (1992)) will be needed and the capability of reasoning from first principles using it. In a way similar to the human counterpart the computer high-level agent should be silent most of the time, just monitoring the overall situation and making small corrections from time to time. It should be fully activated only when the situation requires considerable replanning and deep reasoning for dealing with exceptions for which the low-level agents have no decision.

There is obviously a strong case for implementation of the chairman in a logic programming language. Sammut (1992) has suggested that the low-level agents should *also* be coded in a first order language and constitute a library of primitive actions that the high-level planner could use, setting values for parameters and defining goals.

4.5 Conclusions from autopilot induction

Models extracted as above from decision-data by rule learning are purely heuristic in form. They incorporate no explicit references to time or causality. Yet as reviewed earlier, real-time human problem-solving involves co-operation between two separate kinds of mental model associated with two separate memory systems, updated by separate kinds of learning. The dichotomy of models is recognized in AI under the labels 'heuristic' and 'causal', reflecting the procedural/declarative distinction. The balance in humans is set by the time-constraints imposed by different tasks. A fast situation–action cycle allows time only for executing heuristics and virtually none for reasoning about causes.

As the autopilot experiments demonstrate, complex skills can be built entirely from heuristics. Bears can learn to ride bicycles, and humans can fly combat planes through mission phases which allow no time for analysis. Under such circumstances, everything goes by pattern-invocation. The formal identity between pattern classification and control then stands out clearly. This identity has recently been discussed (Michie, 1991) in connection with a definition of learning which says:

> a learning system uses sample data (the training set) to generate an up-dated basis for improved classification of subsequent data from the same source.

The above-cited discussion continues:

> Notice that the definition, although phrased strictly in terms of classification, logically extends to acquisition of improved performance on tasks which do not look at all like classification. Iterative situation–action tasks come to mind such as riding a bicycle, solving an equation, or parsing a sentence. The extension becomes obvious when for the decision classes we choose names which refer to partitions of the space of situations as 'suitable for action A', 'suitable for action B', etc.

Why should one want, in addition to finding a machine-efficient representation of the above mapping, to construct an operationally redundant superstructure to capture causal relations and to support 'what-if' planning? Answers suggest themselves as soon as one moves to the more demanding definition which animates the characteristically AI approach to learning:

> a learning system uses sample data to generate an up-dated basis for improved classification of subsequent data from the same source and expresses the new basis in intelligible symbolic form.

The requirement for social communication of the 'updated basis' now forces the issue. If synthetic autopilots are to show 'understanding' of flight situations and their own responses, then however necessary heuristic models may continue to be for the sub-structures of skill, insightful performance and explanation at higher and more strategic levels demands causal modelling of a sophisticated kind. It is towards this difficult objective that much work of the kind here reviewed is now turning (see Bratko, 1991).

5 SUMMING UP

In the debate between symbolic and neural-net representations, the two sides have tended to overlook the possibility that different parts of the brain, specialized to address different purposes, employ different representations. Specifically such differing purposes can be broadly grouped under two contrasted main heads:

(1) 'run-time' thinking;
(2) communication of the process and its outcome.

For (1), there is no obvious biological reason to expect symbolic representations to have evolved, in the sense in which 'symbolic' is here used. Indeed there is little evidence that such structures are employed in the brain's real-time problem-solving, some of which is critically supported by varieties of visual and spatial reasoning associated with the brain's right cerebral cortex, and by subcognitive procedures. But the social dissemination of knowledge and thought listed under (2) is of such predominant importance in our species that elaborate symbolic mechanisms have emerged to support the execution of this function. The evolutionary processes have partly been biological and partly cultural. Eccles (in Popper and Eccles, 1987) paints a picture of divergent specialization between the two hemispheres of the brain according to which the 'minor' (usually the right) hemisphere plays roles central to run-time problem-solving, involving pattern-handling and spatial and social orientation. Yet this hemisphere almost wholly lacks capabilities of symbolic reasoning, notably those associated with language and logic. The dominant (usually left) hemisphere, by contrast, not only fluently handles the decipherment of linguistic and logical expressions, but is also the clearing-house for reports on subgoal attainment during problem-solving. Eccles argues that, although 'consciousness' is also manifested by the right hemisphere in the sense of a diffuse awareness, the focussed and organized forms of goal-oriented awareness which we associate with 'self' are functions of the left brain. More recently the possibility has been aired in neurobiological circles (see Benjamin Libet's observations and associated discussion in Behavioural and Brain Science, 1988-89) that the seat of consciousness acts more as a news room than as a planning headquarters, putting a coherent retrospective gloss on the consequences of decision. The decisions themselves, in this model, emanate from activities localized elsewhere. An elaboration of this view has recently been developed by Dennett (1992). Whatever the neural nature of functions (1) and (2) above, modern brain science sees them as operationally and topographically distinct. In such a view, the mechanisms of (2) face a serious problem. Modules specialized to symbolic reporting must interface with dissimilar, even alien, architectures if explanations of the 'self's problem-solving decisions are to be generated. When required to support the more intuitive field of real-time skills, the brain's explanation module tends to fail, or resorts, when pressed by the dialogue-elicitation specialist, to confabulation.

Are we, as engineers of cognition, obliged to burden intelligent artifacts with similar problems? On the contrary, to do so would seem the height of folly. Moreover, from such work as has here been reviewed, an alternative strategy is available. We can treat expert sub–cognition as a 'black box' from which articulate models can be extracted. The product: symbolic models of sub-symbolic behaviour, or, more concretely, machine-executable yet articulate skills derived from 'silent' brains.

Acknowledgments

Thanks are due to Professor Quinlan for active assistance in the use of his C4.5 algorithm. The preparation of the review also benefited from criticisms and suggestions made by Dr. Patricia Smith Churchland, Mrs. J.E. Hayes Michie, Prof Colwyn Trevarthen, and by colleagues in the helicopter division of the Royal Aerospace Establishment, Bedford, UK, and at Advanced Rotorcraft Technology Inc., Mountain View, USA. We were helped by facilities at the Turing Institute, UK, at the Department of Computer Science, University of New South Wales, Australia, at the Department of Statistics, Virginia Tech, USA, and at the

Oxford University Computing Laboratory. Thanks are also due to the Japan Society for Artificial Intelligence, in whose 1991 Proceedings some of the material of this article first appeared. Rui Camacho is supported by a scholarship from Junta Nacional de Investigação Científica e Tecnológica (JNICT) in Portugal.

REFERENCES

Anderson, J.R. (1990). *Cognitive Psychology and its Implications* (Third Edition), W.H. Freeman & Co.

Bratko, I. (1991). *Qualitative modelling: learning and control*. Proc. AI-91, Prague. Copies also available from Electr. Eng. and Comp. Sci., Ljubljana University, Slovenia.

Camacho, R. (1992). *Laboratory notes on experiments with the ACM flight simulator*. Available from the author: Fax no: +44-865-273839. e-mail: camacho@prg.ox.ac.uk.

Chambers, R.A. and Michie, D. (1969). Man–machine co-operation on a learning task. In *Computer Graphics: Techniques and Applications* (eds. R. Parslow, R. Prowse and R. Elliott-Green), London: Plenum, pp. 179–185.

Dennett, D.C. (1992). *Consciousness Explained*. Little, Brown and Co.

Donaldson, P.E.K. (1960). Error decorrelation: a technique for matching a class of functions, in *Proc. Third Internat. Conf. on Medic. Electronics* , pp. 173-178.

Feigenbaum, E.A. and McCorduck, P. (1983). *The Fifth Generation: Artificial Intelligence and Japan's Computer Challenge to the World* , Reading, MA: Addison-Wesley.

French, R.M. (1990). Subcognition and the limits of the Turing Test. In *Mind* , 99, pp.53-65.

Ginsberg, M.L. (ed. 1987). *Readings in Nonmonotonic Reasoning*, Los Altos, CA: Morgan Kaufman.

McCarthy, J. (1959). Programs with common sense. In *Mechanization of Thought Processes Vol. I* , London: Her Majesty's Stationary Office. Reprinted in M. Minsky, ed. (1960), *Semantic Information Processing*, Cambridge, MA: MIT Press.

Makarovic, A. (1987). *Pole-balancing as a benchmark problem for qualitative modelling*. Technical Report DP-4953, Ljubljana: Josef

Stefan Institute. Revised as (1991): A qualitative way of solving the pole-balancing problem. In *Machine Intelligence 12* (eds. J.E. Hayes, D. Michie and E. Tyugu), Oxford University Press.

Michie, D. (1988). The Fifth Generation's unbridged gap. In *A Half-Century of the Universal Turing Machine* (ed. R. Herken), Oxford University Press.

Michie, D. (1989). Problems of computer-aided concept formation. In *Applications of Expert Systems 2* (ed. R. J. Quinlan). Wokingham and Reading, MA: Addison Wesley, pp 310-333.

Michie, D. (1991). Methodologies from machine learning in data analysis and software. *Computer Journal* , 34, 559-565.

Michie, D., Bain, M. and Hayes-Michie, J. E. (1990). Cognitive models from subcognitive skills. In *Knowledge-based Systems in Industrial Control* (eds. Grimble, M., McGhee, S. and Mowforth, P.), Peter Peregrinus.

Michie, D. and Chambers, R.A. (1968) BOXES: an experiment in adaptive control, In *Machine Intelligence 2* (eds. E. Dale and D. Michie), Edinburgh: Edinburgh University Press.

Newell, A. and Simon, H.A. (1976). Computer science as empirical inquiry: symbols and search. *Commun. of the ACM* , 19, 113-126.

Nii, H. P. (1986). Blackboard systems: the blackboard model of problem solving and the evolution of blackboard architectures. In *AI Magazine* , 7, 38-53.

Popper, K.R. and Eccles, J.C. (1977). *The Self and its Brain.* London and New York: Routledge and Kegan Paul.

Posner, M.I. (1973). *Cognition: An Introduction*, Glenview, IL: Scott, Foresman.

Quinlan, J.R. (1987). Generating production rules from decision trees. *Proc. Tenth Internat. Joint Conf. on Art. Intell. (IJCAI-87)* , Los Altos, CA: Kaufmann, pp. 304-307.

Sammut, C. (1988). Experimental results from an evaluation of algorithms that learn to control dynamic systems. In *Proc. Fifth Internat. Conf. on Machine Learning* (ed. J. Laird), San Mateo: Morgan Kaufmann.

Sammut, C. (1992). Automatically constructing control systems by observing human behaviour. In *Proc. of the Second Internat. Workshop on Inductive Logic Programming* , June 6-7, 1992,

Tokyo, Japan.

Sammut, C., Hurst, S., Kedzier, D. and Michie, D. (1992). Learning to fly. In *Proc. Ninth Intern. Machine Learning Conf.* (eds. D.H. Sleeman and P. Edwards), San Mateo, CA: Morgan Kaufman, pp. 385-393.

Sammut, C. and Michie, D. (1991). Controlling a 'black box' simulation of a space craft. *AI Magazine* , 12, (Part 1, Spring), pp. 56-63.

Simon, H.A (in press) Machine as mind, In *Proceedings of the Turing 1990 Colloquium* , 3-6 April 1990, Brighton, UK (ed. Millican, P.) to appear.

Slocombe, S., Moore, K. and Zelouf, M. (1986). *Engineering expert system applications*. Presented at BCS Annual Conference, December 1986.

Squire, L. R. (1987). *Memory and Brain*, Oxford University Press.

Sterling, L. and Shapiro, E. (1986). *The Art of Prolog*, Cambridge, MA: The MIT Press, p.357.

Widrow, B. and Smith, F.W (1964). Pattern recognising control systems, In *Computer and Information Sciences* (eds. Tou, J.T. and Wilcox, R.H.). Clever Hume Press.

16

Learning Perceptually Chunked Macro Operators

M. Suwa and H. Motoda

Advanced Research Laboratory, Hitachi Ltd.,
2520, Hatoyama, Saitama, 350-03, Japan

Abstract

In previous studies search control knowledge has been acquired using explanation-based learning (EBL) techniques. These learn *goal-oriented* control knowledge by explaining how a decision at a control decision node leads to the goal. In the domain of geometry problem-solving, however, this leads to knowledge which is neither sufficiently general nor sufficiently operational. This paper addresses an alternative form of search control knowledge in which the search is controlled at each decision node in such a way that a problem solver can locally recognize relevant 'perceptual chunks'. Previously the effectiveness of perceptual chunks as control knowledge has been reported in a geometry domain. In this paper, we propose a new chunking technique, which acquires, as a chunk, an assembly of diagram elements that can be *recognized and grouped together with the control decision node.* In order to implement this chunking criterion a learner, PCLEARN, employs *recognition rules*, domain-specific knowledge describing necessary conditions for a domain object to be recognizable. Experiments in a geometry domain show that the set of learned knowledge exhibits higher operationality than EBL macro-operators. They also suggest that the PCLEARN chunking technique can be a powerful method for obtaining a small and highly organized set of domain-specific perceptual chunks if augmented with a mechanism for dynamically managing the utility of each chunk.

1 INTRODUCTION

The ability to learn search control knowledge is critically important for problem solvers due to the exponential growth in size of the search spaces they confront. It has been shown that explanation-based learning (EBL), including macro-operator learning by simple EBL techniques (Fikes *et al.* 1972; Minton 1985) and a more sophisticated one that actively selects what to learn (Minton *et al.* 1989), is a powerful technique for learning search control knowledge. This research shares the common view that learners acquire 'goal-oriented'[1] search control knowledge by explaining why a choice taken at a control decision node eventually satisfies the target concept of the problem.

The objective of this paper is to pose and answer the following question. 'Is there any kind of effective search control knowledge which is not goal-oriented?' In the domain of geometry, one of the classical but typical domains with exponential growth in size of search spaces, it has been reported that use of 'perceptual chunks' in the Diagram Configuration (DC) model (Koedinger and Anderson 1990) drastically reduces search spaces. Perceptual chunks are regarded as search control knowledge which is not aimed at achieving a certain goal/subgoal, but at guiding the search process at a control decision node in such a way that problem solvers can recognize the chunks in the problem space. Suwa and Motoda (1989; 1991) have shown that use of 'figure-pattern strategies', a small set of macro-operators whose figurative patterns are the chunks meaningful in geometry domain, enables problem solvers to intelligently select appropriate construction-lines by adding a new point out of an indefinite number of candidate constructions. This research suggests that perceptual chunks also provide critically important search control knowledge in the geometry domain. Here, we pose a sec-

[1]EBL learns from a target concept and produces search control knowledge that is used for accomplishing a unifiable goal in future problems. In this chapter we say that such control knowledge is 'goal-oriented'.

ond question, 'Can the EBL technique simulate acquisition of perceptual chunks in geometry?' Koedinger (1992) suggested that macro-operator-like perceptual chunks in geometry are not primarily organized around the goal-structure explaining target concepts but, rather, are organized around objects or aggregations of objects in the domain of geometry. We will find an answer to the above questions and justify Koedinger's suggestion.

In this chapter, we will address two issues along the lines of the above questions, by illustrating experimental data in the domain of geometry. The first issue is about operationality of EBL macro-operators in the domain of geometry. It is a critical issue because unless learners provide a mechanism of acquiring a small and highly organized set of macro-operators with high operationality, problem solving performance degrades drastically with increasing numbers of macro-operators (Minton 1984, 1985). Operationality of EBL macro-operators depends upon the intrinsic nature of the geometry domain itself concerning whether there is a consistency in goal-structure across many problems or not, because EBL macro-operators can be applied only to those future problems which include the same goal-structure. It is an open empirical question (Koedinger 1992). We will examine it by collecting experimental data on the frequency at which EBL macro-operators are acquired from and applied to many problems, which is a simple measure of the utility of macro-operators (Minton 1985).

The second issue is the proposal of a new learning technique, which is based on another concept different from that of EBL; the learned concept is to be acquired as a chunk, *an assembly of diagram elements that can be recognizable and grouped together with each control decision node.* The learned chunk can be used as control knowledge which guides problem-solving search so that solvers can locally recognize a perceptual-chunk relevant to each of the control decision nodes. The distinguishing point is that there is no notion corresponding to target concepts of EBL. This requires us to provide a criterion for dynamically determining the range of chunking. Suwa and Motoda (1991) proposed the idea of '*recognition rules*', domain-specific rules describing necessary conditions for a domain object to be recognizable, as a guide to determine the area of problem-solving traces to be chunked out. In this chapter, we present a computer program PCLEARN, a domain-independent system of learning perceptual chunks, by use of recognition rules. We also investigate the utility of the rules using experimental data.

In the second section, we characterize the geometry domain and enumerate the problems in applying EBL to this domain. In the third section, we describe the details of the recognition rules themselves and their use in learning perceptual-chunks. In the fourth section, experimental results in the geometry domain are presented and comparisons are made between EBL and the proposed technique in terms of operationality of the learned knowledge. The current limitations of the PCLEARN system and future research issues are discussed in the fifth section.

2 LEARNING SEARCH CONTROL KNOWLEDGE IN GEOMETRY

2.1 Geometry domain

A general characterization of geometric problem-solving is that it is the task of proving a fact holding among an assembly of geometrical objects when a set of other facts are known to hold as given conditions. Domain rules are used for deriving new facts from the set of already given facts. Once a fact is derived, it will never be undone in this domain, because it has been already proved to hold in the given environment. Therefore, the number of facts will increase monotonically as the proving process proceeds. As discussed later, these characteristics are major factors in bringing about difficulties in applying EBL in this domain.

The geometrical objects in this domain are points, segments, directed segments, angles, and triangles. A fact is a nature of an object or a relation between geometrical objects, which is represented in this chapter using the predicate symbols *eqs*, *eqa*, *cong*, *sim*, *para*, *collinear*, *exist*. These express equality of the lengths of two segments, equality of the sizes of two angles, congruence of two triangles, similarity of two triangles, the state of two directed segments being parallel, collinearity of two seg-

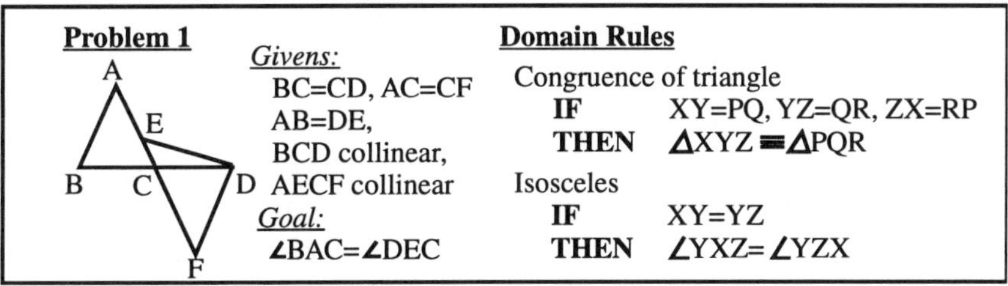

Figure 16.1. Geometry domain: examples of a problem and domain rules

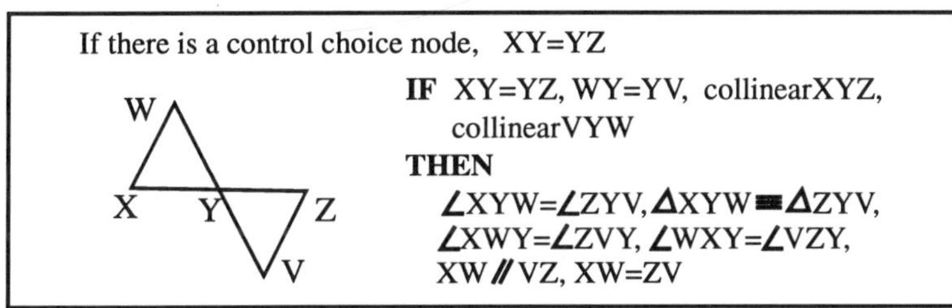

Figure 16.2. An example of perceptual-chunk (which can be useful in solving Problem 1 of Fig.16.1)

ments and the existence of a segment respectively. Domain rules are general knowledge describing the natures of geometrical objects. Figure 16.1 shows an example of a geometry problem as well as two examples of domain rules. Figure 16.2 is an example of a perceptual chunk, which says, 'When there is a fact (control decision node) such as $segment\ XY = segment\ YZ$, try to apply the domain rule of Triangle-Congruence to the fact preferably, and subsequently apply the designated macro-operator if possible.'

2.2 EBL as a learner in geometry

Various versions of EBL systems have been proposed as a technique for learning search control knowledge. Experiments with a STRIPS-like pure EBL technique (Minton, 1985) confirmed that problem solving efficiency degrades remarkably as macro-operators are learned, which is caused by the two limitations of these earlier EBL systems. One is the limitation of the ways of selecting what to learn, i.e. their target concepts were essentially the same as the goals of the problem solving traces (as in

(Fikes *et al.* 1972; Mitchell *et al.* 1983)). The second is the lack or deficiency of utility measures for storing only useful macro-operators. Some methods had no measure (Fikes *et al.* 1972; Minton 1984) and others merely had a simple measure (Minton 1985). The PRODIGY system (Minton *et al.* 1989) addresses these two problems by providing four kinds of meta-level target concepts (i.e. 'succeeds', 'fails', 'sole-alternative' and 'goal-interference') and by evaluating the cost-effective utility of the learned control knowledge (Minton 1990) over a series of experiences of solving other problems.

However, in the domain of geometry problem-solving, learning from 'fails', 'sole-alternative', and 'goal-interference' will not lead to useful knowledge, because there may be no positive reason why a choice leads to a failure, and there may be no problem-solving phenomenon corresponding to sole-alternative and goal-interference in this domain where facts increase monotonically in the problem space as reasoning proceeds. This is unlike task planning where applications of operators successively change the state of the reasoning target. Consequently, it is only 'succeeds' that may work well in the domain of geometry problem-solving. This means again that the target concept is essentially the same as the goal node of the problem. So, the EBL technique cannot go beyond the first limitation mentioned above in this domain.

In the fourth section, we will examine the operationality of the knowledge learned by EBL, based on the experimental data in solving geometry problems.

3 THE PCLEARN SYSTEM

3.1 The learning concept

In order to address the problems mentioned in the previous section for the purpose of learning a useful set of perceptual-chunks from problem-solving traces, we proposed a new learning concept quite different from the 'goal-orientedness' of EBL; PCLEARN acquires, for each control decision node in the problem-solving traces, *an assembly of diagram elements that are visually recognizable and grouped together with the control decision node* as a chunk. It then learns the macro-operator information in-

Input
Training Example : each control decision node
Domain Rules : knowledge represented as production rules
Chunking Criterion : "recognition rules"
Output
The pair of a control decision node and the relevant perceptual-chunk with macro-operator information.

Figure 16.3. The specification of PCLEARN chunking technique

cluded in the chunk as search control knowledge.

For that purpose, PCLEARN has a criterion for determining which portion of the problem diagram is recognizable and grouped together with each control decision node. The criterion is a set of '*recognition rules*', domain-specific knowledge which describes the necessary conditions for a domain object to be recognizable. These take the following form;

$$\text{recognizable}(Obj)\text{:- recognizable}(Obj_1), \ldots, \text{recognizable}(Obj_N), \alpha(Obj_1, \ldots, Obj_N).$$

In order for Obj to be recognizable, all the objects Obj_1 through Obj_N must be recognizable and also an additional condition $\alpha(Obj_1, ..., Obj_N)$ has to hold. $\alpha(Obj_1, ..., Obj_N)$ is a relation between the argument objects and/or the objects composing the argument objects. The precise procedure for acquiring perceptual-chunks by use of these recognition rules is shown in the Section 3.3.

The specification of PCLEARN perceptual chunking is summarized in Figure 16.3. PCLEARN selects training examples from the problem-solving traces according to the definition of control decision node. Control decision nodes are nodes that are members of the successful proof tree for which there is at least one tested domain rule which has been found to be applicable when the other tested ones were not. The output of PCLEARN is the pair of a control decision node and the perceptual chunk relevant to that node with macro-operator information telling what domain rules should be subsequently applied to that node. The learned knowledge can be regarded as a counterpart of 'preference rules' (Minton *et al.* 1989) of the PRODIGY system.

3.2 The overview of PCLEARN

The PCLEARN system includes the following modules;

- **A domain-independent problem solver.** This deals with the task of proving a fact that holds in a given assembly of domain objects when a set of other facts is known to hold, e.g. theorem-proving or diagnosis, using domain rules as well as search control rules. Domain rules are general domain knowledge. They are represented as production rules which have preconditions (sets of facts) in the IF part and a conclusion (a fact) in the THEN part.
- **Chunking facility.** PCLEARN's chunking method is explained in the previous section.

The problem solver's search is conducted by repeating the following decision cycle until the goal node is derived;

1. A node in the search tree is chosen. A node represents a fact which has been given or proved to hold in the problem space.
2. Domain rules (or search control rules) which can be applied to that node in a forward direction are searched for. The domain rule applicable in a forward direction to a node is the one which has an element of the IF part unifiable to that node and whose other elements in the IF part can also be unifiable to the already existing facts.
3. If there is no applicable domain rule, go back to 1 and select another node. If there is one, add a new parent node(s), representing the instantiated fact of the THEN part of the domain rule, whose children are the nodes representing the set of facts in the IF part. Unless the new node is unifiable to the goal node, go back to 1.

3.3 Algorithm for PCLEARN chunking

In creating a perceptual-chunk for a domain rule which is successfully applied to a control decision node, PCLEARN first identifies all the *recognizable* domain objects included in the rule.[2] It then enumerates all the *recognizable* features of these

[2]This rule is denoted as SAR (Successfully Applied Rule) in this paper.

recognizable(X):- recognizable(s(X,Y)).
recognizable(s(X,Y)):- recognizable(a(X,Y,Z)).
recognizable(s(X,Y)):- recognizable(tr(X,Y,Z)).
recognizable(s(X,Y)):- recognizable(X), recognizable(Y), exist(s(X,Y)).
recognizable(s(X,Y)):- recognizable(X), recognizable(Y), collinear(X,Z,Y).
recognizable(a(X,Y,Z)):- recognizable(s(X,Y)), recognizable(s(Y,Z)).
recognizable(tr(X,Y,Z)):-
recognizable(s(X,Y)), recognizable(s(Y,Z)), recognizable(s(Z,X)).

where s(X,Y) -- segment XY, tr(X,Y,Z) -- triangle XYZ, a(X,Y,Z) -- angle XYZ
The literals underlined are additional conditions.

Figure 16.4. The set of recognition rules in geometry

objects. This produces a perceptual chunk which is the assembly of the objects with their features. Recognition rules are used in the first process.

Figure 16.4 is the set of recognition rules in the geometry domain. Points, segments, triangles and angles are the domain objects in this domain. The first rule states that a point X is always recognizable when a segment XY is found to be recognizable because X is a constituent member of XY. In general, when an object is already found to be recognizable and we want to prove the recognizability of another object which is a structural constituent member of the former object, we do not need any additional conditions. The first three rules in Figure 16.4 belong to this category. On the other hand, when we prove the recognizability of an object from the other objects which compose that object, we need some (sometimes no) additional conditions. For example, when we prove the recognizability of segment XY from the recognizabilities of the two points X and Y, an additional condition is needed, i.e. the segment XY actually has to exist in the problem space(corresponding to $exist(s(X,Y))$ in Figure 16.4), or two segments $s(X,Z)$ and $s(Z,Y)$ have to be on the same line for another point Z (corresponding to $collinear(X,Z,Y)$ in Figure 16.4). The last two recognition rules are examples where no additional condition is needed by chance, although they belong to this category.

3.3.1 *Step 1: Picking up recognizable objects.*

The first step is to enumerate all the recognizable objects relevant to a control decision node. The procedures are

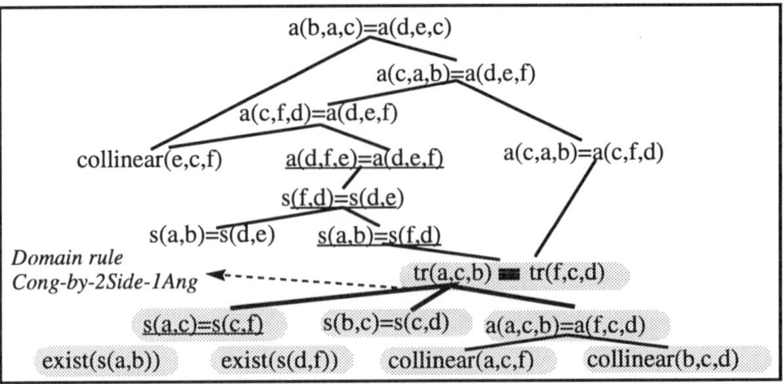

Figure 16.5. The successful proof tree of Problem 1 of Fig.16.1

1. to assert that all the objects which appear as the arguments of the literals in the SAR are recognizable, and
2. to enumerate all the objects which can be proved as recognizable, using **recognition rules.**

Figure 16.5 is a successful proof tree of the Problem 1 in Figure 16.1. The underlined nodes are the control decision nodes. Here, the learning process for the control decision node, $AC = CF$, is illustrated. The SAR for this control decision node is Cong-by-2Side-1Ang. First, the objects appearing in this SAR, $s(b,c)$, $s(a,c)$, $s(c,d)$, $s(f,c)$, $a(b,c,a)$, $a(d,c,f)$, $tr(a,b,c)$ and $tr(f,c,d)$, are asserted to be recognizable. Then, by use of the recognition rules, the following objects, a, b, c, d, f, $s(a,b)$, $s(d,f)$, $s(b,d)$, $s(a,f)$, $a(b,a,c)$, $a(b,a,f)$, $a(a,b,c)$, $a(a,b,d)$, $a(d,f,c)$, $a(d,f,a)$, $a(f,d,c)$, $a(f,d,b)$, $a(b,c,f)$ and $a(a,c,d)$ are justified to be recognizable.

3.3.2 *Step 2: Enumerating recognizable features.*

The second step is to derive from the problem-solving traces all the recognizable features of the above recognizable objects. The procedures are

1. the literals appearing in the SAR are recognizable,
2. the literals of the **additional conditions** which appeared in the recognition rules used successfully for proving the recognizability of objects in Step 1 are recognizable, and

3. all the features that can be derived from the above recognizable literals using domain rules are recognizable.

What we have obtained so far is the derivation tree (the third procedure of Step 2). Note that the derivation tree itself represents a piece of macro-operator information that can be applied to the same control decision node in future problems. The lowest nodes of the tree are the IF-part of the macro-operator and the other nodes are the THEN-part. If we notice that the macro-operator has been derived only from the recognizable features that have been determined by use of recognition rules, the significant role of recognition rules in chunking the macro-operator may be clear.

Let us look at the example case of learning from $AC = CF$ in Figure 16.5. The recognizable literals to be picked up before the derivation process are shown in Fig. 16.5 as the nodes colored grey, out of which the literals that have been incorporated as a result of using recognition rules (the 2nd of Step 2) are $collinear(a, c, f)$, $collinear(b, c, d)$, $exist(s(a, b))$, $exist(s(d, f))$. The first two have been picked up because they appeared in the recognition rules used for proving the recognizability of the object $s(a, f)$ and $s(b, d)$ respectively. Out of these four, the last two will not be used in the derivation process and therefore will be removed from the macro-operator.

Note that owing to the existence of some additional conditions in the set of recognition rules, the learned macro-operator becomes more specific[3] than the SAR itself. In case of the above example, incorporating the two collinearities has been significant in obtaining a perceptual-chunk of the two congruent triangles located in a completely point-symmetry (the one in Figure 16.2), which is more specific than the two merely congruent triangles.

3.3.3 *Step 3: Generalizing.*

The final step is to generalize each node of the acquired derivation tree by dissolving the bindings of the variables of the used

[3]This specificity directly influences the operationality of the learned perceptual-chunks. In this sense, recognition rules play a crucial role in determining the chunked area.

Table 16.1. Assignment of problems to training and test sessions

Sessions	Category 1	Category 2	Category 3
Learning	7	7	6
Test	3	4	3

domain rules. The generalized tree itself represents a macro-operator that has been learned for the control decision node. In the case of the above example, the one in Figure 16.2 is acquired.

We call this sort of macro-operator a perceptually-chunked macro operator because the recognition rules work as a perceptual criterion for determining the area to be chunked out, just as human experts might do visually.

4 EXPERIMENTAL RESULTS

4.1 Method of experimentations

For simplicity, we divided the experiments into two sessions; a training session where problems are solved without using learned search control knowledge and learning is conducted for each problem, and a test session where problems are solved by use of search control knowledge obtained from the training session and no new learning is performed. We selected 30 geometry problems from some reference books on geometry, 20 of which are assigned to the training session and 10 are assigned to the test session. The problems we selected are limited to three problem categories; congruence (and/or similarity) of triangles (Category 1), natures of isosceles and right-angled triangles (Category 2) and natures of quadrilaterals (Category 3). The numbers of the problems selected for each category and assigned to the two sessions are shown in Table 16.1.

In selecting problems, we paid attention mainly to two issues. The first is that the numbers of training problems selected for the three categories should be approximately equal. This is in order to avoid problem selection in terms of categories for the training session which may cause the bias that certain perceptual chunks are learned more frequently, obscuring the issues of consistency in perceptual chunks across problems. The second is that the numbers of the test problems should also be approxi-

Table 16.2. Frequencies of the same perceptual-chunks being learned from many problems

Frequencies of being learned	The number of perceptual-chunks	
	PCLEARN	EBL
1	43	86
2	9	7
3	5	0
4	3	1
more than 4	4	0
total	64	94

mately equal in the three categories because the problems in the three categories should be equally tested using macro-operators.

4.2 Operationality of macro-operators

In this chapter, operationality of macro-operators is measured by the frequency at which each of the macro-operators is acquired from the problems in the learning session and applied to the problems in the test sessions. We investigated it in both cases of the EBL learner which learns from 'succeeds' and the PCLEARN system.

Table 16.2 shows the frequency at which macro-operators with the same diagram configuration are acquired during solving 20 problems in the training session. It seems to be quite a rare case that the EBL learner acquires the same set of perceptual chunks from different problems. On the other hand, PCLEARN learns several kinds of perceptual chunks more frequently in different problems.

Table 16.3 shows the results of the frequency at which those learned macro-operators are successfully applied to problems in the test session. The macro-operators of the EBL learner were applied 15 times, out of which 13 were successful, while the macro-operators by PCLEARN were applied 44 times, out of which 39 applications were successful. Success rate is about the same with both learners but the frequency is much larger in PCLEARN.

Table 16.4 shows the percentage of the nodes which were related to applications of macro-operators against all the nodes

Table 16.3. Frequencies of applications of the learned knowledge

Frequency	PCLEARN	EBL
Total applications	44	15
Successful applications (ratio)	39 (89%)	13 (87%)

Table 16.4. The ratio of the nodes related to applications of macro-operators against all the nodes in a proof-tree (average over all the test problems)

PCLEARN			EBL		
Mean (%)	Max. (%)	Min. (%)	Mean (%)	Max. (%)	Min. (%)
67.5	86.0	38.0	30.0	80.0	0

in the successful proof tree, i.e. a measure of how much macro-operators contribute to constructing a proof-tree in the test session. The shown data (mean, maximum and minimum) are statistics over all the test session problems. The degree of the macro-operators' contributions to constructing proof-trees are larger in using PCLEARN macro-operators.

According to the data on cross-problem learnability (Table 16.2), successful applicability (Table 16.3) and the degree of contribution to proof-trees (Table 16.4), an answer to the empirical question mentioned in the first section is that there is *little* consistency *in goal-structure* across geometry problems while there is indeed cross-problem consistency *in perceptual chunks*, i.e. in the domain of geometry, 'perceptually-chunked' macro-operators have higher operationality than 'goal-oriented' EBL macro operators and hence PCLEARN is more appropriate to this domain than EBL.

Table 16.5 shows an explanation why goal-oriented macro-operators have low operationality. The average size of the applied macro-operators weighted with the frequency of applications is compared with the average size of all the macro-operators learned in the training session[4]. Expert-selected macro-

[4]We define that the size of a macro-operator is the number of its IF part elements, a measure reflecting the ease of finding appropriate instantiations of its preconditions.

Table 16.5. The average sizes of all the learned and applied macro-operators

Range of average	EBL	PCLEARN	Expert-selected
All the learned macros	5.1	3.9	3.2
The applied macros	2.9	2.5	3.1

operators in the third column are the ones which are carefully selected by a geometry expert from among the set of perceptual chunks PCLEARN has acquired. In general, a large difference in both quantities means that a group of macro-operators with a certain size is not applicable. However this may cause considerable costs in testing to apply them in vain. In the case of the EBL macro, the average size of all the macro-operators is much bigger than that of the others which were actually applicable. This is mainly because the EBL learner acquires a chunk from all the paths from each control decision node to the goal of the problem and hence the learned macro-operators tend to be too big in size to be applied to the control decision nodes of future problems. The experimental data suggest that more localized small macro-operators, around control decision nodes which are not always goal-oriented, would have higher operationality in geometry domain.

From all this discussion, we conclude that the PCLEARN chunking module is superior to the typical EBL technique as a method for learning perceptual chunks in the geometry domain, and also that *recognition rules* are effective as an operationality criterion for determining the area of problem-solving traces to be chunked out.

4.3 Learning performance results

Previous experiments have revealed that macro-operator learning has some distinct (both positive and negative) effects on the search process. These reflect two sides of the same coin. The major good effect is referred to as the 're-ordering effect' (Minton 1990); the domain rules encoded as macro-operators are tried before other rules which might be tried first if there were no macro-operators, and consequently unsuccessful search

Table 16.6. Reduction of the explored nodes by use of macro-operators as well as frequencies of macro-operator applications, in four macro modes

Problem No.	Without macros			EBL			PCLEARN			Expert-selected		
	Expl.	Suc.	Tot.	Expl.	Suc.	Tot.	Expl.	Suc.	Tot.	Expl.	Suc.	Tot.
1	33	—	—	33	0	0	14	7	7	12	4	4
2	10	—	—	10	1	1	5	4	4	5	4	4
3	10	—	—	23	2	4	7	2	2	7	2	2
4	19	—	—	18	2	2	15	8	8	19	2	2
5	12	—	—	11	2	2	9	4	4	9	2	2
6	17	—	—	15	2	2	15	3	3	15	2	2
7	17	—	—	17	0	0	7	2	2	7	2	2
8	14	—	—	11	2	2	12	3	6	10	3	3
9	8	—	—	4	2	2	8	2	4	4	1	1
10	15	—	—	15	0	0	8	4	4	7	3	3

Expl. -- The number of the explored nodes in the proof tree
Suc. -- Frequencies of macro-operators being applied successfully
Tot. -- Total frequencies of macro-operators being applied

will be put off later or sometimes left out. This reduces the search space.

Another negative effect is 'increased matching cost' (Minton 1990). As the number of macro-operators increases, the potential frequencies of testing domain knowledge (domain rules and macro-operators) at each control decision node also increases; if no macro-operators are applicable at a control decision node, the problem solver will have to resort to its domain rules, which means that the matching cost in considering the macro-operators was unnecessarily consumed. The number of bindings for each precondition of a domain rule is especially large in domains like geometry. Thus increased matching cost produced by macro-operators severely affects the performance.

The third effect, which also tends to degrade performance, is unsuccessful macro-operator application. The way PCLEARN applies macro-operators is not in goal-oriented search control but in a more local, opportunistic search control. This may sometimes guide the solution search in the wrong direction, which will increase the search space as a whole.

We investigated the above three effects in solving 10 test problems using the following four macro modes; with no macro-operators and with each of the three sets of macro-operators mentioned in Table 16.5. A simple measure of the search space explored by the problem solver is the number of nodes to which the problem solver applied domain knowledge (domain rules and macro-operators). Table 16.6 shows the numbers of the explored nodes in solving 10 problems in each of the four macro modes, together with the statistics about all the applications of macro-operators and successful applications. It is observed that in all three macro-modes the search spaces are reduced (i.e. 're-ordering effect'), compared to no macro mode. There are some exceptional cases of unsuccessful macro-operator applications. Especially the PCLEARN macro-operators contributed much more to reducing the search space than the EBL macro-operators did. In order to reduce the search space considerably, problem solvers need a set of macro-operators with operationality higher than a certain threshold. The PCLEARN macro-operators exhibit a relatively high percentage of success, 89%. The ideal value 100% is seen in the case of Expert-selected-macro mode (see Table 16.6). This shows that the third effect mentioned above is just a minor one in the PCLEARN system.

Table 16.7 shows the experimental data of 'matching costs' (the total cost, the cost of domain rule matchings and the cost of macro-operator matchings) when solving the test problems in the four macro modes. In EBL-macro mode, the cost of macro-operator matchings is extremely large compared to the total cost in No-macro mode. This is mainly because a large number of inapplicable macro-operators with relatively large IF sizes (refer to Table 16.5) were unnecessarily tested.

In PCLEARN-macro mode, the cost of domain rule matchings is smaller than that in No-macro mode, due to the reduction of the search space by the re-ordering effect using PCLEARN macro-operators which have relatively high operationality. However, the cost of macro-operator matchings still exceeds the reduction amount of the cost of domain rule matchings, and hence the total cost does not pay in all the test problems compared to No-macro mode. Expert-selected macro-operators are ones which have been obtained by eliminating some of the PCLEARN macro-operators (as mentioned before) which do not satisfy sim-

Table 16.7. Matching costs in solving the test problems in each of the four macro-operator modes

Problem No.	Without macros			EBL			PCLEARN			Expert-selected		
	Tot.	Rule	Macro	Tot.	Rule	Macro	Tot.	Rule	Macro	Tot.	Rule	Macro
1	477	477	0	3889	841	3048	629	278	351	329	199	130
2	67	67	0	1052	57	995	110	14	96	44	11	31
3	45	45	0	1060	162	898	131	32	99	59	26	33
4	87	87	0	1205	159	1046	198	35	163	267	121	146
5	33	33	0	660	53	607	112	34	78	35	21	14
6	170	170	0	621	100	521	305	75	230	169	67	102
7	81	81	0	541	117	424	103	32	71	57	31	26
8	98	98	0	941	173	768	318	130	188	137	86	51
9	54	54	0	92	5	87	150	52	98	15	6	9
10	106	106	0	1506	175	1331	168	79	89	66	39	27

Tot. -- The total cpu-time cost taken in solving the problem
Rule -- The cpu-time cost taken for domain-rule matchings
Macro -- The cpu-time cost taken for macro-operator matchings (unit: sec)

ple requirements such as successful applicability and cross-problem learnability. This elimination contributed to reducing the cost of macro-operator matchings considerably (ranging from 20% reduction to even 90%) and thereby reducing the cost of domain rule matchings marginally. Consequently, in some of the test problems, the total cost is less than the cost in No-macro mode.

However, there are still some problems in which even the use of the Expert-selected macros does not pay in terms of the total matching cost. This is an issue to be addressed. From the data of Tables 16.6 and 16.7, in all the problems (No. 1, 2, 7, 9, 10) where the use of macro-operators pays, the reduction ratio of search space is more than 50%. This data shows that since the cost of macro-operator matchings is inevitable, the only way to minimize cost using macro-operators is to use those macro-operators which are empirically promising and give great re-ordering effects. For that purpose, we need to manage the utility of each macro-operator dynamically (as in Minton 1990). This must be based on empirical data of how frequently each macro-operator can be successfully applied to problems and in how many nodes of each of the applied macro-operator sequences re-ordering effects are expected.

5 FUTURE WORK

The experiments show that PCLEARN provides a method of learning perceptual chunks with high operationality. However, in order to obtain a small and highly organized set of cost-saving perceptual chunks further study must be made on 1) the mechanism of collecting empirical data of operationality and re-ordering effects for each macro-operator and 2) managing the utility of macro-operators dynamically. PRODIGY addresses this problem (Minton 1990). This is one of the areas we intend to study with a perceptual-chunking learner.

The PCLEARN chunking mechanism must be compared with other goal-structure-based learning methods like SOAR (Laird *et al.* 1987) and compilation of ACT theory (Anderson 1983) in terms of operationality and dynamic utility of learned search control knowledge. The characterization of learning mechanism is that problem solvers learn knowledge of how to satisfy the subgoals the solvers have established during problem solving, i.e. chunking all the lower subgoal structures of the target subgoal. If we applied this method to the domain of geometry, the learning procedure would be as follows. Problem solvers establish a subgoal for finding a domain rule which can be successfully forward applied to a control decision node and then chunk all the necessary conditions for deriving each precondition of the domain rule which was actually applied to the decision node. This mechanism may chunk quite a different range of the problem solving traces from a chunking mechanism which uses 'recognition rules'.

Finally, we have to examine the generality of the PCLEARN perceptual chunking mechanism. Currently, domain independentness is assured if PCLEARN deals with tasks of reasoning within structured objects that satisfy the requirement of monotonicity of the derived facts. A candidate task in which the extension of this method has to be examined might be in design and/or planning tasks where operator applications will change the state or forms of the domain objects, producing domain objects unseen at the initial state of reasonings. The two key requirements that have to be retained even in this extension are the following. Firstly all the kinds of domain objects which will potentially appear in the reasonings have to be listed in advance. Secondly recognition rules holding between those objects can be explicitly described as domain-specific knowledge.

6 CONCLUSION

We proposed a new technique of learning search control knowledge from problem solving episodes by the perceptual chunking mechanism. This approach is quite different from the 'goal-oriented' principle in EBL. Our method learns control knowledge that guides problem-solving search at a control decision so that the solver can recognize locally a perceptual chunk relevant to the node. The learned knowledge consists of chunks which are *assemblies of diagram elements that can be recognizable and grouped together with the control decision node.* In order to implement chunking, PCLEARN employs *recognition rules*, domain-specific knowledge describing the necessary conditions for a domain object to be recognizable.

In this chapter, experimental results of solving and learning from 30 geometry problems were presented for comparing both the goal-oriented EBL technique and the PCLEARN technique in terms of operationality of the learned knowledge and performance improvement by use of them. In the domain of geometry, there is little consistency across many problems in goal structure, but rather a lot of cross-problem consistency in perceptual chunks primarily. Reflecting the intrinsic nature of the geometry domain, perceptually-chunked macro-operators, have higher operationality than goal-oriented EBL ones, which tend to be too large to be applied to problems. In this respect, the EBL technique does not work well in this domain.

'Recognition rules' are useful because they produce search control knowledge with high cross-problem learnability, a high percentage of successful applications, and high contribution to proof-trees construction. The learned knowledge has a re-ordering effect on the search process which reduces the search space considerably. However, the cost of macro-operator matchings, a negative effect, cannot be neglected because some of the learned knowledge is inevitably not operational. So, a key issue in future research is how to obtain a small and highly organized set of perceptual chunks by eliminating ones with low utility and selecting ones with large re-ordering effects. This requires that we add to the current framework a mechanism for empirically measuring the utility of each perceptual chunk.

REFERENCES

Anderson, J. R. (1983). *The Architecture of Cognition*, Harvard University Press, Massachusetts, London, England.

Fikes, R., Hart, P. and Nilsson, N. (1972). Learning and executing generalized robot plans, *Artificial Intelligence, 3*, 251–288.

Koedinger, K. R. (1992). Emergent properties and structural constraints: advantages of diagrammatic representations for reasoning and learning, *Working Notes of the 1992 AAAI Spring Symposium on Reasoning with Diagrammatic Representations*, Stanford Univ., March 27–29.

Koedinger, K. R. and Anderson, J. R. (1990). Abstract planning and perceptual chunks: elements of expertise in geometry, *Cognitive Science, 14*, 511–550.

Laird, J. E., Newell, A. and Rosenbloom, P. S. (1987). SOAR: an architecture for general intelligence, *Artificial Intelligence, 33*, 1–64.

Minton, S. (1984). Constraint-based generalization, *Proceedings AAAI-84*, 251–254.

Minton, S. (1985). Selectively generalizing plans for problem solving, *Proceedings IJCAI-85*, 596–602.

Minton, S. (1990). Quantitative results concerning the utility of explanation-based learning, *Artificial Intelligence, 42*, 363–391.

Minton, S., Carbonell, J. G., Knoblock, C. A., Kuokka, D. R., Etzioni, O. and Gil, Y. (1989). Explanation-based learning: a problem solving perspective, *Artificial Intelligence, 40*, 63–118.

Mitchell, T., Utgoff, P. and Banerji, R. (1983). Learning by experimentation: acquiring and refining problem-solving heuristics, *Michalski, R. S., Carbonell, J. G. and Mitchell, T. M., eds., Machine Learning: An Artificial Intelligence Approach.* Tioga, Palo Alto, CA, 163–190.

Suwa, M. and Motoda, H. (1989). Acquisition of associative knowledge by the frustration-based learning method in an auxiliary-line problem, *Knowledge Acquisition, 1*, 113–137.

Suwa, M. and Motoda, H. (1991). Learning abductive strategies from an example, *Working Notes of AAAI-91 workshop on Towards Domain-independent Strategies for Abduction; also Tech. Report 91-JJ-WORKSHOP, Department of Computer and Information Science, The Ohio State University*, 72–79.

17

Inductively Speeding Up Logic Programs

M. Numao, T. Maruoka, and M. Shimura

Department of Computer Science
Tokyo Institute of Technology, Tokyo

Abstract

This chapter presents a speed-up learning method for logic programs, which accelerates a program by composing macro clauses based on partial structures of proof trees. Many systems have been proposed for composing useful macros, e.g., some of them select macros that connect two peaks in a heuristic function. Another employs heuristics that select useful macros. Although they work well in some domains, such methods depend on domain-dependent heuristics that have to be exploited by their users.

We propose a heuristic-independent mechanism by detecting backtracking. The method uses a dead-end path as a negative explanation tree, compares it with positive one, and finds a first different node to remove its corresponding rule by composing a macro. Repeated substructures in such a macro are then combined by applying the generalize-number technique and by sharing common substructures.

Experimental results in STRIPS domain show that, by selecting an appropriate set of macros, 1) backtracking in solving training examples are suppressed, 2) its problem solving efficiency does not deteriorate even after learning a number of examples, 3) after learning 30 training examples, no backtracking occurs in solving 100 test examples different from the training examples. In conclusion, the proposed method speeds up the problem solving from 10 to 100 times.

1 INTRODUCTION

Explanation-Based Learning (EBL) has been used for speed-up learning in problem solving. Since there are many combinations of macros in each explanation, EBL systems need a selective learning mechanism of macros. Some systems select macros that connect two peaks in a heuristic function (Iba 1985; Minton 1985). Another system employs heuristics that select useful macros (Yamada and Tsuji 1989). Although they work well in some domains, such methods depend on domain-dependent heuristics that have to be exploited by their users.

This chapter presents a heuristic-independent mechanism by detecting backtracking. The method uses a dead-end path as a negative explanation tree, compares it with positive one, and finds a first different node to remove its corresponding rule by composing a macro. Experimental results in STRIPS domain show that, after learning 30 training examples, no backtracking occurs in solving 100 test examples, and that the problem solving is speeded up from 10 to 100 times.

2 OUR APPROACH

Symbol-Level Learning (SLL) improves a rule set based on training examples. Figure 17.1 shows some classes of SLL. A rule set is better than the others if it works faster and occupies smaller memory space, i.e., i) its reasoning speed is faster, and ii) its size is smaller. Reasoning speed depends on the cost of unification and backtracking. Optimization by the recent Prolog compilers removes most of the unification processes, while they do not optimize backtracking that depends on the query and occurs dynamically. AI programs are usually declarative and nondeterministic, and thus cause backtracking. They have had to be hand-rewritten to efficient ones.

Composing macros is a classical method of speed-up learning, which decreases both the cost of backtracking and unification. However, composing too large or too many macros worsens reasoning speed and causes the utility problem (Minton 1988). Another method controls reasoning by employing a meta-interpreter. It is analysed very well (Minton 1988), though such

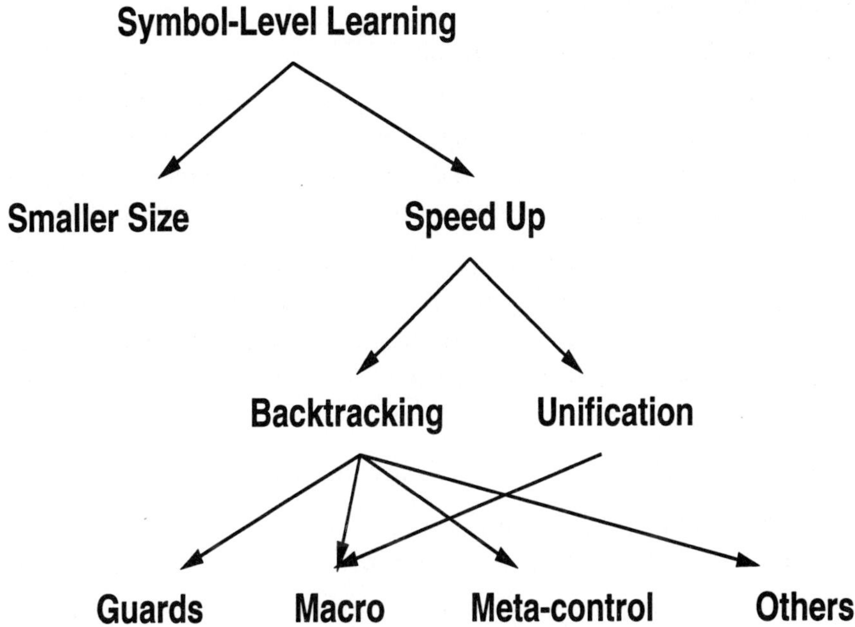

Figure 17.1. Classes of Symbol-Level Learning.

an interpreter usually slows down reasoning. A rather new technique is to attach *guards* to each clause and to select an appropriate clause (Cohen 1990; Zelle and Mooney 1992). This works well for suppressing backtracking, if the guards are kept simple and checked fast.

On the other hand, the authors proposed a learning method based on partial structures of explanations, which selectively learns macro operators (Numao and Shimura 1989, 1990). The method learns at the knowledge level in the field of translation and logical circuit design, and outputs term-rewriting rules for a production system *Nat*. In this chapter, we apply this method to the learning of Horn-clauses, and investigate a method to suppress backtracking by assuming a failure as a negative instance.

To reduce the size of a learned rule grown during the learning process, we will develop a method for folding a repetition in each macro and extracting a common subsequence in macros. It makes the subsequence shared to save the memory space and the cost of unification.

3 LEARNING ALGORITHM

3.1 Generalization and Specialization

In prolog, a computation progresses via *goal reduction*[1]. In each stage of the computation, there exists a resolvent, i.e., a conjunction of goals to be proved. To suppress backtracking, we construct a learning mechanism for avoiding failing paths.

Definition 17.1 (Instance) An *instance* is a pair of a query and a resolvent: (query, resolvent). A clause set *derives* an instance iff the query in the instance derives the resolvent. An instance is *positive* if its resolvent is reduced to be empty, and *negative* otherwise.

Example 17.2 Consider the following logic program:

```
p :- q(X), q(Y), r(X,Y).
q(b).  q(c).  r(a,b).  r(b,c).
```

A computation of a query '`?- p.`' progresses via the following goal reduction:

p	▷	`q(X),q(Y),r(X,Y)`
	▷	`q(Y),r(b,Y)`
	▷	`r(b,b)` (failure)
(backtracking)	▷	`r(b,c)`
	▷	□ (success)

A pair of `p` and each resolvent above is an instance. The resolvents in the following instances are reduced to be empty:

(`p`, (`q(X),q(Y),r(X,Y)`))
(`p`, (`q(Y),r(b,Y)`))
(`p`, `r(b,c)`)
(`p`, □)

Thus, they are positive instances. The following instance is negative since its resolvent fails:

(`p`, `r(b,b)`)

□

[1]See 1.8 (p.12) and 4.2 (p.72) in (Sterling and Shapiro 1986).

Definition 17.3 (Generalization and Specialization of a Clause Set) Let I_1 and I_2 be instance sets derived from clause sets S_1 and S_2, respectively. S_1 is a *generalization* of S_2, and S_2 is a *specialization* of S_1, denoted by $S_1 \preceq S_2$, iff $I_1 \supseteq I_2$.

The following two theorems show that Explanation-Based Generalization (EBG) (Mitchell *et al.* 1986; Kedar-Cabelli and McCarty 1987) specializes a clause set:

Theorem 17.4 Let c be a macro generated by EBG from an explanation including clauses $c_1, \ldots, c_n$. Then, $\{c_1, \ldots, c_n\} \preceq c$.

Proof. Let i be any instance derived from $\{c\}$. Since c is generated from $c_1, \ldots, c_n$, i is also derived from $\{c_1, \ldots, c_n\}$. Therefore, $\{c_1, \ldots, c_n\} \preceq \{c\}$. In general, $\{c_1, \ldots, c_n\} \neq \{c\}$, since some instances derived from $c_1, \ldots, c_n$ are not derived from c.□

Theorem 17.5 If $S_1 \preceq S_2$ and $S_3 \preceq S_4$ then, $S_1 \cup S_3 \preceq S_2 \cup S_4$.

Proof. Suppose $S_2 \cup S_4$ derives i. Consider a clause in S_2 applied at a step in the derivation. Since $S_1 \preceq S_2$, S_1 derives the same step. Similarly when a clause in S_4 is applied, S_3 derives the same step. Since $S_1 \cup S_3$ derives each step of the derivation, $S_1 \cup S_3$ derives i, by which we conclude that $(S_1 \cup S_3) \preceq (S_2 \cup S_4)$.

□

These theorems proclaim that a clause set is specialized when $c_1, \ldots, c_n$ is replaced by c. We speed up a program by making a specialization that avoids backtracking.

3.2 An Algorithm for Speed-Up Learning

Figure 17.2 shows an algorithm to suppress backtracking based on positive and negative instances. A *proof tree* is an ordered tree, each of whose nodes specifies a clause. Children of a node specify clauses applied to its subgoals in their order. The function *clause*(x) gives a clause applied in a node x. The algorithm traverses a proof tree, finds an inappropriate application of a clause, and replaces the clause by a macro to suppress backtracking.

The algorithm only needs part of the instances given in Definition 17.1. A *positive instance* P_i, is: *(instantiatedQuery,*$\{\}$*)*

```
begin
  clauseSet := domain_theory;
  EP_i := proof tree of each positive instance P_i (i = 1, ..., n);
  for i := 1 to n do
    while clauseSet derives any negative instance do begin
      N := the derived negative instance;
      EN := the proof tree of N;
      Compare each node in EN with its corresponding node
      in EP_i in depth-first order;
      h := the first different node;
      clauseSet := {};
      for i = 1 to n do
        for each p ∈ EP_i do
          if clause(p) = clause(h) then begin
            q := the parent or a child of p;     ...(*)
            Make a macro c based on clause(p) and clause(q);
            clauseSet := clauseSet ∪ {c}
          end
          else clauseSet := clauseSet ∪ {clause(p)}
    end
end.
```

Figure 17.2. Learning Algorithm.

where *instantiatedQuery* is a query with output variables instantiated correctly. In the fifth line, the algorithm checks whether a query with output variables uninstantiated causes backtracking or not. If it does, the *derived negative instance* N in the sixth line is a pair: $(query, resolvents)$ where *resolvents* is a set of resolvents at the point of failure. EP_i and EN are proof trees of P_i and N, respectively. The algorithm finds the first incorrectly applied clause in EN by comparing both the proof trees, and replaces it by a macro.

Let us suppress backtracking in Example 17.2 by using this algorithm. Positive and negative instances P_1 and N are $(\mathtt{p}, \Box)$ and $(\mathtt{p}, \mathtt{r(b,b)})$, respectively. Their proof trees EP_1 and EN are shown in Figure 17.3. In each node, *predicate_place* specifies

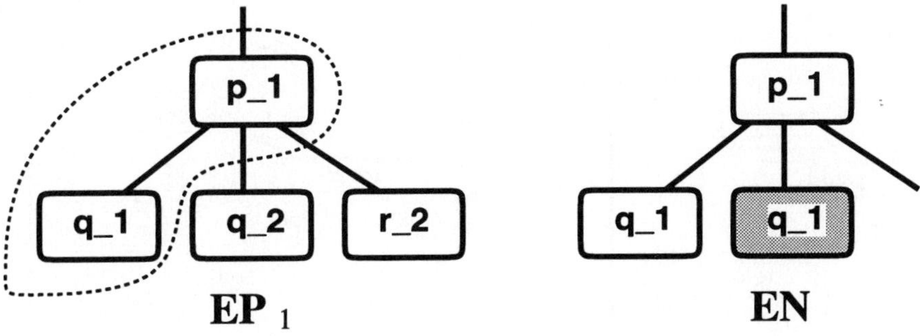

Figure 17.3. Proof Trees of the Positive and Negative Instances.

a clause, where *predicate* and *place* are a predicate and a place in its definition, respectively; e.g., **q_2** is the second clause in the definition of `q`, i.e., the fact ' `q(c)`'. The first different node h between EP_1 and EN is the shadowed node **q_1**. While traversing EP_1 in the **for** loop, $clause(p) = clause(h)$ when $p =$ **q_1**. Thus, the algorithm makes a macro c based on $clause$(**p_1**) and $clause$(**q_1**) encircled with a dotted line, and generates the following deterministic $clauseSet$:

```
p :- q(Y),r(b,Y).
q(c).
r(b,c).
```

We need only one macro for removing backtracking above. In general, the **while** loop in the fifth line invokes iterative composition of a larger macro while the *clauseSet* causes backtracking.

A clause usually checks their applying conditions in the leftmost part of its body as follows:[2]

head :- *conditions*, ... , *goals*, ...

We assume such *conditions* are operational and ignore backtracking there.

EBG generates a macro c based on p and its parent or child q at (*) in the algorithm. Nodes p and q should be a node and its first child; e.g., **p_1** and **q_1** in Figure 17.3. A macro based on the node and its other children does not suppress backtrack-

[2]The planning program for the following experiment consists of such clauses.

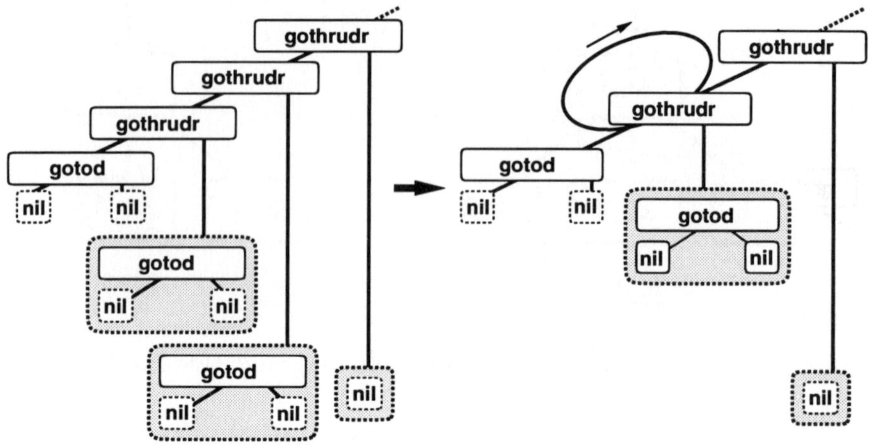

Figure 17.4. Generalizing Number.

ing, since the generated clause has conditions and goals aligned alternatively.

3.3 Generalizing Number and Sharing a Common Sequence

The presented algorithm tends to generate a large macro, which occupies much memory space and slows down its unification. Since such a large macro usually contains a repeated sequence, we apply the *generalizing-number* technique (Shavlik and DeJong 1987a, 1987b) to the extracted macros, which integrates the different number of repetitions. Figure 17.4 shows an example of a generalizing number. Such a generalized repetition is implemented as a group of clauses internally calling one another.

After the generalizing number, if two sequences have a common structure, such as the left trees in Figure 17.5, make the sequence shared as shown in the right tree. This transformation saves memory space and the cost of unification.

Sharing a common sequence may cause over-generalization, i.e., the left trees do not include the combination of **nil** and *subtree C*, while the right tree does. In the experiment to be presented in the next section, this causes no problem since the selection between *subtree B* and *subtree C* does not depend on **gothrudr** nor **nil**, though in general we need more investigation.

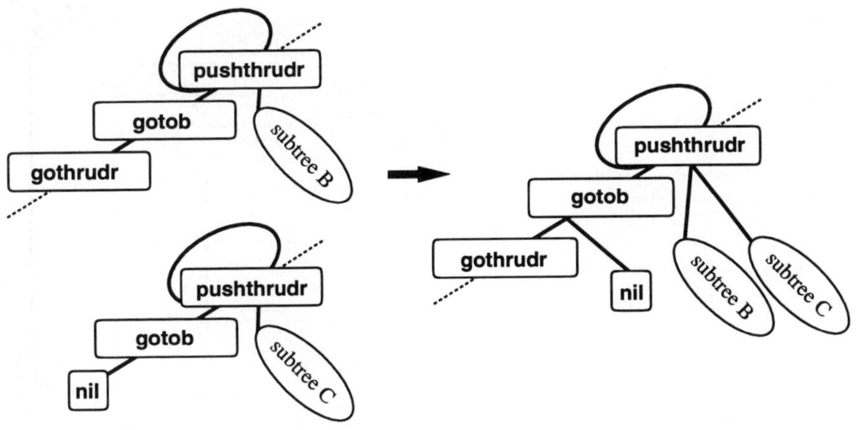

Figure 17.5. Sharing a Common Sequence.

4 EVALUATION OF THE METHOD

4.1 STRIPS in Prolog

We evaluate the proposed method based on STRIPS planning system (Fikes et al. 1971, 1972). A system written in Prolog is needed to apply the proposed learning mechanism. The authors rewrote a simple planner in (Sterling and Shapiro 1986)[3]. It searches a plan in a depth-first manner prevented from an infinite loop by checking formerly occurred models and by limiting the search depth.

Consider forming a plan for achieving the goal state in Figure 17.6 from the initial state. The proof tree in the figure shows its planning process. The solid line shows the proof tree of the positive instance (EP_1). Each dotted line partially describes a proof tree of a negative instance (EN).

The following is a query for the planning procedure:

```
?- transform(
     [type(room1,room),...,pushable(box1),
      connects(door1,room1,room2)],                 ··· fact
     [inroom(box1,room1),inroom(robot,room1),
      status(door1,closed)],                        ··· initial state
     [inroom(robot,room2),inroom(box1,room2)],      ··· goal state
     Plan).
```

[3]Program 14.11 (p.222).

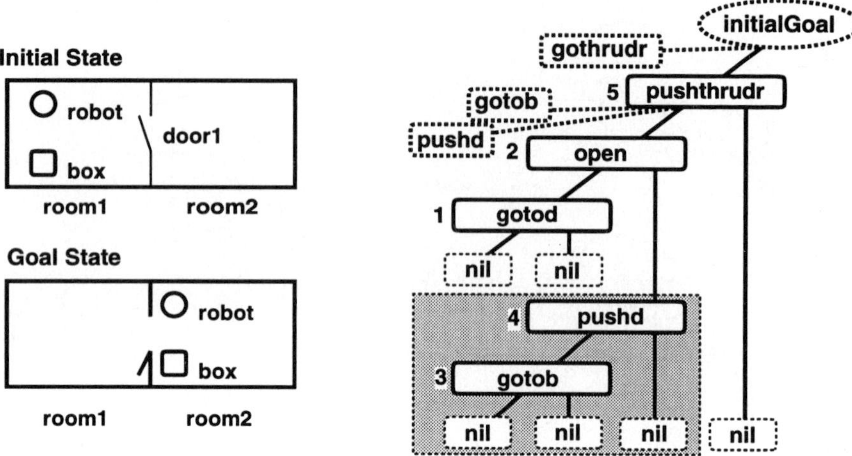

Figure 17.6. A Planning Example and its Proof Tree.

This query invokes the following clause, represented by **initial-Goal** in the proof tree:

```
transform(Fact,Model,Goal,Plan):-
    diff(Model,Goal,Diff),
    transform(Fact,Model,Goal,Diff,Ac,[],Model,Actions,N,0),
    reverse(Actions,Plan).
```

where `diff(Model,Goal,Diff)` is an operational predicate that calculates the difference between `Model` and `Goal`. If no difference is detected, the following clause terminates the recursion:

```
transform(Fact,Model,G,[],Na,Vg,Vm,[],Model,_).
```

It is indicated by **nil** in the proof tree.

Each of the other nodes indicates a planning clause selecting an operator. The system has the following eight operators:

- *gotob(BX)* Go to object *BX*.
- *gotod(DX)* Go to door *DX*.
- *pushb(BX,BY)* Push *BX* to object *BY*.
- *pushd(BX,DX)* Push *BX* to door *DX*.
- *gothrudr(DX,RX)* Go through door *DX* into room *RX*.
- *pushthrudr(BX,DX,RX)* Push *BX* through door *DX* into room *RX*.
- *open(DX)* Open door *DX*.
- *close(DX)* Close door *DX*.

As an example of the planning clauses, one selecting *pushthrudr* operator is as follows:

```
transform(Fact,Model,Goal,Diff,pushthrudr(BX,DX,RX),
          Vgoal,Vmodel,Plan,New_model,M) :-
  conditions(Fact,Model,Goal,pushthrudr(BX,DX,RX),Diff), ...,
  transform(Fact,Model,Sub_goal,Diff1,Action,[Effect|Vgoal],
           Vmodel,Actions,Nmodel,NM), ...,
  transform(Fact,Update_model,Goal,Diff2,Next_ac,
           [Effect|Vgoal],[Update_model|Vmodel],Acts,
           New_model,NM),
  append(Acts,[pushthrudr(BX,DX,RX,RY)|Actions],Plan).
```

where conditions(_,_,_,_,_) is an operational predicate that checks conditions for applying the operator. The first subgoal transform(Fact, Model, ···) makes a subplan required before the application of the operator. The second subgoal transform(Fact, Update_model, ···) makes a subplan after the application.

The differences between the initial state and the goal state in Figure 17.6 are the places of the robot and the box. If the planner selects the latter, the operator is pushthrudr(box1, door1, room1, room2) (node 5 in the figure), whose precondition is:

```
inroom(box1,room1), inroom(robot,room1),
nextto(robot,box1), nextto(box1,door1), status(door1,open),
```

out of which a subplan has to reduce:

```
nextto(robot,box1), nextto(box1,door1), status(door1,open).
```

If the planner determines to reduce status(door1,open), the operator is open(door1) (node 2). nextto(robot,door1) is reduced by the directly applicable operator gotod(door1) (node 1). After open(door1) is applied, the subtree whose root is node 4 reduces other differences for the application of pushthrudr(box1, door1, room1, room2). The result is the inorder sequence of operators in the tree, indicated by numbers in the figure, as follows:

gotod(door1)
open(door1)
gotob(box1)
pushd(box1,door1)
pushthrudr(box1,door1,room2)

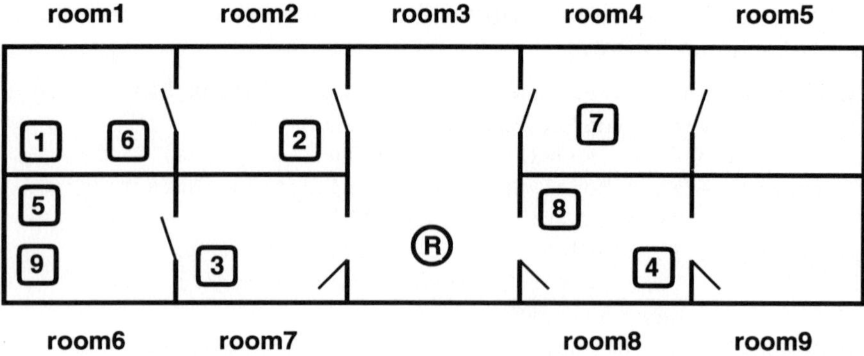

Figure 17.7. A State.

Backtracking occurs when the planner determines to reduce other differences during the above process. If it selects the place of the robot instead of the box, the operator is `gothrudr(door1, room2)` as shown in a dotted line in the figure. As such, the operator becomes `gotob(box1)` and `pushd(box1, door1)`, if the planner tries to reduce `nexto(robot, box1)` and `nextto(box1, door1)`, respectively, instead of `status(open)`. To suppress backtracking, the algorithm in Figure 17.2 composes a macro based on nodes 3 and 4, and removes **gothrudr**, **gotob** and **pushd**. Since the composed macro is applied instead of node 2, a larger macro shadowed in the figure is composed. Operators **pushthrudr**, **open**, **gotod**, **nil** and the last macro make the plan without backtracking.

4.2 Experiments

The authors performed experiments on planning in a robot domain shown in Figure 17.7. They generated the initial and the goal states of planning by randomly choosing:

- the number of boxes (0–9),
- the locations of boxes,
- whether each box contacts another box or not,
- whether each door is open or not, and
- the location of robot.

1000 pairs of such states are stored, out of which training and test examples were sampled for each experiment. The length of plans for the pairs are from 10 to 120 steps.

By solving the training examples with learned clauses, the authors first verified that our method suppresses backtracking:

> **Experiment A** After learning 20 examples, compare CPU time for the non-learned and the learned clauses to solve the same examples.

If the system extracts superfluous macros, problem solving efficiency deteriorates after learning a number of examples. This phenomenon is checked by the following experiment:

> **Experiment B** Perform the same experiment for 30 examples by adding 10 examples to Experiment A.

Figure 17.8 shows the result of Experiments A and B. The CPU time is for SICStus Prolog version 0.7 on Sparc Station 2. No backtracking occurred after learning in Experiment A. The learning accelerated the problem solving from 10 to 100 times, and reduced the number of searched nodes from 1/50 to 1/1000 in both the experiments.

Although learned sets of clauses in both the experiments are different, their efficiency is almost the same as shown in Figure 17.8, exhibiting that correct macros are extracted. When the system learns more than 30 examples, it composes no more combinations as a macro, i.e., the result is the same even if the system learns more than 30 examples.

After learning 30 examples, the system composes 5 or 6 macros. Each macro is constructed from some clauses, which implement the generalizing number and share a common sequence. The number of clauses constructing the macros sums up to 104, occupying rather much memory space, though their reasoning speed is fast.

To test generality of extracted macros, the system solved a different set of examples:

> **Experiment C** After learning 30 training examples, solve the other 100 test examples.

Figure 17.9 shows the result of Experiment C. The learned macro solved the 100 examples without any backtracking, and

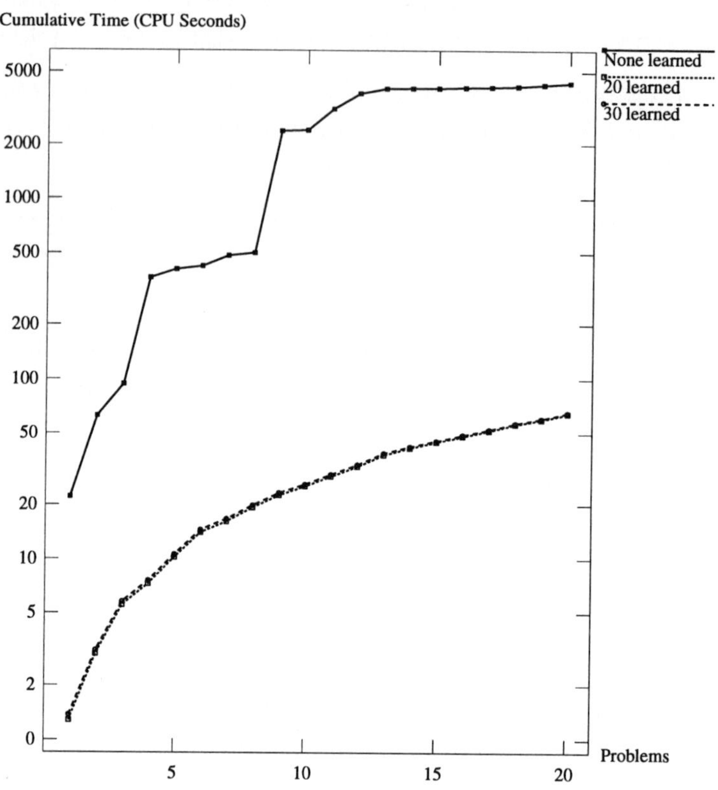

Figure 17.8. The Result of Experiments A and B.

improved the problem solving similarly to Experiment B. This result shows that the system learned general and useful macros.

The problem in the current system is that it takes much time for learning, e.g., about 10 hours for learning 30 examples. Therefore, the method is appropriate for speeding up frequently used programs, though it should be improved in a future implementation.

5 RELATED WORK

Iba (1985) and Minton (1985) proposed the *peak-to-peak* heuristics that composes a macro connecting two peaks in an eval-

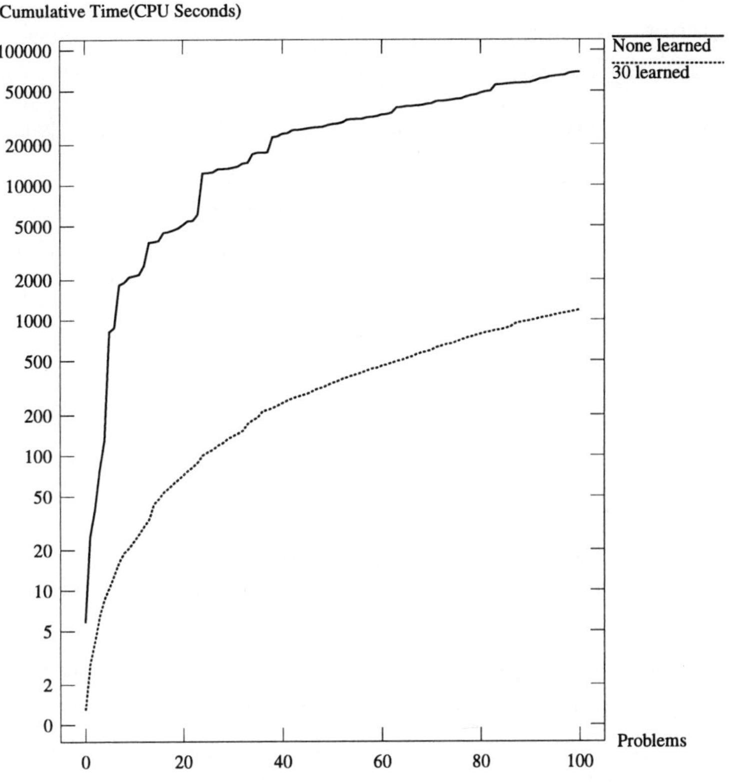

Figure 17.9. The Result of Experiment C.

uation function. However, it is usually hard to construct an appropriate heuristic function for evaluating goals in logic programs.

Minton (1985), Yamamura and Kobayashi (1991) presented methods to extract common substructures in explanations. This is useful for reducing the number of macros, though it may not accelerate problem solving. Our method concentrates on suppressing backtracking, after which it reduces the number of macros by applying the generalizing-number technique and making common structures shared.

Yamada and Tsuji (1989) proposed a heuristics called the

perfect causality for selecting appropriate macros. Although it works well in many problems, it may slow down problem solving in others. By suppressing backtracking, our method straightforwardly speeds up problem solving.

Minton (1988) presented *learning from failure*. It is a similar method to ours in the sense that both the methods are based on failures. The difference is in their representation of learned knowledge. Minton's method learns control heuristics for the *Prodigy* system, and thus needs a meta-interpreter for application to logic programs. Such an interpreter runs slower than compiled Prolog programs. In contrast, our method learns macros, efficiently executable by Prolog processors.

In the field of Combining Empirical and Explanation-based Learning, some researchers pursue methods to extract partial structures of explanations based on positive and negative examples (Numao and Shimura 1989, 1990), and to search a macro on a generalization hierarchy of common explanation structures (Kobayashi et al. 1991). In contrast to our research focusing on learning at the symbol level, they evaluate macros at the knowledge level.

Cohen (1990), Zelle and Mooney (1992) proposed to add some guards to each clause for controlling and speeding up a program. Such guards are synthesized by using the techniques of Inductive Logic Programming. Instead of introducing such additional guards, our method composes macros, which are decomposed into simpler clauses by generalizing numbers and sharing common structures.

6 CONCLUSION

We have presented a method for extracting macros based on backtracking, and evaluated it by using planning examples in STRIPS. This method is widely applicable to automatic improvement of logic programs and knowledge bases.

ACKNOWLEDGMENT

The author would like to thank Dr. Seiji Yamada, who sent us data of his experiment (Yamada and Tsuji 1989). Tatsuya Kawamoto has assisted us in using SICStus Prolog.

REFERENCES

Cohen, W. (1990). Learning approximate control rules of high utility. *Proc. the Seventh International Conference on Machine Learning*, 268–276, Morgan Kaufmann, San Mateo.

Fikes, R. E., Hart, P. E., and Nilsson, N. J. (1972). Learning and executing generalized robot plans. *Artificial Intelligence*, *3*, 251–288.

Fikes R. E. and Nilsson N. J. (1971). STRIPS: a new approach to the application of theorem proving to problem solving. *Artificial Intelligence*, *2*, 189–208.

Iba G. A. (1985). Learning by discovering macros in puzzle solving. *Proceedings of the Ninth International Joint Conference on Artificial Intelligence*, 640–642, Morgan Kaufmann.

Kedar-Cabelli S. T. and McCarty. L. T. (1987). Explanation-based generalization as resolution theorem proving. *Proceedings of the Fourth International Workshop on Machine Learning*, 383–389, Morgan Kaufmann.

Kobayashi, S., Shirai, Y., and Yamamura, M. (1991). Acquiring valid macro rules under the imperfect domain theory by top-down search on a generalization hierarchy of common explanation structures (in Japanese). *Journal of Japanese Society for Artificial Intelligence*, *6(3)*, 416–425.

Minton. S. (1985). Selectively generalizing plans for problem-solving. *Proceedings of the Ninth International Joint Conference on Artificial Intelligence*, 596–599, Morgan Kaufmann.

Minton, S. (1988). *Learning Search Control Knowledge.* Kluwer Academic Publishers, Boston/Dordrecht/London.

Mitchell, T. M., Keller, R. M., and Kedar-Cabelli, S. T. (1986). Explanation-based generalization: a unifying view. *Machine Learning*, *1*, 47–80.

Numao, M. and Shimura, M. (1989). A learning method based on partial structures of explanations (in Japanese). *The Transac-*

tions of the Institute of Electronics, Information and Communication Engineers, J72–D–II(2), 263–270.

Numao, M. and Shimura, M. (1990). A knowledge-level analysis of explanation-based learning. *Proceedings of the Third International Conference on Industrial & Engineering Applications of Artificial Intelligence & Expert Systems*, 959–967, ACM Press, New York.

Shavlik, J. W. and DeJong, G. F. (1987a). BAGGER: an EBL system that extends and generalizes explanations. *Proceedings of the Sixth National Conference on Artificial Intelligence*, Morgan Kaufmann.

Shavlik, J. W. and DeJong, G. F. (1987b). An explanation-based approach to generalizing number. *Proceedings of the Tenth International Joint Conference on Artificial Intelligence*, Morgan Kaufmann.

Sterling L. and Shapiro. E. (1986). *The Art of Prolog.* The MIT Press, Cambridge.

Yamada, S. and Tsuji, S. (1989). Selective learning of macro-operators with perfect causality. *IJCAI89*, 603–608, Morgan Kaufmann, San Mateo.

Yamamura, M. and Kobayashi, S. (1991). An augmented EBL and its application to the utility problem. *IJCAI91*, 623–629, Morgan Kaufmann, San Mateo.

Zelle J. M. and Mooney, R. J. (1992). Speeding-up logic programs by combining EBG and FOIL. *Proc. ML92 Workshop on Knowledge Compilation & Speedup Learning.*

INDEX

Note: illustrations and captions are indicated by *italic page numbers*, footnotes by suffix 'n'

AC2 algorithm 324
  performance on various datasets 338, 341, *352–5*, *357–9*
ACE computer 9–10, 11, 40–2
ACT theory 437
agents
  coordination using blackboard 411
  derivation by inductive learning 411–12
  low-level skills implemented by 411
aircraft flight simulators 400–1
  inductive rule-learning experiments 404–8
    results 406, 408, *409–10*
ALGOL 14
algorithm, use by Turing 7
algorithmic information theory, hypothesis evaluation using 90, 91
Alloc80 algorithm 317
  performance on various datasets 335, 336, 339, *352–9*
α-helix formation 101n, 200
  chemical principles 200, *212*
ambient vision 258
Angluin's learning algorithm
  compared with Yokomori's algorithm 182–4, *185*
  example operation 183, *185*
animal learning 213–28
  basic procedures 214–15
  perspectives 215–16
anticipation 225–7, *245–6*
AQ algorithm 325
  see also CN2 algorithm
artificial intelligence
  and logic 6, 21–3
  Turing's contribution 9, 24
  von Neumann's contribution 24
artificial neuron concept 4, 12, 28
  see also neural networks
assembler program 23
associative transfer (in conditioning) 219–20
atoms, Boolean algebra 146
attribute generalization 139
attribute noise 124
  CAFE's learning ability affected by 135, 136, *136*
  COBWEB's learning ability affected by 134, *136*
attribute-oriented induction 77–9
  application of minimal multiple generalization 79–83
automated hypothesis-formation 87, 88
automatic programming 13–14, 23
automatization, skill-learning 389
autopilots 406, 408
  degree of control achieved 408, *409–10*
  heuristic vs causal models 413–14
Babbage's Analytical Engine 42
background knowledge
  chess endgames 298, 299
  drug design 196, 201
  encoding of 94–7
  protein structure 199–200

representation of 93, 196, 200, 201
back-propagation 326–7
compared with other systems 313–14, *352–9*
performance on various datasets 336, 337, 338, 339, 341, *352–9*
Back-propagation Multi-Layer Perceptron 326–7
biological interpretation 327
backtracking 442, *443*
occurrence in planning 452
suppression of 443, 444, 446–7, 453
bang-bang control 364, 396, 398
batch processing, inductive inference from frequency data 159–60
Bayesian metworks 320–1
see also CASTLE algorithm
Bayesian posterior variance, as significance measure 102–3, *104*
Bayes rule algorithm 316
performance on various datasets 335, 337, 339, 341, *352–9*
behavioural field, mechanism for generating 251, 252, 255
beta-reduction rule 16, 20
bimolecular modelling, aims 194
bindings
meaning of term 286
minimum number in universal engineering knowledge base 286, 287
blackboard-like model, coordination of agents using 411–12
BMT expert system 390, *391*
Boolean algebra
atoms 146
construction of Euclidean space 146, 148
Hasse diagram *147*
Boswell (patient with no episodic memory) 390, 392
BOXES algorithm 364–9, 396
cleaning up *379*, 379–80
comparison of variants *370*
Cribb–Sammut variant 367–8
Cribb's version 367
decisions on action 365–9
Law–Sammut variant 368–9
properties 369, 370–3
Law's version 368
local merit criterion introduced 367
original Michie–Chambers algorithm 366–7, 396
patterns discovered 379–80
readability of code 379, 381–2
task addressed 363–4
brain
left hemisphere *257*
activities associated 258, 414
right hemisphere *257*
activities associated 258, 414
brain development, embryo-to-young-child 250–5
brain stem systems, activities controlled 258–9
brain systems 248, 249
British Museum algorithm 30
brittleness bottleneck (problem in knowledge bases) 278–9
C4.5 algorithm 324
aircraft flight simulation 405, 406
performance on various datasets *352–7*, *359*
pole-and-cart system control rules 401
CAFE concept formation system
algorithm *131*
characteristics 125, 129–30

classification and learning by 129–30
compared with COBWEB 129–30, *134*, 135–6, 137–8
compared with LABYRINTH 125, 126, 129, 138–9
concept-predictability used 139–40
evaluation of system 134–9
artificial domain 134–6
natural domain 137–9
knowledge representation formalism 126–9
learning curves *134*, *136*, *138*
Cal5 algorithm 324–5
discrimination measure used 325
performance on various datasets *352–9*
canonical discriminant correlations (of StatLog project datasets) 346, *351*
canonical discriminants, variation (for StatLog project datasets) 346–7, *351*
cardiology 306–7, 332–3
see also heart disease dataset; KARDIO system
CART system 321–3
compared with ID3-type algorithms 322, 323
compared with neural networks 314
impurity measure used 322
performance on various datasets 333, 338, *352–3*, *355–8*
pruning used 323
versions available 323
see also INDCART
CASTLE algorithm 320–1
characterization measures for 334, 342
performance on various datasets 335, 336, 338, 339, 341–2, *352–9*
category utility 130
causal networks 320
approach to learning 321
determination of structure 320–1
see also CASTLE
Central Limit Theorem 336
cerebral cortex
development 248
links between two halves 255, *257*
specialized areas 260
Cessna aircraft flight simulator 404–6
flight plan described 404
variables sampled 405
CFIX concept formation system 125
checkers (draughts)
knowledge acquisition 392
playing programs 37, 46, 47
chess endgames
learning classification rules 292
learning optimal strategies 291–307
materials used 293–6
method used 296–8
results 298–306, 307
see also KRK chess endgame
chess playing machines, Turing's discussions 39, 41
chess program 291–2
child development, comparison with machine learning 45
choice-point encoding 98–9
Church, Alonzo 10
solution of Decision Problem 10, 38
Church machines 18–20
Church's lambda calculus 5
beta-reduction rule 16, 20
mathematical interpretation 14
see also lambda calculus

classical conditioning 215
  with feedback 216
classical logic, new (Boolean algebra) model 150
classification, meaning of term 311, 312
classification algorithms
  comparative study 311–59
  representative collection 313
classification noise model 101, 113
CLASSIT conceptual clustering system 333
clause sets, specialization of 445
clean-up effects
  aircraft flight simulator control 406, 408, *409–10*
  BOXES algorithm *379*, 380
  pole-balancing control system *379*, 380, 402–3, *403*
closed-world specialization (CWS) 107, 297
  algorithm *108*
  clauses produced, compression-based selection used *112*
CN2 algorithm 325–6
  performance on various datasets 338, *352–5*, *357–9*
COBWEB 124, 125
  compared with CAFE 129–30, *134*, 135–6, 137–8
  evaluation function used 130, 132–4
  learning curves *134*, *136*, *138*
Colmerauer, Alain 16, 21
commissurotomy 255–6
  effect on speech 258
  effect on visual information 256, 258
common sequence, sharing in learning algorithm 448, *449*
common substructures, extraction in explanations 455
communication mechanisms, human infants 261
compactness, with respect to containment 63–4
competitive learning (in conditioning) 220
compiler program 23
compression
  estimation from theory *111*
  meaning of term 91
  as noise-meter 104–6
  as significance measure 100–4, 115
    compared with other measures 102–3, *104*
  see also database compression
compression-based clause selection 107–13
  CWS-produced clauses *112*
compression-guided learning
  case study 106–15
    empirical evaluation 113–15
compression model
  characteristics 117–18
  effect of noise 114–15
computers, and logic 4–6, 6–12
concept formation
  structure information utilized 123–41
  see also CAFE; COBWEB; LABYRINTH
concept hierarchy 77, 79
  generation of 124, 129
  inductive learning of 124
concept-predictability, CAFE using 125, 129–30, 132, 139–40
concept-predictability function 132, 141
  examples of calculation *133*
conditioned stimulus/response 215
conditioning 215
  associative transfer in 219–20, *239*
  CSPUS interval studied 218, *236*

facilitation of remote associations 219, *237*
first-order 220, *239*
impairment of proximal associations 219, *238*
representation in 221–3
second-order 220, *239*
serial conditioned stimuli 219
summation in 221–2, *240*
SuttonPBarto model 220–1
confirmation theories 90
conformity (for knowledge base) 277–8
Connection Machine 31
consciousness, genesis of 247–50, 414
consistent max trees 67
constraint networks 273, *274*
context-free grammars, learnability 170
context-free transformations with flat base 72–5
inference algorithm for 74
operation of 73–4
contradiction backtracking 173, 176
control decision nodes 425, *428*
successfully applied rule 428
correlation coefficients (for StatLog project datasets) 345–6, *351*
cost matrices, use with classification algorithms 337, 341
counterexamples, learning using *171*, 173
credit assignment 217–21, 364
basic principles 219–20
credit risk dataset (StatLog project) 333–4
characteristics *351*
performance of various algorithms 338–9, *358*
Cribb–Sammut algorithm (for pole-balancing problem) 367–8
compared with other variants of BOXES algorithm *370*
asymmetric pushing 373, *374*
Cribb's local merit criterion 367
critical feedback 216
curriculum construction, design of knowledge base similar to 276, 284
CWS *see* closed-world specialization
cyclopean *252*
with jointed member *253*
watching another cyclopean *254*
CYC project 276
database compression 307
databases, discovery of knowledge in, minimal multiple generalization algorithm used 77–83
datasets (StatLog project) 328–35
characterization of datasets 334–5, 340–1, *351*
measures described 345–7
performance of various algorithms 335–40, *352–9*
see also credit risk...; handwritten digits...; head injury...; heart disease...; Karhunen–Loeve digits...; satellite image...; Space Shuttle control...; vehicle silhouettes dataset
Decision Problem 7
solutions 8, 10, 38
decision rule algorithm 325
see also CN2
decision tree algorithms 321–2, 323, 324
advantages 322
see also AC2; C4.5; Cal5; CART; ID3; INDCART; NewID
decision tree induction 307
patterns discovered using 380

decision trees, aircraft simulation data 405, 406
declarative knowledge 388–9
compared with procedural knowledge 393–4
declarative memory, contrasted with procedural memory 388
deductive knowledge base 25, 26
Desch, Joseph 9n
descriptive approach to programming 14–15, 16, 21
deterministic finite automata (DFAs)
learning using 169
compared with NFA learning *184*
see also non-deterministic finite automata
device malfunction data *165*
description length of propositions obtained 164
Diagram Configuration (DC) model, perceptual chunks used 420
dialogic closure 264
dialogue (knowledge) elicitation methods 390, 392, 393
digits datasets (StatLog project) 329–30
characteristics *351*
performance of various algorithms 336–7, *353–4*
see also handwritten...; Karhunen–Loeve digits dataset
dihydrofolate reductase (DHFR)
binding of trimethoprim analogues 195–8
favourable interactions 197–8, *207*
Discrim algorithm 317–18
performance on various datasets 338, 340, *352–9*
discriminant analysis *see* linear...; quadratic discriminant analysis
DNA promoters data set 137
evaluation of CAFE and other systems 137–9
draughts *see* checkers
drug design 100–1, 195–8
Golem used
compared with Hansch equation 196–8, *208–11*
rules obtained for favourable drugs 197–8
significance measures compared *104*
dual personality, human brain born with 264
EBL systems 117
see also explanation-based learning
ECG interpretation 306–7
Eckert, J.P. 4, 11
Eckert–Mauchly computers 11
EDSAC computer 11
EDVAC computer 11
embryo brain 250
emotional referencing 263
emotions, infant–mother 260–5
engineering, role in computer design 12
engineering knowledge
classification 284–6
general 284, 285
specific 284–5, 285–6
see also nonspecific knowledge
engineering knowledge base
clusters of knowledge *283*
segmentation of knowledge 285–6
size 286, 287
ENIAC computer 11
entropy gain 323
entropy gain ratio 324
episodic memory 388
equivalence queries, learning using 169, 175
error rate

data for protein structure prediction 101, *102*
reduction/correction 110
Escherichia coli dihydrofolate reductase
modelling of trimethoprim analogues binding to 195–6
see also dihydrofolate reductase (DHFR)
Euclidean space, construction using Boolean algebra 146, 148, 151–3
evaluation of logical theories
encoding scheme for 93–100
comparison with FOIL's scheme 99–100
input tape encoding 93–9
background knowledge and hypothesis encoding 94–7
example 96
proof encoding 97–9
output tape encoding 99
exclusive-OR (XOR) problem, solving of 221
experimental machine learning 289–359
expert systems 26–7
knowledge acquisition 392
typical rule-based systems *391*
Explanation-Based Generalization (EBG) 445, 447
explanation-based learning (EBL) 420
compared with PCLEARN 431–6
in geometry domain 423–4
macro-operators 421
operationality in geometry domain 421, 431–3
use in speed-up learning 442
eyelid closure
conditioning in rabbit 217, 226, *234–5*, *245*
real-time models 226–7, *246*
as protective measure 217, 226
F-16 aircraft flight simulator 406–8, *409–10*
clean-up effect due to autopilot *409–10*
flight plan described 407
variables sampled 407–8
failure, learning from 456
familiar objects, time taken for recognition 393
feature construction technique (in CAFE) 124, 125, 141
Fifth Generation Project 27, 51, 388
figure-pattern strategies 420
finite elasticity 71
finite thickness 71
fire-protection equipment configuration, expert system for *391*
see also BMT
first-order conditioning 220, *239*
flight control simulations
blackboard-like model for coordination among agents 411–12
Cessna aircraft 404–6
F-16 combat aircraft 406–10
flowchart diagram
early use 13
learning algorithms (for NFAs) *179*, *187*
FOIL learning system 51
compared with neural network approach 314
encoding scheme 99–100
forebrain commissurotomy 258
see also commissurotomy
Friedman, William 9
functional constraints *272*
see also higher-order constraint network
functional programming 143
GASOIL expert system 390, *391*
GCWS algorithm 297

chess endgame Prolog programs learned by 293
generalization relation 79
generalized database 80
with conceptual hierarchy 79
LCHR algorithm used 77–8
generalized linear models (GLIMs) 318
generalizing-number technique 448
geometry domain 422–3
EBL techniques used 423–4, 431–6
experimental results using EBL and PCLEARN 430–6
learning search control knowledge in 422–4
PCLEARN used 427, 431–6
recognition rules *427*
GINI criterion 322
Glasgow Outcome Scale 331–2
Gluck's category utility 130
goal-oriented search control knowledge, learning 420
goal reduction, Prolog computation via 444
Gödel, Kurt, as first programmer 7
Gödel's descriptive recursive function formalism 6–7, 15
Gödel's Incompleteness Theorem 6–7, 38, 39
Turing's attempt to circumvent 38
Golem learning program 194–5
algorithm used 194–5
applications 50
bimolecular modelling 193–212
drug design 196–8, *208–10*
protein structure prediction 101n, 199–201, *212*
approach 51, 193, *206*
compared with Hansch equation 196–8, *208–11*
inputs 194, 196–7, 199–200
RLGGs implemented 297, 307
Good, Jack 9, 39
Grisin's logic 151
guards, speed-up learning using 443, 456
handwritten digits dataset (StatLog project) 329–30
characteristics *351*
performance of various algorithms 336–7, *353*
Hansch linear regression equation 195
compared with Golem 196–8, *208–11*
Hasse diagram 146, *147*
head injury dataset (StatLog project) 331–2
characteristics *351*
performance of various algorithms 337–8, *356*
heart disease dataset (StatLog project) 332–3
characteristics *351*
performance of various algorithms 337, 338, *357*
helicoptor control 404
heterogeneity of knowledge 277
Hierarchical Image Processing System (HIPS) 331
hierarchy tree of concepts *128*
higher-level computing machines 31
higher-order constraint network 273, *274*
Hilbert Program 6, 7, 10, 38
historical perspectives 1–52
homogeneity of covariances (of StatLog project datasets) 345, *351*
Horn-clause-resolution predicate calculus 5, 16
implication on Church machines 20
Huffman coding 94, 96

human cognition, regulation of 247–65
human experts
  non-determinism used 171
  rule-formulation not liked 392, 394
human mind, working of 28–9
human reasoning, logic rarely used 24, 25
human short-term memory capacity 297
hydrocarbon separation system configuration, expert system for *391*
  see also GASOIL
hypotheses
  automated formation 88
  construction 89–90
  evaluation 90–2
  see also incremental hypothesis
IAS computer 11
  similar machines listed 11
ID3 algorithm
  application to KRKN chess endgame 292–3
  compared with neural networks 313–14
  other algorithms based on 323, 324
  patterns discovered using 380
image analysis datasets (StatLog project) 328–31
  performance of various algorithms 335–7, *352–5*
  see also handwritten digits...; Karhunen–Loeve digits...; satellite image...; vehicle silhouettes dataset
imperative approach to programming 15, 21
incremental freezing 376–7
  difficulties with BOXES algorithm 377, *378*
incremental hypothesis
  construction 89–90
  evaluation 90–2
incremental learning 44
  frequency data 161
INDCART system 323
  performance on various datasets *352–9*
  see also CART system
induced rule-sets, performance 402–3
inductive inference 57–189
  from frequency data 159–65
    algorithm used 159–61
    batch processing 159–60
    cloud/rain example 145, 161
    incremental processing 161
  see also concept formation
inductive logic programming (ILP) 45–6, 47–9
  applications 50
    chess endgame 291–307
  minimal multiple generalization used 69–77
  theory 49–50
  see also Golem
inductive skill-grafting
  applications 400–10
  see also skill-grafting experiments
infant consciousness, development 260–5
inference algorithm 70
information of probability distribution 155
inheritance calculus, use in language translation for large knowledge bases 279
inner product, definition 151
inner product space *see* Euclidean space
input tape encoding (for evaluation of logical theories) 93–9
instance, definition 444
instance-based prediction 333

Institute for New Computer Technology (ICOT) 27, 31
instructive feedback 216
instrumental learning 215
integrity, knowledge bases 278
intelligence, theories of regulation 249
interpreter, use in machine learning 42, 48
intuitive motherese (babytalk) 261
Inverse Resolution 49
Ishizaka's learning algorithm 187
ITrule algorithm 326
  performance on various datasets *353–8*
justification of logical theories 87–118
KARDIO system 306–7
Karhunen–Loeve (KL) digits dataset (StatLog project) 330
  characteristics *351*
  performance of various algorithms 336, *354*
Kemeny, John 14
kernel density estimation method 317
  see also ALLOC80 algorithm
k-minimal multiple generalization 63
  algorithm for 59
  application to attribute-oriented induction 79–83
    compared with use of LCHR algorithm 82–3
k-nearest-neighbour (K-N-N) algorithm 316
  performance on various datasets 335, 336, 337, 339, *352–9*
knowledge
  types 387–8
  see also declarative...; procedural knowledge
knowledge acquisition, support of 276–7
knowledge-based programming environments, knowledge representation in 270
knowledge bases
  applications 276
  building analogous to teaching humans 276, 284, 287
  changeability 278
  conformity 277–8
  engineering knowledge base
    clusters of knowledge *283*
    segmentation of knowledge 285–6
    size 286, 287
  extendability 278
  integrity 278
  large knowledge bases 269–87
  openness 278
  processing users' queries 282–3
  stratification of knowledge 280
  uniting into large knowledge bases 276, 287
knowledge engineers, aims 387
knowledge representation
  CAFE system 126–9
  large knowledge bases 271–5, 279–80
knowledge system, formalization of 280
Knuth, Donald 14
Kohonen net 328
  performance on various datasets 336, *352–3*, *355*, *357–8*
Kowalski, Robert 16, 17, 21
k-pivot 66
KRK chess endgame
  BTM (black-to-move) WFW (won-for-white) positions *300*, *301*, *303–4*
    depth 0 induced Prolog definition *299*, 300–1

depth 1 induced Prolog definition 301, *302*, 305–6
number of legal positions 295–6, *295*
canonical positions *294*
databases 291, 293–6
depths of win 293
number of positions available *295*
illegality induction task 101
significance measures compared *104*
legality learning task
compression-guided CWS procedure used 113–15
GCWS algorithm used 297–306
symmetrical translation *294*, 295
KRKN chess endgame 292
Kullback–Leibler measure 335, 342
kurtosis (of StatLog project datasets) 347, *351*
LABYRINTH 125, 126
attribute generalization used 139
compared with CAFE 125, 126, 129, 138–9
representation of instance *127*
lambda calculus
on Landin machine 20
see also Church's lambda calculus
Landin, Peter 20
LandSat image data 328–9
see also satellite image dataset
Laplace's Law of Succession 325
large knowledge bases 269–87
amount of work needed to build 287
architecture 279–83
contents 283–6
knowledge representation for 271–5
operational parts described 280–2
referential part 282
requirements 276–9
size 278, 286, 287
Law–Sammut algorithm (for pole-balancing problem) 368–9
compared with other variants of BOXES algorithm *370*
asymmetric pushing 373, *374*
properties 369, 370–3
Law's version of BOXES algorithm 368
layered network model 222–3, *241*
simulated learning curves using 223, 225, *242*, *244*
LCHR algorithm 77
generalized database obtained using 78
compared with use of *k*-mmg algorithm 82–3
leap-year example 109
learning, definitions 413
learning algorithm
LA (for NFAs) 177–8
correctness 178, 180–1
flowchart diagram *179*
time analysis 181–2
speed-up learning 445–8
learning control 361–456
learning curves
animal learning *235*, *237–40*
simulations on layered network *242–4*
CAFE system *134*, *136*, *138*
COBWEB *134*, *136*, *138*
LABYRINTH *138*
Tsukimoto–Morita algorithm *163–4*
learning from failure 456
learning to learn 224–5
demonstration *243*
simulation on layered network 225, *244*

least general generalization 60
applications 69
generalization of 59–83
meaning of term 62
least Hebrand model (of program) 70, 72
least-mean square rule 223
linear discriminant analysis 317–18
performance on various datasets 337, 338, 340, 341
see also Discrim algorithm
linear regression methods, compared with Golem 197–8, *208–11*
linguistic learning 263–4
LISP 16
living organisms, growth 247
local merit criterion 367
logic
and artificial intelligence 6, 21–3
and computers 4–6, 6–12
and neurocomputation 27–30
and programming 12–13
and programming languages 16–21
logical entropy 154
and information of probability distribution 155
logical function, information of 154
logical theories
evaluation, encoding scheme for 93–100
justification 87–118
logical vector 146, 149–53
definitions used 149
interpretation of nonclassical logical vectors 156, *157*
meaning of term 153
relationship to probability vector 148, 153–9
correspondences 156–8
transformation from probability vector 146, 158–9
Logic Theorist 29–30
logistic regression (LogReg) method 318–19
performance on various datasets 337, 338, 340, *352–9*
McCarthy, John 16, 21
Minsky's opinion on his approach to AI 24, 25
McCarthy's logical deductive knowledg base paradigm 24, 25, 26, 388–9
McCullogh, W.S. 4, 12, 27
machine learning
first program developed 37, 46–7
first shown by imitation of trained human 402
integral role in computers 51
memory size required 41
Turing's ideas 41, 51
see also concept formation; inductive learning
machine learning algorithms 321–6
performance on various datasets 340, *352–9*
time taken 341
see also AC2...; C4.5...; Cal5...; CART...; CN2...; Golem; ID3...; INDCART...; ITrule...; NewID algorithm
macro-operators
average sizes *433*
increased matching cost 434
matching costs compared 435–6
operationality
experimental results 431–3
factors affecting 421
re-ordering effect 433–4, 435
unsuccessful application 434, 435
see also PCLEARN; perceptual chunks

Makarovic's rules (for pole-and-cart system) 380, 398
Martin-Löf's Intuitionistic Type Theory 5
Marvin program 44–5
Mauchly, J.W. 4, 11
MAXT algorithm 68
max trees *see* consistent max trees
medical datasets (StatLog project) 331–3
  performance of various algorithms 337–8, *356–7*
  see also head injury...; heart disease dataset
medical diagnosis expert system *391*
membership queries, learning using 169, 175
memory
  accessibility to 388
  declarative 388
  procedural 388, 394
metacontrol (in split brain) 259
meta-interpreter 442–3, 456
metaknowledge 282
Michie, Donald, discussions with Turing 39, 40
Michie–Chambers algorithm (for pole-balancing problem) 366–7, 396
  compared with other variants of BOXES algorithm *370*
microtheories, knowledge representation using 284
minimally adequate teacher (MAT) learning model 169
  interactive learning in *170*
  learnability of NFAs in polynomial time 171–2
minimal multiple generalization (MMG)
  algorithm for 59, 65
  applications
    in inductive logic programming 69–77
    in knowledge discovery in databases 77–83
Minimum Description Length (MDL) criteria 99, 144
  proposition obtained by Tsukimoto–Morita algorithm 162, 164
Minsky, Marvin 21
  on McCarthy's approach to AI 24, 25
MMG algorithm 65
modularity of knowledge 277
molecular modelling, Golem used 50
mother love 261
multigroup discriminant analysis 317
  see also Alloc80 algorithm
multimodality measures 335, 339
mutual induction, use by CAFE 125, 129, 141
MYCIN expert system *391*
naive Bayes algorithm 316
  performance on various datasets 335, 337, 341, *352–9*
NASA
  LandSat image data 328–9
  Space Shuttle data 334
National Cash Register (NCR) Company 9n
National Physical Laboratory (NPL), Turing's work 9–10, 40, 43
natural intelligence (NI) 25
natural numbers, encoded using prefix codes 95
nearest-neighbour algorithms 316, 340, 341
  see also *k*-nearest-neighbour (K-N-N) algorithm
negative patterning 222, *240*
  simulation by layered network model 223, *242*
neocortical commissurotomy 255

see also commissurotomy
nervous system 247
neural networks 326–8
compared with ID3 algorithm 313–14
compared with other algorithms and systems 200, 201, 313–14, *352–9*
drug design using 195
lack of diagnostic help criticized 342
performance on various datasets 336, 337, 340–1, *352–9*
problems solved 223
protein structure prediction using 199, 200
see also back-propagation...; Kohonen algorithm...; layered network model...; radial basis functions
neurocomputation 27–30
neurological data *see* head injury dataset
neurone activity 250–1
newborn baby's brain 249, 250
NewID algorithm 323
performance on various datasets 338, *352–9*
see also ID3 algorithm
Nim game 37, 46–7
noise
effect in learning 44, 101
compressive theories 114, *115*
estimation using compression 104–6
noisy data, discovery of propositions in 143–66
nonclassical logical vectors, interpretation 156, *157*
nonclassical logics 27
non-determinism, use by human experts 171
non-deterministic finite automata (NFAs)
application in pattern matching *171*
query learning by 172–88
algorithm used 177–8
compared with DFA learning *184*
construction of new candidate rules 176
diagnosis of set of transition rules 176–7
introduction of new states 175
learning protocol used 175
notation used 174–5
usefullness 170–1
non-monotonic inductive inference, closed-world specialization used *108*
nonmonotonicity 278
nonspecific knowledge 284, 285
norm, definition 152
NTgrowth 333
NUT programming environment 275
Occam's razor 90
operant conditioning 215
with feedback 216
optimal chess strategies, learning 291–307
orthonormal systems 152–3
output tape encoding (for evaluation of logical theories) 99
overfitting 144
pruning to avoid 144, 323
PAC-learning 114, 117
parallel computation, Church machines used 20
partial structures of explanations, learning methods based on 443, 456
pattern-matching problem
DFA compared with DFA *171*
non-determinism used 170–1
Pavlovian conditioning 215

PCLEARN system 424–30
chunking facility 426
algorithm used 426–30
method used 425
specification *425*
compared with EBL learner 431–6
experimental results
learning performance results 433–6
methods used 430–1
operationality of macro-operators measured 431–3
future work proposed 437–8
input 425
learning concept 424–5
output 425
overview 426
problem-solver module 426
peak-to-peak heuristics 454–5
perceptual chunks 420
learning based on 421–2
see also PCLEARN
perceptually chunked macro-operators 430
perfect causality 456
physical symbol system hypothesis *386*, 386–7
Pitts, W. 4, 12, 27
planning
example 449–50
proof tree for *450*
experiments in robot domain 452–4
Plotkin, Gordon 37
pole-balancing problem 363–4
asymmetric pushing problem 373
comparison of various algorithms *374*
BOXES algorithm used 364–9, 396
clean-up effects *379*, 380
C4.5 algorithm used 401, 402
clean-up effects 402–3, *403*
determination of *DK* *372*
determination of *K* 369, *371*
induction of rules from behaviour 402
line-and-pointers representation 401
order in which dimensions checked 398
reliable controllers 375–8
role of problem representation 401
pole-and-cart system 363–4, 396, *402*
see also pole-balancing problem
polynomial time *k*-mmg algorithm 62–9
polynomial update time inference algorithm 70
polytree algorithms 320, 342
see also CASTLE algorithm
polytrees 321
pooling of learning results 376
predicate calculus 46, 48
use in Golem 195
see also Inductive Logic Programming
predicate invention 50, 307
prefix codes 94
natural numbers encoded using 95
prefunctional morphogenesis 248
primitive attribute 126
primitive Prologs 75–7
definition 75
inference algorithms for 76–7
probability vector
relationship to logical vector 148, 153–9
correspondences 156–8
transformation to logical vector 146, 158–9
problem-solving, involvement of both declaraive and procedu-

ral knowledge 412
procedural knowledge 389–93
compared with declarative knowledge 393–4
procedural memory 394
contrasted with declarative memory 388
PRODIGY system 424, 425, 437, 456
productivity 223–5
programming
descriptive approach 14–15
imperative approach 15
and logic 12–13
Turing's attitude 13–14
see also automatic programming
programming languages
development 5, 16
and logic 5, 16–21
program-translation programs 23–4
projection pursuit regression 319–20
see also SMART algorithm
Prolog 16
computation via goal reduction 444
as ideal language for symbolic learning 48
proof encoding 97–9
example 97–8
proof trees 445
geometry problem *428*
planning example *450*
positive/negative instances *447*
propositional calculus theorem proving program 29
propositions
acquisition of 144
discovery in noisy data 143–66
protein docking problem 201
protein secondary structure prediction task 100, 198–9
chemical rules governing 200, *212*
error rate data 101, *102*
Golem used 199–201
significance measures compared *104*
protoconversations (baby–mother) 261, *262*
proximal associations, impairment in conditioning 219, *238*
psychological development 249
psychology, in AI 24–5, 401
PVM algorithm, compared with backpropagation 314
quadratic discriminant analysis 318
performance on various datasets 336, 337, 338, 340, 341, *352–9*
quantitative structure–activity relationship (QSAR) modelling 195–6
rules 202–3
query learning 169–88
see also minimally adequate teacher (MAT) learning model
rabbit eyeblink, conditioning of 216–17
radial basis functions 327–8
performance on various datasets 335, 336, 337, 339, *352–8*
real-time skills, building symbolic representations 385–415
reclassification error rate 101
recognize–act cycle, frequency 389–90
REDUCED algorithm 66
operation of 65–6, 67
reinforcement learning, combined with induction 380–2
relational attributes 126
value distribution 129, 139–40

relative least general generalizations (RLGGs) 49, 297, 307
remote associations, facilitation in conditioning 219, *237*
RescorlaPWagner rule 223
resubstitution error rate 101
retrograde analysis method, chess endgame databases generated using 293–5
rewriting rules 18–19
Robinson's resolution logic 21
  inversion 49
    see also Inverse Resolution
rote learning programs 37, 47
rotorcraft control systems 404
rule-based control, experimental study 395–400
rule-conjecture-and-test procedure 394
Runge–Kutta algorithm 396–7
run-time thinking 414
Samuel, A.L. 37, 47
satellite fault diagnosis, ILP approach used 307
satellite image dataset (StatLog project) 328–9
  characteristics *351*
  performance of various algorithms 335–6, *352*
scientific domains 191–287
Scott, Dana 14
search control knowledge
  learning 420
    in geometry domain 422–4
SECD machine 20–1
second-order conditioning 220, *239*
self, origin in brain 414–15
self-reproduction 28
semantic memory 388
Shapiro's contradiction backtracking algorithm 176
Shapiro's debugging system 42, 51
Shapiro's Model Inference System 41–2
Shuttle... *see* Space Shuttle...
skewness (of StatLog project datasets) 347, *351*
skill, meaning of term 389
skill acquisition, postulates for 393–4
skill-grafting experiments 400–14
  aircraft flight 404–10
  pole-balancing 401–4
SMART algorithm 319–20
  performance on various datasets 337, 338, *352–9*
SOAR learning method 437
software reusability 269–71
somatotopic maps 251, 255
spacecraft attitude control problem 382, 395–400
  black-box simulator 396–7
  constraints 395–6
  decision arrays used 399
  result of control strategy 398–400
  rules used 397–8, 399
Space Shuttle control dataset (StatLog project) 334
  characteristics *351*
  performance of various algorithms 339–40, *359*
speed-up learning 40, 441–56
  algorithm for 445–8
split brain
  anatomy *257*
  levels and directions of consciousness in 255–60
state transition diagram, BOXES algorithm *376*
statistical algorithms 316–21
  performance on various datasets 338, 340, *352–9*
  time taken 341
  see also Alloc80...; Bayes...; CASTLE...; Discrim...; K-N-N...; LogReg...; Quadra...; SMART algorithm

statistical analysis methods, protein structure prediction using 199
StatLog project
- aim 311, 312
- algorithms studied 316–28
  - machine learning algorithms 321–6, *352–9*
  - neural network algorithms 326–8, *352–9*
  - statistical algorithms 316–21, *352–9*
  - time taken 341
- datasets studied
  - characterization of datasets 334–5, 340–1, *351*
  - credit risk data 333–4, 338–9, *358*
  - handwritten digits data 329–30, 336–7, *353*
  - head injury data 331–2, 337–8, *356*
  - heart disease data 332–3, 337, 338, *357*
  - image analysis data 328–31
  - Karhunen–Loeve (KL) digits data 330, 336, *354*
  - medical data 331–3, 337–8, *356–7*
  - satellite image data 328–9, 335–6, *352*
  - Space Shuttle control data 334, 339–40, *359*
  - vehicle silhouettes data 330–1, 337, *355*
- key personnel listed 343–5
- objectives 312
- previous studies commented on 314–15
- results 335–40, *352–9*
  - credit risk dataset 338–9, *358*
  - image analysis datasets 335–7, *352–5*
  - medical datasets 337–8, *356–7*
  - Space Shuttle dataset 339–40, *358*
- testing methodology described 315–16

stimulus compounding 221, *240*
- simulation by layered network model 223, *242*

Strachey, Christopher 13, 14, 37, 46–7
STRIPS planning program 449–52
strong AI 26
- criteria summarized *386*
- logico-neural version 27
- supporters of thesis 26, *386*

structure information, utilization in concept formation 123–41
supervised learning 124, 140
- with instructive feedback 216
- pruning methods used 144

support, knowledge acquisition 276–7
SuttonPBarto model 220–1
symbol description header, components 94
symbolic propositional algorithms 321
- performance on various datasets 340, *352–9*
- *see also* machine learning algorithms

symbolic representation of intuitive processes 387
- compared with neural-net representations 414–15

Symbol-Level Learning (SLL) 442
- classes *443*

teaching of humans, building of knowledge bases analogous to 276, 284, 287
thought, and knowledge 387–94
tightening process 65

touch-typing, declarative memory disabled during 392–3
training set accuracy/covery, as significance measure 102, 103, *104*
transfer of training 224
 see also learning to learn
transparency of knowledge 277
tree pattern language 62
trial-and-error learning 215
trimethoprim analogues *207*
 binding with dihydrofolate reductase 195–8, *207*
  observed vs predicted activity *208–9*
Tsukimoto–Morita algorithm 145–6, 159–61
 description length of proposition obtained 162, 164
 experimental analyses 161–2, *163–4*
 incremental processing version 161
Turing, Alan
 attitude to programming 13–14
 building of universal machine 8, 9–10
 child development compared with machine learning 45
 contributions to AI 24, 37, 43
 on Hilbert's decidability question 7, 8, 38
 on incremental learning 44
 invention of universal machine 1–2, 6, 7–8
 at Manchester 12
 on neural-net vs logic-based learning 42
 at NPL 9–10, 40, 43
 at Princeton 10
 and von Neumann 2, 3n, 8–9
 war-time work 8–9, 39–40
Turing machine 7–8
 invention 1–2, 6
 model for
  evaluating logic programs 91
  incremental hypothesis evaluation 92
 von Neumann's practical version 11, 30
Turing's imperative notation 15
Turing Test 24, 26
tying shoelaces, as example of procedural knowledge 390
unconditioned response 215
unconditioned stimulus 215
unit clause programs, inference algorithm for 71–2
univariate skewness (of StatLog project datasets) 347, *351*
universal computing machine 38
 implications of concept 3–4, 8
 invention by Alan Turing 1–2, 6, 7–8, 18, 38
 practical exploitation 5, 8, 9–10, 30–1
unsupervised learning 124–5, 140
VAX computer configuration, expert system for *391*
 see also XCON
vehicle silhouettes dataset (StatLog project) 330–1
 characteristics *351*
 performance of various algorithms 337, *355*
virtual other 264
visual information uptake 256
visual input (to brain hemispheres) 256
von Neumann, John
 contribution to AI 24, 27–8
 practical development of Turing machine 11, 30
 on practical implications of universal machine 3–4

propensity to develop other people's ideas 2, 4
and Turing 2, 3n, 8–9, 10
von Neumann architecture 2, 11
voting, use in learning 376
weak AI, criteria summarized *386*
weather (rain/cloud) example 145, 161
Widrow-Hoff rule 223
Winograd, T. 21
Womersley, J.R. 4n
XCON expert system *391*
Yokomori's learning algorithm 177–8
compared with Angluin's algorithm 182–4, *185*
correctness 178, 180–1
example operation 183, *185*
flowchart diagram *179*
practical variant 186–8
flowchart diagram *187*
time analysis 181–2